Microwave Power Amplifier Design
with MMIC Modules

For a listing of recent titles in the
Artech House *Microwave Library*,
turn to the back of this book.

Microwave Power Amplifier Design with MMIC Modules

Howard Hausman

ARTECH HOUSE

BOSTON | LONDON

artechhouse.com

Library of Congress Cataloging-in-Publication Data
A catalog record for this book is available from the U.S. Library of Congress

British Library Cataloguing in Publication Data
A catalog record for this book is available from the British Library.

ISBN 13: 978-1-63081-346-8

Cover design by John Gomes

© 2018 Artech House
685 Canton Street
Norwood, MA

10 9 8 7 6 5 4 3 2 1

To my wife Gale,
our children, and our grandchildren

Contents

CHAPTER 3

CHAPTER 9

System Cascade and Dynamic Range Analysis

CHAPTER 12

MMIC Amplifier Modules for Use in Parallel Combining Circuits

CHAPTER 15

Power Amplifier Chain Analysis   235

Preface

Solid state power amplifiers (SSPA) are key parts of many microwave systems. The design of these parts has evolved from designing the individual amplifier stages to integrating microwave monolithic integrated circuits (MMICs) as building blocks with known key parameters. This technique cuts the design cycle, lowers the engineering cost, and ultimately results in a more consistent and reliable product. This text focuses on the theory and technology needed to select, integrate, and optimize the design of an SSPA using MMIC modules configured on microstrip transmission lines. The text is separated into three parts plus an introduction.

Introduction

The introduction, Chapter 1, focuses on SSPA applications, configurations, specifications, and documentation used to start the product design.

Part I: Useful Microwave Design Concepts

Part I, starting with Chapter 2, contains eight chapters of useful information on general microwave theory relating to the design of microwave power amplifiers using MMIC modules. The concepts discussed are applicable to the design of many other similar microwave devices, assemblies, subsystems, and systems. The chapters cover: lumped components in RF and microwave circuitry (Chapter 2), transmission lines (Chapter 3), S-parameters (Chapter 4), microstrip transmission lines (Chapter 5), circuit matching and voltage standing wave ratio (VSWR) (Chapter 6), noise in microwave circuits (Chapter 7), nonlinear signal distortion (Chapter 8), and system cascade and dynamic range analysis (Chapter 9). The generality of these chapters makes them useful as a means of introducing students or engineers to basic microwave concepts and design techniques.

Part II: Designing the Power Amplifier

Part II, starting with Chapter 10, contains six chapters focusing on the design of power amplifiers using MMIC modules. The chapters cover: defining the output power requirements for a communication link and other wireless systems (Chapter 10), parallel amplifier topology-enhancing SSPA performance (Chapter 11), MMIC amplifier modules for use in parallel combining circuits (Chapter 12), measuring and matching the impedance of high-power MMIC amplifier modules (Chapter 13), power dividers and combiners used in parallel amplifier SSPAs (Chapter 14), and power amplifier chain analysis (Chapter 15). The information presented is applicable to the design of many types of integrated microwave assembly (IMA), containing active components such as amplifiers, switches, mixers, and frequency converters. A course for students or engineers can be built around the chapters contained in this part.

Part III: Designing the Power Amplifier System

Part III, starting with Chapter 16, contains five chapters focusing on the interface of an SSPA with other system components including issues with DC power supplies, monitoring circuits and electromagnetic interference compatibility (EMC). The chapters cover: RF signal monitoring circuits (Chapter 16), DC power interface with the RF signal path (Chapter 17), SSPA DC voltage and current (Chapter 18), thermal design and reliability (Chapter 19), and electromagnetic interference (EMI) (Chapter 20). These issues focus on the SSPA devices but the concepts are applicable and critical to ensuring system compatibility of many other amplifier and nonamplifier subsystem functions.

Summary

The primary purpose of this text is to instruct engineers and engineering students on the issues pertaining to the design of SSPAs, an import and growing segment of the microwave component and system market. The material presented is applicable to many facets of microwave industry, making this book useful as a primary or secondary text for a course focused on the design of SSPAs or a course focused more generally on microwave engineering.

CHAPTER 1

Introduction

1.1   Introduction to Designing Microwave Solid State Power Amplifiers

Power amplifiers are a critical part of many microwave systems. Designing these devices with power amplifier monolithic microwave integrated circuits (MMICs) has changed the focus from the individual semiconductor device to the next higher level, that is, a power amplifier module. The intention of this text is to provide the power amplifier engineer information to: (1) select the best power amplifier module for a particular application, (2) interface the selected module with other power amplifier modules in the system, and (3) identify and mitigate peripheral issues concerning the power amplifier module, the solid state power amplifier (SSPA) or the microwave system.

The text is separated into three parts:

- Part I, Useful Microwave Design Concepts: Microwave theory and concepts helpful to design the SSPA and other microwave devices;
- Part II, Designing the Power Amplifier: Design of the SSPA includes selection of MMIC modules, configuring the modules to provide the desired result and interfacing the modules to optimize the final design;
- Part III, Designing the Power Amplifier System: An SSPA is a primary part of any microwave system. Designing and optimizing the interface of this part with the applicable microwave system are essential to a successful project. The theory and techniques of some major system compatibility topics of are discussed and analyzed.

Each section of this text is totally independent and can be used as a reference for the respective titled subject matter. With regard to the use of the terms RF (radio frequency), microwave, and RF/microwave, in general, the term RF refers to signals at a frequency above baseband. The term microwave typically refers to signals at a frequency above 1 GHz. In this text the terms RF, microwave, and RF/microwave are used interchangeably, denoting the frequency of the signals being amplified in the SSPA [1]. With regard to the term IF (intermediate frequency), IF signal frequencies are interim conversion frequencies between RF and baseband, and there no specific frequency range for IF signals.

1.2   Applications of SSPAs

SSPAs increase low-level signals to higher-power signals terminated usually in 50Ω [2]. Typically, the high-power signals are used in applications such as missile guidance systems, radars, electronic warfare (EW) systems, communications systems, and microwave test systems. The operational platforms can be spacecraft, aircraft, ships, and ground-based systems operating in relatively benign to very severe environments.

With regard to SSPA design concepts, the variety of SSPAs is so diverse that addressing a particular application is deemed too narrow of a focus. The concepts presented are generally applicable to most SSPA requirements. The intention of the text is to give the engineer the ability to adapt the knowledge gained to a specific application.

1.3   A Typical SSPA Configuration

Amplifier power levels are usually classified as low-power, medium-power, and high-power with very few definitions of the actual levels of demarcation. Many times, the high-power amplifier is distinguished from other amplifiers by its unique output stage. To boost the output power, multiple high-power amplifier modules are configured in parallel (see Figure 1.1).

Sometimes parallel amplifier stages are designed into a single MMIC module and then the modules are placed in parallel. The number of parallel stages can be well into the double digital range and higher.

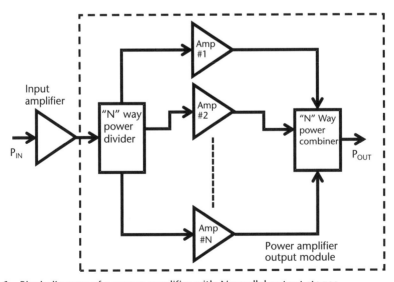

Figure 1.1   Block diagram of a power amplifier with *N* parallel output stages.

1.4   Typical Documents Starting a Project

Starting an SSPA project or any subassembly of a larger system generally requires three documents: a contract, a statement of work (SOW), and a specification control drawing or a source control drawing (SCD).

1. *Contract:* Usually contains the item description, the delivery date, the applicable specification, price, general terms and conditions, and other contractual items as required. Note that in the case of a conflict, the contract usually takes precedence over all other documents.
2. *SOW:* The SOW is a document that states the general requirements of the project. It defines (either directly or by reference to other documents) all work that must be performed by the contractor to fulfill the requirements of the contract [3]. Note that management requirements such as schedule, design reviews, part selection, part derating [4], and reliability are usually contain in an SOW.
3. *SCD:* The SCD contains all the specifications that must be met to ensure that the item being produced will work in the next higher assembly. This document also defines the tests that must be performed successfully before the unit is accepted.
   The following are notes on SCD:
   a. A source control drawing is similar to a specification control drawing except that it identifies specific contractors that have been qualified to produce the item.
   b. A specification control drawing might or might not suggest contractors that may be qualified to produce the item.
   c. A specification control drawing describes the form, fit, and functional design criteria necessary to ensure that the item will meet the intended purpose [5].
   d. It is generally understood that all specifications in the SCD must be met even if they are not included in the testing requirements.

The details of an SOW, contract, and portions of the SCD are beyond the scope of this text.

1.5   General Format of the SCD

The design phase of a project starts with an SCD, whether it be formalized or an informal document. The specification control drawing has many formats, but generally follows the following major paragraph headings [6]:

1. Scope;
2. Applicable documents;
3. Requirements;
4. Verification;

5. Packaging;
6. Notes.

1.5.1   Paragraph 1.0: Scope

The scope or introduction is usually a short paragraph describing the program and the item specified in the SCD.

1.5.2   Paragraph 2.0: Applicable Documents

This section is a list all the ancillary documents that are part of the SCD as applicable to the respective parts of the specification.

1.5.3   Paragraph 3.0: Requirements

The requirements section lists all of the specifications that ensure that item functions properly in the next higher assembly. There are usually several subparagraphs each critical to the design of the respective item. Some of the subparagraphs typically found in an SSPA SCD are: electrical requirements, mechanical requirements, and environmental requirements.

1.5.4   Paragraph 4.0: Verification

This paragraph describes the testing to be performed and passed to verify that the end item is acceptable to be installed in the next higher assembly. There may be multiple subparagraphs describing a different set of tests to be performed such as: qualification testing, random or regular periodic production testing, and testing of all production units (100% testing).

1.5.5   Paragraph 5.0: Packaging

This includes any specialized end item packing instructions.

1.5.6   Paragraph 6.0: Notes

This is a miscellaneous section. Many times, the suggested or qualified source's information is included in this paragraph.

1.6   Requirements Section of an SCD

The requirements section focuses on the electrical, mechanical, and environmental requirements of the SSPA. The requirements listed are those necessary to ensure form, fit, and function such that the item will perform as expected in the next higher assembly. All parameters of the device may or may not be listed. In the paragraphs below is a sample of typical specifications that may be seen in the requirements section of the SCD. Actual numerical values are applicable to a specific item and are therefore not included in this paragraph.

A drawing of the item is usually shown or referenced in the requirements section of the SCD. Figure 1.2 is a typical simplified functional drawing of an SSPA.

The functional diagram in Figure 1.2 omits some information that might be stated in an actual SCD. Some of these items are:

1. Actual location of all connectors, that is, the actual location of the RF connectors, DC power terminals, or connectors as applicable, and signal terminals or connectors as applicable;
2. Location and types of mounting connections;
3. Tolerance of all dimensions.

Note that, on the thermal mounting surface, the thermal mounting surface defining the heat path away from the unit is critical information to SSPA design and all modules dissipating a significant amount of power should be located on the surface that is designed to dissipate that power.

1.6.1   Electrical Requirements

The RF/microwave SSPA specifications or design goals are usually broken into functional parts and subparts. Shown in this paragraph is a sample of expected requirements. Typical applicable units are shown in parentheses:

Electrical characteristics:

1. RF requirements:
   a. Input requirements (as applicable):
      i.   Operating frequency (MHz or GHz);

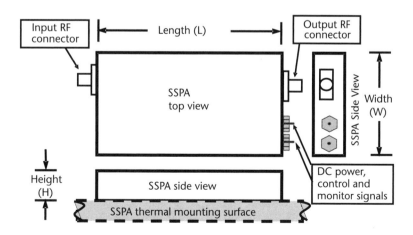

Figure 1.2   Simplified functional drawing of an SSPA that might be included in the SCD.

      ii. Input power level (dBm);

     iii. Input return loss (dB) or VSWR;

     iv. Nominal input impedance (Ω);

      v. Signal modulation type (CW, pulsed, or FM, as applicable).

b. Output requirements (as applicable):

      i. Output power level (dBm or dBW), either or both of saturated pulsed SSPA [Psat, Saturated output power (dBm or dBW), pulse width (seconds), and pulse repetition frequency (Hz)] and linear CW SSPA [P1dB, 1-dB gain compression (dBm or dBW) and intermodulation distortion (dBc) at a specific output power level (dBm or dBW)];

      ii. Output return loss (dB) or VSWR;

     iii. Nominal output impedance (Ω);

     iv. Spurious (dBc or dBm).

c. Transfer requirements:

      i. Gain (dB);

      ii. Level flatness ($\pm$ dB);

     iii. Group delay distortion (nanoseconds).

2. Control and monitor signal requirements;

3. DC power:

    a. DC voltage range;

    b. Total DC power.

4. Power-added efficiency (PAE) [7]:

$$PAE = \frac{P_{\mathrm{OUT}} - P_{\mathrm{IN}}}{P_{\mathrm{DC}}} \qquad (1.1)$$

where PAE is the power-added efficiency, usually expressed as percentage, P_{OUT} is the RF output power (watts), P_{IN} is the RF input power (watts), and P_{DC} is the dissipated DC power (watts).

1.6.2 Mechanical Requirements

The mechanical requirements are as follows: size, weight, mounting configuration, RF connectors, signal connectors, and DC connectors.

1.6.3 Environmental Requirements

The environmental requirements are as follows: operating and storage temperature range, mechanical stress (vibration or shock), altitude, and humidity.

1.6.4 Other Design Criteria

Other design criteria include:

1. Reliability: Minimum anticipated mean time between failure (MTBF);
2. Electromagnetic interference: Specify the limits for electromagnetic compatibility (EMC).

The requirements of reliability may be contained in the SOW.

References

[1] Golio, M., *The RF and Microwave Handbook*, Boca Raton, FL: CRC Press, 2008.

[2] Cripps, S., *RF Power Amplifiers for Wireless Communications*, Norwood, MA: Artech House, 2006.

[3] U.S. Department of Defense, "Handbook for Preparation of Statement of Work (SOW)," MIL-HDBK-245D APRIL 19965

[4] Sahu, K., "EEE-INST-002: Instructions for EEE Parts Selection, Screening, Qualification, and Derating," National Aeronautics and Space Administration Goddard Space Flight Center, May 2003.

[5] Watts, F., *Engineering Documentation Control Handbook*, New York: Elsevier, 2012.

[6] U.S. Department of Defense, "Defense and Program-Unique Specifications Format and Content," MIL-STD-961E w/CHANGE 1, April 2008.

[7] Jacob, M., "Optimizing RF Power Amplifier System Efficiency Using DC-DC Converters," Texas Instruments Incorporated, 2011.

Part I
Useful Microwave Design Concepts

CHAPTER 2

Lumped Components in RF and Microwave Circuitry

2.1 Applicability of Lumped Element Analysis

Transmission line theory deals with the problem of signal transfer between components. Lumped circuit theory is a specific case of transmission line theory where it is assumed that the distance between components is zero and therefore the signal emitted from a component is identical to the signal received by the connecting component [1]. If the signal changes between components due to the distance traveled, the change can be analyzed using transmission line theory.

The line of transition between when lumped circuit theory can be used and when transmission line theory should be used is not distinct and approximated by many rule-of-thumb assumptions. The actual line of demarcation when to use transmission theory and when lumped circuit can be used is application-related. A quantitative way of looking at this problem is to assume that the applicable signal is sinusoidal with a peak amplitude at zero degrees (cosine wave). If the signal travels one-twentieth of a wavelength ($\lambda/20 \rightarrow 18°$) between the emitting and receiving component, the signal amplitude changes 5%. A tenth of a wavelength ($\lambda/10 \rightarrow 36°$) distance between components translates to a 20% change in amplitude. Typically lumped constant theory can be used when the distance between components is less than one-twentieth of a wavelength ($\lambda/20 \rightarrow 18°$) and transmission line theory should be used when the distance between components is greater than one-tenth of a wavelength ($\lambda/10 \rightarrow 36°$). Component separation between lambda over 20 ($\lambda/20$) and lambda over 10 ($\lambda/10$) is an application-related judgment decision.

2.1.1 Calculating Wavelengths

The relationship between frequency and wavelength is:

$$f \cdot \lambda = \frac{c}{\sqrt{\varepsilon_r}} \tag{2.1}$$

where f is the frequency of the signal, λ is the wavelength of the signal, c is the velocity of light in free space (2.99792458E+10 centimeters/second), and ε_r is the relative dielectric constant of the material ($\varepsilon_r = 1$ for free space)

$$\lambda = \frac{c}{f \cdot \sqrt{\varepsilon_r}} \tag{2.2}$$

2.1.2 Example: Calculating Wavelengths for Lumped Circuit Analysis

For the lumped circuit theory, if it is desired to have the distance between components (l) less than one-tenth of a wavelength

$$l \leq 0.1 \cdot \frac{c}{f \cdot \sqrt{\varepsilon_{\mathrm{r}}}} \tag{2.3}$$

For $f = 1$ GHz in a dielectric material, $\varepsilon_{\mathrm{r}} = 4$. The maximum distance between components (l) would be:

$$l \leq 0.1 \cdot \lambda \approx 1.5 \text{ cm or } \approx 0.59 \text{ inch in free space}$$

2.2 Capacitor Characteristics at High Frequencies

Capacitors are typically represented by a nominal value of capacitance (C) in farads (F) that is usually measured at frequencies below the point that parasitic effects significantly alter the measurement results. As the operating frequency increases, parasitic inductance and resistance change the measured impedance.

2.2.1 Single-Layer and Multilayer Capacitor Construction

2.2.1.1 Single-Layer Capacitor

A capacitor in its simplest form (see Figure 2.1) is two parallel plates separated by a dielectric material as shown in Figure 2.2.

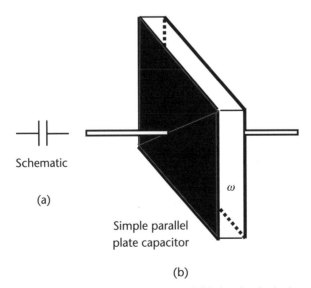

Schematic

(a)

Simple parallel
plate capacitor

ω

(b)

Figure 2.1 A parallel plate capacitor: (a) schematic and (b) simple physical model.

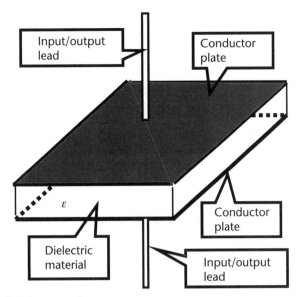

Figure 2.2   A parallel plate capacitor.

The capacitance is given [2] in (2.4)

$$C = \varepsilon_r \cdot \varepsilon_0 \cdot \frac{A}{d} \qquad (2.4)$$

where C is the capacitance in farads, A is the area of each parallel plate (meters^2), d is the distance between the plates (meters), ε_0 is the dielectric constant of free space ($\varepsilon_0 = 8.854187817 \cdot 10^{-12}$ farads/meter), and ε_r is the relative dielectric constant of the material between the plates.

From (2.4), it is seen that larger areas (A), greater relative dielectric constants (ε_r) and smaller distances between the plates (d) leads to higher capacitance. Typical capacitor ranges for single layer capacitors range from a fraction of a picofarad to thousands of picofarads.

2.2.1.2   Multilayer Capacitors

To increase the capacitance per unit volume manufactures, use multiplate devices fed like interlacing fingers from caps on the ends of the capacitor (see Figure 2.3). As shown, the plate area is multiplied many times and the distance between plates is significantly decreased. Very high relative dielectric materials (ε_r) are also used to further increase the device capacitance.

Note that some devices are manufactured with larger values of capacitance in very small volumes using very high dielectric materials, many plate layers and very small distances between layers. Under these conditions, the capacitor may be less resistant to stress resulting in possibly higher failure rates. As the dielectric material between plates (d) decreases, the ability of the dielectric material to with stand

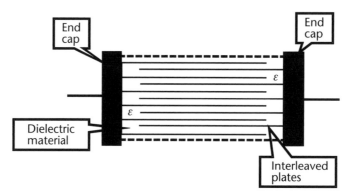

Figure 2.3   Multilayer capacitor.

voltage breakdown is decreased. There is a trade-off between capacitance in a small volume and voltage breakdown of the device.

2.2.2   High-Frequency Capacitor Models

A capacitor is fundamentally a dielectric material separated by two plates as shown in Figure 2.2, but at RF and microwave frequencies, the second-order parasitic effects of these materials can seriously affect the performance of the basic device. One parasitic effect is the resistance of the dielectric material proportionally increasing as the distance between the conductor plates increases. This is modeled as a resistor in parallel with the capacitor. A second effect pertains to the capacitor leads or end-cap terminations shown in Figures 2.2 and 2.3. These second-order effects are modeled as a series resistor and series inductor on either side of the capacitor. The combined primary RF/microwave model of the capacitor is shown in Figure 2.4 [3]. In Figure 2.5, the series resistance and series inductance are combined and shown on one side of the capacitor to simplify the analysis.

The parasitic effects shown in Figure 2.5 are the combined series resistance (Rs) and the combined series inductance (Ls) of the input and output conductors. The resistor (Rp) in parallel with the basic capacitor (C) is the effective resistance of the dielectric material [1].

2.2.3   Capacitor Losses (Q)

The parallel resistance (Rp) affects the loss in the capacitor, which can be expressed as the parallel Qp of the capacitor:

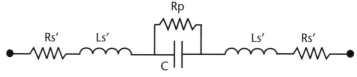

Figure 2.4   Model of a capacitor showing the primary parasitic effects that relate to the RF and microwave performance.

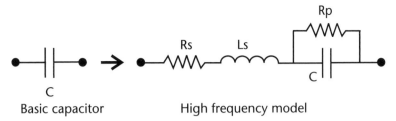

Figure 2.5   The RF/microwave model of a capacitor showing the combined series resistance (*Rs*) and series inductance (*Ls*).

$$Qp = Rp \cdot 2 \cdot \pi \cdot F \cdot C \qquad (2.5)$$

where F is the signal frequency (Hz), C is the capacitance in (F), and Rp is the parallel resistance (Ω). At any single frequency, the parallel circuit ($Rp//C$) can be converted to a capacitor and a series resistor (Rp_{equiv}) [4]; see Figure 2.6.

The equivalent series resistance (Rp_{equiv}) of the capacitor gives better insight into the expected capacitor loss characteristic and allows an easier interpretation of the relationship of capacitor resistance to conductor resistance (Rs). The value of the equivalent series resistance (Rp_{equiv}) is an approximation based on $Qp > 10$. Under this condition, the Q of the parallel circuit (Qp) approximately equals the Q of an equivalent series circuit (Qs) at any single frequency of interest ($Qs \approx Qp$).

$$Qs = \frac{1}{Rp_{\text{equiv}} \cdot 2 \cdot \pi \cdot F \cdot C} \approx Qp = Rp \cdot 2 \cdot \pi \cdot F \cdot C \ \text{ for } Qs \text{ and } Qp > 10 \qquad (2.6)$$

Solving for Rp_{equiv}

$$Rp_{\text{equiv}} \approx \frac{1}{Rp \cdot (2 \cdot \pi \cdot F \cdot C)^2} \qquad (2.7)$$

Higher Qs usually mean less loss in the circuit. From (2.6), it is obvious that Q degrades (1) as frequency increases and (2) as Rp_{equiv} increases or Rp decreases, as shown in (2.7).

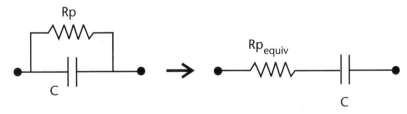

Figure 2.6   At any single frequency the parallel capacitor model can be converted to a series model.

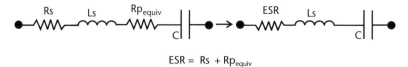

$$ESR = Rs + Rp_{equiv}$$

Figure 2.7 The narrowband equivalent circuit of a capacitor. The equivalent series resistance (ESR) equals the effective resistance of the parallel circuit (Rp_{equiv}) plus the conductor resistance (Rs).

Figure 2.7 is a narrowband equivalent circuit of a capacitor in which:

1. Rs is the resistance of the conductor leads emanating from the parallel plates. Rs increases with frequency due to the skin effect. The skin effect is a phenomenon in which the signal penetrates the conductor less at higher frequencies, effectively decreasing the conductor area (which increases the conductor resistance) with respect to the signal of interest.
2. Ls is the inductance of the conductor leads.
3. ESR is the effective series resistance, which is a function of the resistance of the dielectric material ($Rp \rightarrow Rp_{equiv}$) plus the conductor resistance (Rs).
4. C is the device capacitance calculated from the area of the parallel plates, the distance between the plates, and the relative dielectric constant of the material between the plates.

2.2.4 Capacitor Resonance

From the high-frequency model of the capacitor shown in Figure 2.7, it is seen that the capacitor (C) and inductance (Ls) form a resonant circuit with a resonant frequency (f_R) [5] given by

$$f_R = \frac{1}{2 \cdot \pi \cdot \sqrt{Ls \cdot C}} \tag{2.8}$$

where f_R is the resonant frequency (Hz), Ls is the series inductance in the capacitance model (H), and C is the capacitance (F).

Assume that the resistance ESR is small and the frequency of operation is much less than F_R the reactance of the capacitor (X_C) is

$$X_C = \frac{1}{2 \cdot \pi \cdot F \cdot C} \tag{2.9}$$

where F is the operating frequency.

As f approaches f_R, the reactance of the capacitor becomes

$$X = X_C - X_L \tag{2.10}$$

where

$$X_L = 2 \cdot \pi \cdot f \cdot Ls \qquad (2.11)$$

when the capacitor is used in a filter circuit where the actual value of the reaction is critical to the operation of the device must be at a frequency that ensures $X_L <<$ X_C or $f << f_R$, the resonant frequency.

If the capacitor is used in a bypass or coupling mode where minimum reactance (X) is the only critical criterion, the operating frequency (f) can go up to or sometimes a little greater than the resonant frequency (f_R).

2.3   Resistor Characteristics at High Frequencies

Resistors like other devices can be modeled to explain the devices behavior as the operating frequency increases. A resistor has many parasitic effects that influence its performance at high frequencies. Figure 2.8 shows the dominant parasitic model, an ideal resistance (R) in series with a parasitic inductance (Ls), and a parasitic parallel shunt capacitance (Cs) [6]. The inductance (Ls) is associated with distance between the resistor and the conductor leads attaching the unit to the substrate and capacitance (Cs) is created by the two conductor plates that are typically at the end of the resistor. Mounting the device on a substrate can introduce additional capacitance with respect to the body of the component and add additional series inductance due to the traces or leads used to attach the component. Lead resistance is usually neglected when compared with the device resistance (R). Factors that affect the magnitude of the parasitic elements $(Ls$ and $Cs)$ are the resistor base material, the length of the resistor, the width of the resistor, the distance from the resistor to the interconnection, and the size and proximity of the interconnection conductor. At RF and microwave frequencies, most resistors are surface mount devices as shown in Figure 2.9.

2.3.1   High-Frequency Surface Mount Resistors

The parasitic capacitance (Cs) and inductance (Ls) are functions of the mechanical dimensions of the device. Table 2.1 lists the dimensions and typical parasitic inductance and capacitance for a few standard size resistor surface mount packages, and Figure 2.10 shows the relationship between the parameters in Table 2.1 to the physical package and the high frequency model of the device [7].

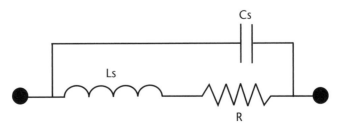

Figure 2.8 High-frequency model of a resistor.

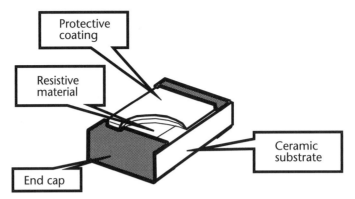

Figure 2.9   Surface mount resistor.

The parameters listed in Table 2.1 suggest that the smaller chip sizes perform better at higher frequencies.

Table 2.1   Typical Parasitic Inductance and Capacitance for Some Standard Surface Mount Packages

	Wrap Chips			Typical Performance		
Case Size	Length Inches	Width Inches		Capacitance Cs (pF)	Inductance Ls (nH)	Resonance GHz
0402	0.04	0.02		0.0392	0.1209	73.108
0603	0.06	0.03		0.25	0.85	10.918

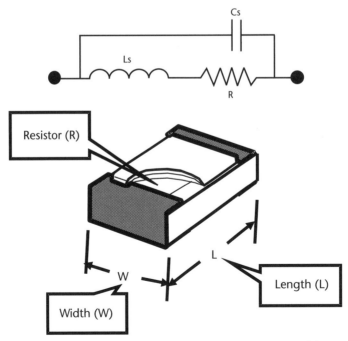

Figure 2.10   Standard chip resistor with its associated high-frequency model.

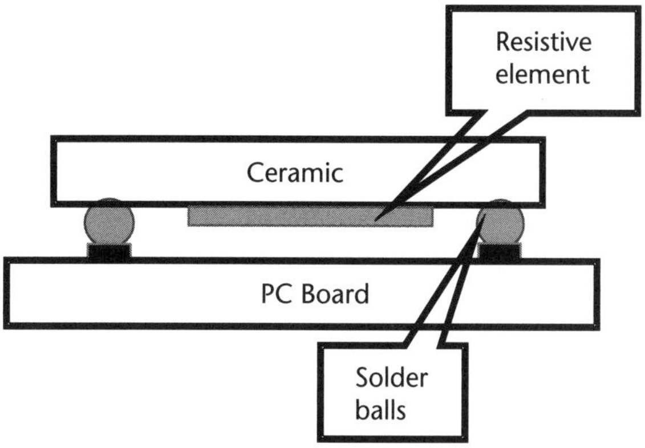

Figure 2.11   Flip-chip surface mount resistor.

2.3.2   Flip-Chip Surface Mount Resistors

The high-frequency performance of a surface mount resistor can be improved by inverting (flipping) the chip mounting. The flip-chip technique brings the resistive element closer to the microwave substrate reducing the parasitic effects (see Figure 2.11).

In the flip-chip configuration, the resistive element is below the ceramic substrate [8]. This configuration shortens the distance from the resistor to the printed circuit (PC) connection and lower value of the parasitic inductance (Ls) and allows a shorter interface connection. The solder balls instead of the end caps' lower parasitic capacitance (Cs). Table 2.2 lists some typical values of Ls and Cs for the same standard surface mount chip sizes as in Table 2.1 plus an additional smaller size resistor chip (0201). The two values of parasitic inductance (Ls) and capacitance (Cs) for chip size 0603 represent the data from different manufacturers and show the variability of the available parasitic data.

With regard to flip-chip devices, the flip-chip configuration utilizes the same model as the standard surface mount devices, shown in Figures 2.8 and 2.10. The parasitic inductance (Ls) and capacitance (Cs) are smaller than those of standard surface-mount devices. The resonance frequency of the flip-chip devices is much higher due to the lower parasitic inductance (Ls) and capacitance (Cs). Flip-chip technology utilizes blind connections (i.e., the connecting balls are below the device).

Table 2.2   Typical Parasitic Inductance and Capacitance for Some Flip-Chip Surface Mount Packages

Case	Flip Chips		Typical Performance		Resonance
	Length	Width	Capacitance	Inductance	
Size	Inches	Inches	Cs (pF)	Ls (nH)	GHz
0201	0.02	0.01	0.0206	1.73E-05	8430.697
0402	0.04	0.02	0.0262	1.89E-03	715.218
0603	0.06	0.03	0.0404	0.1209	72.014
0603	0.06	0.03	0.1	0.15	41.094

Inspection of the connection integrity is very difficult. Assembly of flip-chip devices can be more difficult than standard surface-mount devices.

2.3.3   Thick-Film and Thin-Film Surface-Mount Resistors

Surface-mount technologies are selected for their small size and minimal parasitic effects. Thick film and thin film are alternate surface-mount manufacturing techniques used to manufacture surface-mount resistors. They are commonly used in RF and microwave devices; see Figure 2.12 for a typical device cross section. Thick film is formed by screen printing metal particle resin composites and thin film is deposited by a vacuum process such as sputtering.

Thin film is generally superior in performance than thick film, and thick film is usually considered more reliable than thin film. See Table 2.3 for a comparison between the two types of resistor construction.

2.3.4   High-Frequency Effects of Thick-Film and Thin-Film Resistors

The parasitic elements (Ls and Cs) shown in Figure 2.8 are package-related but the resistive material (R) also has a frequency-related characteristic. As frequency increases, the electric field exhibits less penetration into a conductor (i.e., current flows nearer to the surface of the conductive material). This phenomenon is known as skin effect and it decreases the effective depth of the conductor decreasing the cross-sectional area by the signal which increases the effective conductor resistance. Thin-film resistors create the required resistance value with a very thin cross-sectional area. If the depth of the initial cross-sectional area is less than the skin depth at the maximum frequency of interest, the resistance value will be unaffected up to the frequency of interest because the field penetration into the conduction remains constant.

In Figure 2.13(a) the resistive material and the flow of the electromagnetic fields lines under low-frequency conditions are shown. The resistance of the device (R) is calculated from the material resistivity (R_{mat}) in ohm-meters; see (2.12).

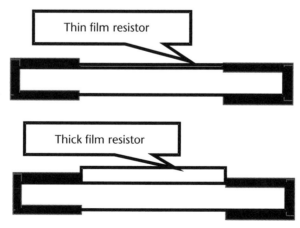

Figure 2.12   Cross section of a thick-film and thin-film surface mount resistor.

Table 2.3 A Comparison Between Thick-Film and Thin-Film Surface-Mount Resistors

Parameter	Thick Film	Thin Film
Cost	Lower	Higher
Handling power surges	Better	Worse
Tolerance	Worse	Better
Temperature effects	Worse	Better
High frequency parasitics	More	Less
Skin effect	Worse	Better
Thickness	Micrometers	Nanometers
Reliability	Better	Worse
Performance	Worse	Better

$$\text{Resistance} = R = R_{\text{mat}} \cdot \frac{\text{length}}{\text{area}} \tag{2.12}$$

As the frequency increases higher, the depth of signal penetration decreases, effectively decreasing the cross-sectional area and increasing the resistance of the device; see Figure 2.13(b). Under these conditions, the resistance increases as frequency increases.

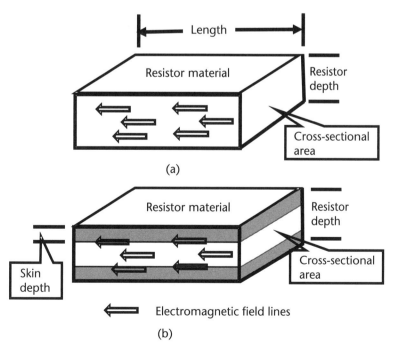

Figure 2.13 (a, b) Thick-film resistor model showing the effects of skin depth.

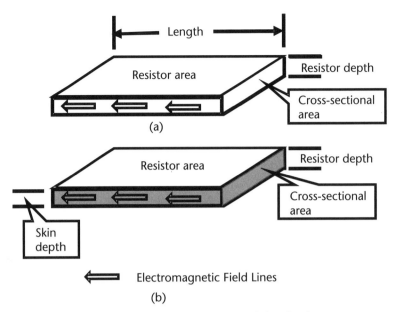

Figure 2.14 Thin-film resistor model showing the effects of skin depth.

Figure 2.14 depicts a thin-film resistor where the calculated depth of the cross-sectional area is muchly smaller than that of the thick-film device, such that at the same frequency depicted in Figure 2.13 the depth of signal penetration is less than the thin-film resistor device depth. Under these conditions, the effective resistance is unchanged up to the frequency of interest.

2.3.5   Notes on Thin-Film Resistors

Thin films can produce a higher resistance per given area. The stability of the thin film is affected by elevated temperatures. The thin-film aging and stabilization process varies depending upon the film thickness and, therefore, is variable for different values of resistance. Thin-film resistors are more susceptible to self-etching in the presence of moisture because of the small mass of metal. A high-value thin-film resistor has greater deterioration rates due to the thinner deposition.

2.3.6   Notes on Thick-Film Resistors

Thick film relies on particle-to-particle contact. The contact points can be interrupted by thermal strain during service. Because there are many parallel paths, the resistor does not open, but continually increases in value with time and temperature. Thick films are less stable (time, temperature, and power) than other resistor technologies. The granular structure used in thick-film manufacturing accounts for the high noise due to the bunch-and-release electron charge movement through the structure. The higher the resistance value for a given size, the less the metal content, the greater the noise, and the less stability. The glass content of the

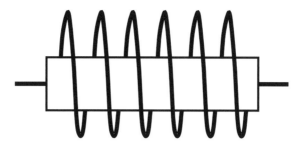

Figure 2.15   Typical inductor

thick-film structure gives the thick-film resistor greater moisture resistance than the thin-film resistor.

2.4   Inductors

The basic inductor is a coil of wire, but there are many other construction techniques each accomplishing a goal related to a specific application. At lower frequencies, inductors may be wound on cores made of nonmagnetic materials or with materials that have magnetic properties, such as ferrites (see Figure 2.15).

2.4.1   Calculating Inductance of a Cylindrical Coil of Wire

The inductance of a cylindrical coil of wire is shown in Figure 2.16. The coil has N turns closely wound around diameter D. The wire size is such that N times the diameter of the wire is length l. The inductance is found using (2.13).

$$L = \frac{N^2 \mu A}{\ell} = \frac{N^2 \mu_r \mu_0 A}{\ell} \text{ and } l \approx N \cdot d_l \text{ (coil is closely wound)} \quad (2.13)$$

where L is the inductance (henry), N is the number of turns, A is the area as shown in Figure 2.16 (meters squared), l is the length of the coil (meters), d_l is the diameter of the wire, μ_r is the relative permeability ($\mu_r = 1$ for free space), and μ_0 is the permeability of free space ($4 \cdot \pi \cdot 10^{-7}$ H/m).

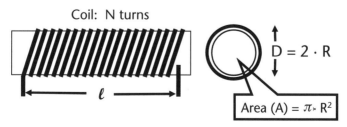

Figure 2.16   Cylindrical coil.

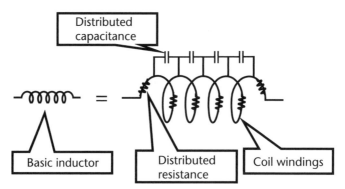

Figure 2.17   The high-frequency parasitic effects of an inductor.

2.4.2   Inductors at High Frequencies

At microwave frequencies, the inductor has many parasitic effects that must be considered to better understand its performance in a circuit design [9]. The wire creating the inductor introduces a series distributed resistance, and the individual windings of the coil in such proximity to one another introduce a parallel distributed capacitance modeled as shown in Figure 2.17.

The high-frequency circuit model of the inductor in Figure 2.18 shows the distributed resistance (Rs) in series and the distributed capacitance (Cp) in parallel with the ideal inductor (L).

The inductor wire losses are much higher than the dielectric losses in a capacitor so the Q of the inductor is usually significantly lower than that of a capacitor. Equation (2.14) is used to calculate the unloaded Q (Qu) of the inductor. The unloaded Q (Qu) of a device is the Q not affected by a surrounding circuit.

$$Qu = \frac{X_l}{Rs} = \frac{2 \cdot \pi \cdot F \cdot L}{Rs} \tag{2.14}$$

where X_l is the reactance of the inductor (ohms), Rs is the parasitic distributed series resistance (ohms), F is the signal frequency (hertz), and L is the inductance of the induction (henries).

Equation (2.14) suggests that the unloaded Q (Qu) increases with frequency, but in reality the series resistance Rs also increases with frequency due to the skin effect phenomenon cancelling the effective increase in the inductor reactance (X_l).

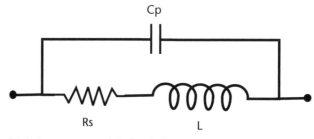

Figure 2.18   The high-frequency model of an inductor.

2.4.3 Inductors at Resonance

To simplify the high-frequency analysis of the inductor, the physical equivalent circuit of the inductor shown in Figure 2.19(a) can be represented as a parallel resonant circuit as shown in Figure 2.19(b) where the value of parallel resistance (Rp) is a function of frequency. The basis for the circuit conversion of an inductor (L) and a series resistor (Rs) from Figure 2.19(a) to the equivalent circuit in Figure 2.19(b) is that the Q of the series circuit (Qs) shown in Figure 2.19(a) equals the Q of the parallel circuit (Qp) shown in Figure 2.19(b).

The value of resistor Rp the resistance of the equivalent circuit at resonance is found in (2.15).

$$Qs = \frac{2 \cdot \pi \cdot F \cdot L}{Rs} \approx Qp = \frac{Rp}{2 \cdot \pi \cdot F \cdot L} \quad \text{for } Qs \text{ and } Qp > 10 \quad (2.15)$$

Solving for Rp

$$Rp \approx \frac{(2 \cdot \pi \cdot F \cdot L)^2}{Rs} \quad (2.16)$$

It is apparent from the analysis of the equivalent circuit in Figure 2.19(b) that the magnitude of the impedance of the parallel resonant circuit is at a maximum when the inductive reactance (X_l) is equals reactance of the parasitic capacitance (Xc). The frequency where the positive reactance of the inductor (L) equals the negative reactance of the parasitic capacitor (Cp) is called the self-resonant frequency (Fp). At this frequency, the impedance is at a maximum equal to Rp. The value of the self-resonant frequency is given by (2.17)

$$Fp \approx \frac{1}{2 \cdot \pi \cdot \sqrt{L \cdot Cp}} \quad (2.17)$$

where L is the inductance of the induction (henries), Cp is the parasitic capacitance (farads), and Fp is the resonant frequency (hertz).

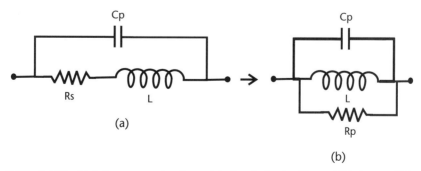

Figure 2.19 (a) The high frequency circuit model of an inductor (L), and (b) the parallel equivalent high frequency circuit model of an inductor (L) where the parallel resistance (Rp) is a function of frequency.

When an inductor is used as a choke (an RF high impedance device) higher frequency operation, higher up the self-resonant frequency (Fp) is acceptable, but when the inductor is used for circuit matching or as part of a filter, the maximum operating frequency should be significantly below the self-resonant frequency (Fp).

2.4.4   Inductance of a Straight Wire

Every length of wire has an associated inductance. The inductance of a straight wire [10] is given in (2.18).

$$L = \left[2 \cdot l \cdot \left\{ \ln\left(\left(\frac{2 \cdot l}{d} \right) \cdot [1 + k] \right) - k + \frac{d}{2 \cdot l} \right\} + \frac{l}{2} \right] \cdot 10^{-3} \cdot \frac{\mu H}{cm} \qquad (2.18)$$

$$k = \sqrt{1 + \left(\frac{d}{2 \cdot l} \right)^2} \qquad (2.19)$$

where l is the length of the wire (cm), d is the diameter of the wire (cm), and L is the inductance (μH).

Example:

d = 0.051 cm (wire size AWG 16);

l = 2.54 cm (1 inch);

L = 0.023 μH.

2.4.5   Planar Spiral Inductors

Designing components into the printed circuit substrate reduces costs and improves performance because it reduces parts count and device interconnections. Small-value inductors such as planar spirals shown in Figure 2.20 can be reliably etched directly onto the substrate. The basic structure is a circular spiral, but straight line segments are often used to emulate that configuration [11]. Figure 2.20 shows a circular spiral, a spiral using a square topology, and a spiral using a hexagonal topology. A hexagonal configuration (not shown) is also commonly used as a planar spiral inductor. The simplest configuration to draw is the square and therefore used most often even though higher-order polygons and ultimately the circular spiral configurations usually exhibit a better defined performance.

For each topology, the inductor is specified by the number of turns (N), the conductor width (W), the spacing between turns (S), and the inner diameter (D_{in}) or the outer diameter (D_{out}). The outer diameter (D_{out}) is:

$$D_{out} = D_{in} + 2 \cdot N \cdot W + 2 \cdot (N - 1) \cdot S \qquad (2.20)$$

Knowing the outer diameter, the inner diameter can be calculated as shown in (2.21).

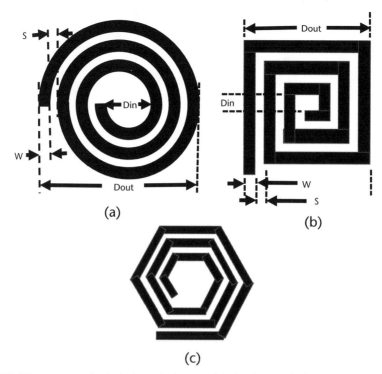

Figure 2.20 Three types of spiral planar inductors: (a) circular spiral, (b) square spiral, and (c) hexagonal.

$$D_{in} = D_{out} - (2 \cdot N \cdot W + 2 \cdot (N - 1) \cdot S) \tag{2.21}$$

Two dependent factors used in the calculation of inductance are the average diameter (D_{ave}) and the fill factor (F_{fill}), which can be calculated from the parameters in (2.20) and (2.21). The conductor thickness can usually be neglected because the skin depth at higher frequencies is usually less than the conductor thickness (t).

$$D_{ave} = \frac{D_{out} + D_{in}}{2} \tag{2.22}$$

$$F_{fill} = \frac{D_{out} - D_{in}}{D_{out} + D_{in}} \qquad D_{in} \geq S \tag{2.23}$$

The analysis of the inductance of a spiral conductor is very complex and best determined using an electromagnetic simulation program. There are several closed-form equations that give accuracies better than 15% considering all of the substrate and manufacturing variables.

2.4.5.1   Current Sheet Approximation of Planar Spiral Inductors

One method calculating the value of a planar spiral inductor (2.24) is what is commonly called the current sheet approximation.

Table 2.4 Spiral Inductor Evaluation Constants for the Current Sheet Approximation

Spiral Topology	K1	K2	K3	K4
Square	1.27	2.07	0.18	0.13
Hexagonal	1.09	2.23	0.00	0.17
Octagonal	1.07	2.29	0.00	0.19
Circle	1.00	2.46	0.00	0.20

Note: The accuracy of (2.24) degrades as the ratio of conductor spacing to conductor width (S/W) increases. Typically, spiral inductors are built with spacing (S) divided by width (W) less than 3 ($S/W \le 3$).

$$L = \frac{\mu_0 \cdot N^2 \cdot D_{ave} \cdot K1}{2} \cdot \left\{ \ln\left(\frac{K2}{F_{fill}}\right) + K3 \cdot F_{fill} + K4 \cdot F_{fill}^2 \right\} \qquad (2.24)$$

where L is the inductance in henries, μ_0 is the permeability constant $= 4 \cdot \pi \cdot 10^{-6}$ henries/meter, N is the number of turns in the spiral, D_{ave} and F_{fill} are given by (2.22) and (2.23), respectively, and K1, K2, K3, and K4 are constants related to the spiral topology, given in Table 2.4.

2.4.5.2 Example Using the Current Sheet Approximation for the Square Topology

Assume the following baseline conditions: $S = 75 \ \mu m$, $W = 100 \ \mu m$, $N = 3$, and $D_{out} = 1,200 \ \mu m$. The calculated values are: $D_{in} = 300 \ \mu m$, $D_{ave} = 750 \ \mu m$, and $F_{fill} = 0.6$. Using the coefficients for the square topology, L = 7.504 nH. Using the coefficients for the hexagonal topology, L = 6.352 nH.

Note that the current sheet approximation for a square planar spiral inductor only considers the mutual inductance between opposite sides of the square since the adjacent sides are orthogonal.

2.4.5.3 Comparison of Planar Spiral Induction Calculation Methods and Topologies

As a reference to possible method of calculation errors, Table 2.5 lists results using different methods and different topologies.

As expected, the results using the square topology gives the largest error. Square topology is used because it is the easiest to lay out on a printed circuit board (PCB).

2.4.6 Conical Inductors

Conical inductors are designed for broadband and high frequency applications. They have a geometry that broadens the bandwidth of the coil and, because of their small size and fine wire, keeps the parasitic capacitance is low. Their main application is

Table 2.5 Evaluation of Inductance Using the Same Parameters (D_{in}, D_{out}, S, N, and W) Applied to Different Calculation Methods and Different Topologies

Analysis Technique	Topology			
	Square	Hexagonal	Octagonal	Circular
Modified wheeler	7.49 nH	6.004 nH	6.098 nH	6.161 nH
Current sheet	7.504 nH	6.352 nH	6.389 nH	6.29 nH
Monomial fit	7.636 nH	6.238 nH	6.332 nH	6.332 nH

for use in DC decoupling applications, isolating the power source from the active device. The topology of a conical coil is essentially a three-dimensional spiral (see Figure 2.21), which has an inherent high resonant frequency.

2.4.7 Inductance of Via Holes

Via holes are vertical connections between layers of a printed circuit board [12]. They are made by drilling a small hole and lining it or filling it with conductive material (see Figure 2.22).

Between the top and bottom of the via holes, there is an effective inductance [13] approximately conforming to the results of (2.25).

$$L_{via} = \frac{\mu_0}{2 \cdot \pi} \cdot \left\{ h \cdot \ln\left(\frac{h + \sqrt{r^2 + h^2}}{r} \right) + \frac{3}{2} \cdot \left(r - \sqrt{r^2 + h^2} \right) \right\} \qquad (2.25)$$

The diameter of the via hole (D) and the height of the via hole (h) are shown in Figure 2.23. The radius of the via hole (r) is the diameter divided by two (r = D/2).

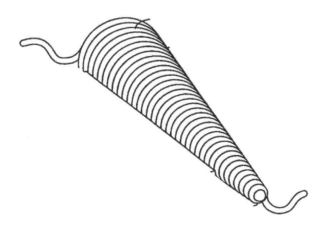

Figure 2.21 Conical coil.

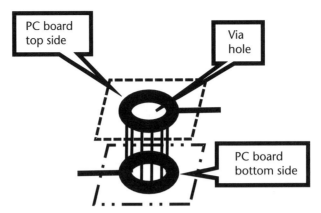

Figure 2.22 Via hole.

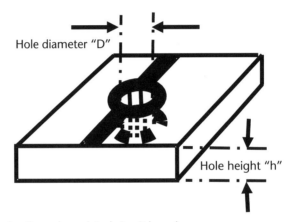

Figure 2.23 Via holes through a printed circuit board.

2.4.7.1 Example: Calculation of the Inductance of a Via Hole

Assume:

$h = 500\ \mu m\ (500 \cdot 10^{-6}\ \text{meters}) = 0.019685\ \text{inch.}$

$D = 250\ \mu m\ (250 \cdot 10^{-6}\ \text{meters}) = 0.009843\ \text{inch.}$

$r = D/2 = 125\ \mu m\ (125 \cdot 10^{-6}\ \text{meters}) = 0.0049215\ \text{inch.}$

L_{via} is the effective inductance of a via hole.

$L_{via} = 9.235 \cdot 10^{-11}\ \text{henries} = 92.35\ \text{picohenries} = 92.35\ \text{pH.}$

2.4.8 Inductance of Bond Wire

Wire bonding is typically used as a connection between an integrated circuit (IC) die and its substrate but can be used to connect a module to a printed circuit board (PCB). Bond wires are usually made of gold, but sometimes aluminum and copper are used. The inductance of a straight bond wire [14] is:

$$L = \frac{\mu_0}{2 \cdot \pi} \cdot l \cdot \left[\ln\left(\frac{1}{r} + \sqrt{1 + \frac{l^2}{r^2}} \right) - \sqrt{1 + \frac{r^2}{l^2}} + \frac{r}{l} + \frac{1}{4} \right] \tag{2.26}$$

where r is the bond wire radius, l is bond wire length, and μ_0 is the permeability of free space ($1.2566 \cdot 10^{-6}$ H/m)

For the bond wire length (l) >> the bond wire radius (r),

$$L \approx \frac{\mu_0}{2 \cdot \pi} \cdot l \cdot \left[\ln\left(\frac{2 \cdot l}{r} \right) - 0.75 \right] \quad \text{for } l >> r \tag{2.27}$$

2.4.9   Inductance of Flat or Ribbon Wire

Ribbon connections have a certain advantage over round wire connections. Thin ribbons have of the relatively larger surface area for the same cross-sectional area of circular wire. At a high frequency, the loss is related to surface area and therefore connections could have a lower loss [15].

The inductance of a wire with a rectangular cross section (flat or ribbon wire) is found using (2.28).

$$LR(\text{nH}) = 5.08 \cdot 10^{-3} \cdot L \cdot \left[2.303 \cdot \text{Log}\left(\frac{2 \cdot L}{W + T} \right) + .5 + .2235 \cdot \frac{W + T}{L} \right] \tag{2.28}$$

where L is the ribbon length (inches), W is the ribbon width (inches), and T is the ribbon thickness (inches).

References

[1] Voldman, J., "Lumped-Element Modeling with Equivalent Circuits," Massachusetts Institute of Technology, Lecture 8, Spring 2007.

[2] Rahulkumar, S., "Measurement and Modeling of Passive Surface Mount Devices on FR4 Substrates," Portland State University, Dissertations and Theses, Summer 2012.

[3] Schmitz, T., and M. Wong, "Choosing and Using Bypass Capacitors," Intersil Application note 1325, 2011.

[4] Agilent, *Impedance Measurement Handbook, A guide to measurement technology and techniques, 4th ed.*, 2013.

[5] Niknejad, A., and J. Dunsmore, *High Frequency Passive Components*, University of California, Berkeley, 2013.

[6] Vishay Intertechnology, Inc., "Frequency Response of Thin Film Chip Resistors," Technical Note, 2009.

[7] Aeroflex/KDI-Resistor Products, Surface Mounted Terminations, Resistors and Attenuators, Application Note.

[8] Seams, J., "All Terminators Are Not Created Equal," *RF Design*, November 2005.

[9] Breed, G., "Fundamentals of Passive Component Behavior at High Frequencies," *High Frequency Electronics*, June 2006.

[10] Patel, P., "Calculation of Total Inductance of a Straight Conductor of Finite Length," *Physics Education*, July-September 2009.

[11] Mohan, S., et al., "Simple Accurate Expressions for Planar Spiral Inductances," *IEEE Journal of Solid-State Circuits*, Vol. 34, No. 10, October 1999.

[12] LaMeres, B., "Characterization of a Printed Circuit Board Via," Montana State University, Technical Report, 2000.

[13] Jahn, S., "Microstrip Via Hole," 2007 from IEEE paper "Modeling Via Grounds in Microstrip," *IEEE Microwave and Guided Wave Letters*, Vol. 1, No. 6, June 1991.

[14] Berkner, J., and K. Pieper, "Bond Wire Inductance Calculation and Measurement," Infineon Technologies, 2014.

[15] Gilardoni, R., "Ribbon Bonding for High Frequency Applications Advantages of Ribbon and the Impact on the Microwave Market," Hesse & Knipps Semiconductor GmbH, Germany, .

CHAPTER 3

Transmission Lines

3.1 Introduction to Transmission Line Theory

An electromagnetic field is a combination of electric and magnetic force vectors containing a defined amount of energy. In a microwave circuit, this energy is guided from device to device in a transmission line. Transmission line theory deals with the problem of signal transfer between components in a circuit when the phase of the signal at the source is not the same as the phase of the signal at the destination due to the frequency of the signal and the path delay. Theoretically, this condition always exists, but when the distance (d) between at the source and destination are close enough as an example when the distance $d << \lambda$ where λ is the wavelength of the signal, the circuit can be analyzed using the lumped circuit theory, which assumes that the distance between components is approximately zero. Lumped circuit theory is a special case of transmission line theory [1].

An important factor in the design of transmission lines is to minimize the loss between devices. The primary loss factors are energy dissipated in the transmission line (i.e., resistive losses), energy radiated off the transmission line, and energy reflected from the destination load.

3.2 Common Transmission Line Topologies

At microwave frequencies, power in a circuit is transmitted through electric and magnetic fields guided in a physical structure, a transmission line [2] (see Figure 3.1).

There are many types of transmission lines, each configuration developed to satisfy a set of desirable characteristics such as impedance, insertion loss, cost, signal operating frequency, and mechanical design. Some common transmission line configurations are shown in Figure 3.2.

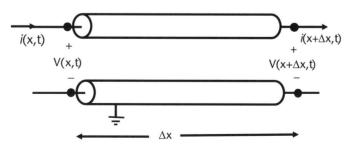

Figure 3.1 Transmission line guiding electromagnetic energy.

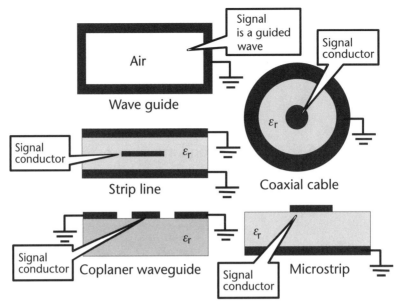

Figure 3.2 Some commonly used transmission lines.

All the transmission lines shown are filled with a dielectric material with relative dielectric constant ε_r except the waveguide, which is usually not filled with any material. Waveguide with no dielectric medium is the lowest loss transmission line available. A vacuum in free space is theoretically the lowest loss medium, but in free space the wave being unguided is dispersed and cannot be practically completely collected. Each type of transmission line has characteristics that make it useful for selected applications. Table 3.1 lists some particularly useful properties for each transmission line configuration [3] (shown in Figure 3.2).

Table 3.1 Advantages and Disadvantages of Some Common Transmission Line Topologies

Transmission Line Configuration	Advantage	Disadvantage
Waveguide	Low loss	Rigid: reconfiguring is very difficult
Coaxial cable	Flexible structure	Higher loss than wave guide
Stripline	Fields are confined between the ground planes	Difficult to tune, not convenient to mount discrete devices, used mainly in passive device applications
Coplanar waveguide	Ground plane is on top of the printed circuit board making an easier interface to coaxial connectors	Greater loss; radiates more than microstrip
Microstrip	Easy to tune, suitable to mount hybrid and monolithic circuits	Some signals radiate into the air

3.3   Transmission Line Characteristics Using Lumped Circuit Elements

Modeling transmission lines as a distributed circuit gives insight into how a signal is affected traveling between devices. The model uses a network with a distributed series resistance, a distributed series inductance, a distributed shunt (parallel) capacitance, and a distributed shunt (parallel) conductance.

3.3.1   Distributed Lumped Constant Model

Figure 3.3 shows the distributed passive components modeling the transmission line. The passive elements, $R(\Delta x)$, $L(\Delta x)$, $G(\Delta x)$, and $C(\Delta x)$, are not necessarily a linear function of distance (Δx) [1]. The series and shunt distributed passive components modeling a transmission line are a function of distance (Δx) and the signals, current and voltage, are a function of distance and time (Δx, Δt), all with respect to the initial condition at point (x, t).

3.3.2   Modeling a Microstrip Transmission Line with Distributed Lumped Elements

The distributed passive network elements in the Figure 3.3 model most transmission line topologies [4] as well as the microstrip topology shown in Figure 3.4. The microstrip transmission line shown has a length x, width w, and thickness t. The

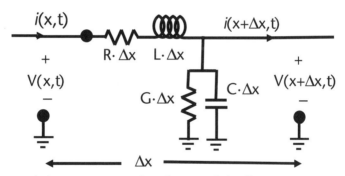

Figure 3.3   Lumped-element representation of a transmission line.

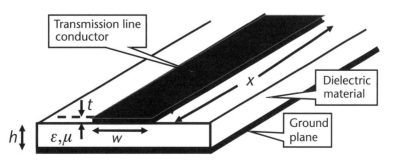

Figure 3.4   A typical microstrip transmission line.

conductor is sitting on a dielectric material width, a permeability μ, a permittivity (dielectric constant) ε, and loss tangent δ [tan(δ) ≈ δ]. The transmission line conductor is height h above the ground plane.

3.3.2.1 Distributed Series Components: Resistance *R* and Inductance *L*

R is the series resistance which is a function of length, in Ω/(unit length), determined by the resistance of the conductor material, the conductor width (w), the thickness of the conductor (t), and the signal frequency. The resistance R increases as frequency increases.

L is the series inductance which is a function of length, in nH/(unit length), affected by the width of the conductor (w), the height of the conductor above the ground plane (h), and the permeability constant (μ) of the conducting material.

3.3.2.2 Distributed Shunt Components: Conductance *G* and Capacitance *C*

G is the shunt conductance which is a function of length, in S/(unit length), ($S = 1/\Omega$), determined by the conductor width (w), the conductor height (h) above the ground plane (the height of the dielectric material), the permittivity (ε) of the dielectric material, and the loss tangent of the dielectric material (tan [δ]). The inverse of conductance ($1/G$) is the parallel resistance (Rp in ohms · [unit length]).

C is the shunt capacitance that is a function of length, in pF/(unit length), determined by the width of the conductor (w), the permittivity of the dielectric material (ε), and the thickness of the dielectric material (h).

3.3.3 Characteristic Impedance of Transmission Line from the Lumped Circuit Model

The characteristic impedance (Z_0) of a transmission line is the ratio of the voltage (V^+) divided by the current (I^+) when the voltage and current are in the forward direction [designated by the plus sign (+) in the exponent], transformed into the frequency domain; see (3.1).

$$Z_0(j\omega) = \frac{V^+(j\omega)}{I^+(j\omega)} \tag{3.1}$$

Equation (3.1) assumes there is a single sinusoidal wave propagating along an infinite line with no discontinuities (i.e., the electromagnetic wave is traveling in one direction, has no reflections, and is independent of the transmission line length).

The characteristic impedance (Z_0) is determined by the geometry and materials of the transmission line and approximated using distributed lumped constant passive elements. Equation (3.2) is the transmission line characteristic impedance from the schematic in Figure 3.3.

The characteristic impedance of the transmission line [5] is:

$$Z_0 = \sqrt{\frac{(R + j \cdot \omega \cdot L)}{(G + j \cdot \omega \cdot C)}} \tag{3.2}$$

where Z_0 is the characteristic impedance of the transmission line: ohms (Ω), ω is the signal frequency in radians/second: $\omega = 2 \cdot \pi \cdot f$ (f is the signal frequency in hertz), R is the distributed series resistance [$R(\Delta x)$] in Figure 3.3, L is the distributed series inductance [$L(\Delta x)$] in Figure 3.3, C is the distributed parallel capacitance [$C(\Delta x)$] in Figure 3.3, and G is the distributed parallel conductance [$G(\Delta x)$] in Figure 3.3.

3.4   Lossless Transmission Line

An ideal lossless transmission line has only reactive components:

Series resistance $R \rightarrow 0$ or ($j \cdot \omega \cdot L >> R$);
Parallel resistance ($Rp = 1/G$) goes to infinity;
$Rp \rightarrow \infty$;
$G \rightarrow 0$ or ($j \cdot \omega \cdot C >> G$).

Under these lossless conditions, ($j \cdot \omega \cdot L >> R$ and $j \cdot \omega \cdot C >> G$):

$$Z_0 = \sqrt{\frac{(R + j \cdot \omega \cdot L)}{(G + j \cdot \omega \cdot C)}} \approx \sqrt{\frac{(j \cdot \omega \cdot L)}{(j \cdot \omega \cdot C)}} = \sqrt{\frac{L}{C}} \qquad (3.3)$$

Equation (3.3) describes an impedance Z_0 that has only real parts since the imaginary terms and the frequency-dependent terms canceled out.

3.5   Characteristics of a Signal Traveling Through an Infinite Transmission Line

Figure 3.5 shows an electromagnetic wave traveling through an infinite transmission line. The wave is driven from a source (V_0) with an impedance Z_0 at point $x = 0$. The transmission line is modeled as having an infinite length to limit the discussion to a forward wave with no reflections.

The signal traveling through the transmission line in Figure 3.5 modeled as shown in Figure 3.3 can be described by (3.4).

$$V^+(x,t) = V_0(t) \cdot e^{-\gamma \cdot x} \qquad (3.4)$$

where V_0 is the signal at $x = 0$ and is a function of time [$V_0(t)$], V^+ is the voltage wave in the positive x direction, V^+ is a function of time and distance, [$V^+(x, t)$], x

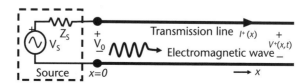

Figure 3.5   An electromagnetic wave in a transmission line traveling in the positive x direction.

is the distance that the wave traveled with respect to the start of the transmission line, $x = 0$, and γ is the propagation constant in (3.4) and (3.5). In general, γ has real and imaginary parts.

$$\gamma = \sqrt{(R + j \cdot \omega \cdot L) \cdot (G + j \cdot \omega \cdot C)} \tag{3.5}$$

L, R, C, and G are distributed lumped element parameters modeling the transmission line. The other parameters j and ω are: j is an imaginary number constant $j = \sqrt{-1}$ and ω is the signal frequency in radians per second, where f is the frequency in hertz and $\omega = 2 \cdot \pi \cdot f$. Equation (3.5) can be expanded into two factors, a real part (α) and an imaginary part (β) [6].

$$\gamma = \alpha + j \cdot \beta \tag{3.6}$$

where

$$V^+ = V_0 \cdot e^{-\gamma \cdot x} = V_0 \cdot e^{-(\alpha + j \cdot \beta) \cdot x} = V_0 \cdot e^{-\alpha \cdot x} \cdot e^{-j \cdot \beta \cdot x} \tag{3.7}$$

and V^+ is the voltage wave traveling down an infinitely long transmission line in the x direction (i.e., no reflections) [7].

Solving and redistributing (3.7),

$$V^+(x,t) = V_0 \cdot e^{-\alpha \cdot x} \cdot \cos(\omega \cdot t - \beta \cdot x + \Phi) \tag{3.8}$$

where V_0 is the peak value of the wave, $V_0 \cdot \cos(\omega \cdot t + \Phi)$ is the sinusoidal wave function with an arbitrary starting phase Φ, α is the attenuation constant, x is the distance that the voltage wave travels down the transmission line, and β is the delay function in radian per unit distance.

3.5.1 Attenuation Constant α

$e^{-\alpha \cdot x}$ models the signal attenuation (signal amplitude decreases exponentially as x increases). The value of α relates to:

- The loss tangent of the dielectric material: Tan $(\delta) \approx G/(\omega \cdot C)$ [G and C are distributed conductions and distributed capacitance, respectively, and ω is the radian frequency of the signal ($\omega = 2 \cdot \pi \cdot f$, where f is the signal frequency in hertz)].
- Conductor loss R.

In Figure 3.6, V_0 is a sinusoidal wave propagating down a transmission line shown at a distance x where x is increasing and the attenuation increasing at a rate $e^{-\alpha \cdot x}$.

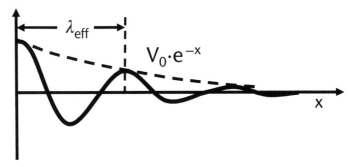

Figure 3.6 Showing sinusoidal signal propagating down a transmission line at distance x. The attenuation in this figure is over a very long distance to show the exponential effects.

When α is given in nepers/meter, the attenuation in decibels (dB) per meter is calculated in (3.9):

$$\text{Attenuation} \left(\frac{\text{dB}}{\text{meter}} \right) = -20 \cdot \text{Log}_{10} \left(e^{-\alpha \cdot (x=1)} \right) = 8.686 \cdot \alpha = A_{\text{dB}} \qquad (3.9)$$

$$\alpha = \frac{A_{\text{dB}}}{8.686} \qquad (3.10)$$

α or A_{dB} is per-unit-length constants not necessarily in meters. The length units in the value of x must be consistent with the length units of α and A_{dB}.

3.5.2 Phase Constant β

The constant β is a function of the signal wavelength (λ_s); see (3.11):

$$\beta = \frac{2 \cdot \pi}{\lambda_s} \qquad (3.11)$$

The $\beta \cdot x$ term in the sinusoidal part of the wave equation (cos ($\omega \cdot t - \beta \cdot x + \Phi$)) represents the phase shift in radians when x is expressed in terms the number of wavelengths (λ_s). For example, if $x = 0.5$ wavelength, ($\beta \cdot x$) is π radians $\rightarrow 180°$. Therefore, the wavelength shifted $180°$ for each distance x traveled.

With regard to the lumped circuit model, the lumped circuit model in Figure 3.3 does not fully describe the microstrip transmission line in Figure 3.4. The actual characteristics are more complicated when higher-order effects are considered. A level of accuracy required by most designs requires considering higher-order effects (i.e., using a more complex model). The performance of a microstrip transmission line is best determined by sophisticated microwave simulation programs.

3.6   50Ω Transmission Lines

Critical to interconnecting microwave devices with maximum power transfer is matching the impedance of the devices to the impedance of the transmission line. Most microwave devices are designed with 50Ω input and output impedances.

Fifty ohms (50Ω) was established as a quasi-standard for microwave systems as a compromise between minimum loss (77Ω) and maximum power handling (30Ω) for air dielectric coaxial cables. These cables were used extensively in multikilowatt radar transmitters around the World War II time frame; see Figure 3.7.

Designing 50Ω impedance transmission lines is important to connect microwave devices, but that is not the only application. Transmission lines other than 50Ω at specific lengths are used to design passive microwave devices such as power dividers, transformers, and filters. When designing these devices, the lengths are usually expressed as fractions of wavelength (λ) (i.e., $\lambda/2$, $\lambda/8$, and so forth) where the wavelength refers to the designed frequency for optimal operation.

3.7   Example of a Passive Microwave Circuit Using Transmission Lines at Different Impedances: Wilkinson Power Divider

The Wilkinson power divider splits an input signal into two identical outputs ideally (no loss), each having an output level at half the input signal level (−3 dB). All input and output ports are matched to the characteristic impedance (Z_0) of the input and output transmission lines [8].

This passive microwave device is commonly etched directly on a microstrip substrate schematically shown in Figure 3.8. The power divider requires different impedances at specific wavelengths, where the wavelength is related to the signal frequency (f) of optimum performance.

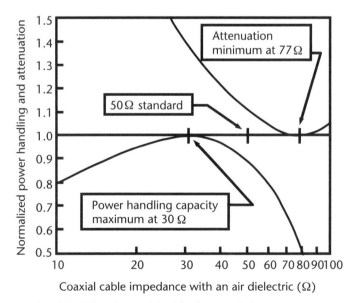

Figure 3.7   Normalized air dielectric coaxial cable characteristics.

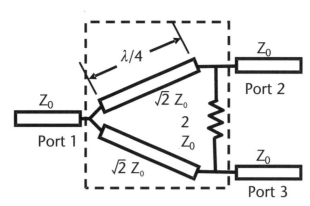

Figure 3.8 Wilkinson power divider.

The power divider structure consists of two quarter-wave ($\lambda/4$) transmission lines attached at the input port 1. Each transmission line has an impedance equal to $\sqrt{2} \cdot Z_0$, terminating at the output ports 2 and 3. The output ports are connected together through a $2 \cdot Z_0$ resistor. The input at port 1 and two equal outputs at ports 2 and 3 each matched to Z_0 optimized at frequency (f) equal to the velocity of the electromagnetic signal (v) divided by the wavelength of that signal (λ). The characteristic impedance of the interconnecting transmission line (Z_0) is typically 50Ω ($Z_0 = 50\Omega$) and it operation depends on the each of the outputs terminated in 50Ω.

The Wilkinson power divider is a bidirectional device, it can be used as a power combiner with the inputs at ports 2 and 3 and the combined output at port 1 [9]. If the input signals at ports 2 and 3 are coherent, have matched output impedances (Z_0), and are in phase, the output at port 1 into a characteristic load impedance (Z_0) will be lossless (assuming lossless transmission lines).

References

[1]   Adam, S., *Microwave Theory and Applications*, Agilent Technologies, Inc.

[2]   Pozar, D., *Microwave Engineering*, 4th ed., New York: John Wiley & Sons, 2011.

[3]   Farahmand, F., "Introduction to Transmission Lines," Lecture slides, University of California, Santa Barbara, 2012.

[4]   El-Banna, A., "Microstrip Lines," Lecture slides, Benha University, 2014.

[5]   Amanogawa, "Transmission Line Equations," Digital Maestro Series, 2006.

[6]   Niknejad, A., Lecture 6: Lossy Transmission Lines and the Smith Chart, University of California, Berkeley.

[7]   Wu, T., "Transmission Line Basics," Lecture slides, Nanyang Technological University, Singapore, 2012.

[8]   Berens, L., "Design, Analysis, and Construction of an Equal Split Wilkinson Power Divider," Marquette University, 2012.

[9]   Grebennikov, A., "Power Combiners, Impedance Transformers and Directional Couplers: Part II," *High Frequency Electronics*, January 2008.

CHAPTER 4

S-Parameters

4.1 Introduction

The expected performance of a linear two-port microwave device can be determined by measuring the devices' response to signals applied to the input and output ports, characterizing the two-port network without knowledge of the internal circuitry. There are many two-port network parameters that will accomplish this task, but most measure voltage and current requiring measurements to be taken with one port shorted (voltage is equal to zero) or one port looking into an open circuit (current is equal to zero). At RF and microwave frequencies, creating an ideal short-circuit is difficult because the distance between the network and the short would have to be so close so that the phase change is insignificant. This criterion is very hard to implement, especially as the frequency increases. The other issue is creating an open circuit because microwave signals tend to radiate out of an open port, creating a current when zero current is necessary to measure the intended parameter.

With respect to microwave devices, the preferred two-port network parameters are called scattering parameters, or S-parameters [1]. S-parameters have the most success accomplishing the goal of measuring the required two-port parameters because they utilize the properties of electromagnetic waves traveling through a transmission line with a defined source and load impedance (Z_0). When the wave encounters a device that could be as simple as a disturbance in the transmission path, a passive device such as a filter, an active device such as an amplifier, or a complex device such as a multifunction microwave network, a portion of the incident wave (forward wave, V_i) is typically reflected back (reflected wave, V_r) into the input transmission line with the remainder of the signal passing through and possible changed by the microwave device. The input port is driven by a voltage source through the characteristic impedance (Z_0) and output port is terminated in the same characteristic impedance (Z_0).

The necessary data is obtained over the frequency range of interest by measuring the amplitude and phase characteristics of the forward and reflected voltage waves when a signal is applied to port 1 through a characteristic impedance Z_0 and

Figure 4.1   Microwave two-port network; port 1 is the input and port 2 is the output.

Figure 4.2 Microwave two-port network; port 2 is the input and port 1 is the output.

observed at port 2 terminated in the same characteristic impedance Z_0 (see Figure 4.1). Similar data must also be taken with the voltage wave incident on port 2 and an observed reflected wave at port 2 and a wave transmitted to port 1 with the same input and output characteristic impedance Z_0 (see Figure 4.2). The input and output forward and reflected waves are voltage vectors with an associated magnitude ($|\mathbf{V}|$) and phase ($\angle\mathbf{V}$). Typically, the magnitude of these waves is expressed as power equal to the square of the magnitude of the voltage ($|\mathbf{V}|^2$) with respect to the characteristic impedance (Z_0).

4.2 The *S*-Parameter Matrix

S-parameters for a two-port network are a set of two equations describing the characteristics of a wave through the network and reflected by the network when the input and output impedance is equal (Z_0), usually 50Ω. One set of equations describes the results when port 1 is the input port (see Figure 4.3). The other set of equations describes the results when port 2 is the input port (see Figure 4.4).

The *S*-parameter matrix [2] in one equation relates a microwave voltage signal (magnitude and phase) incident to port 1 ($V1_i$) to a signal coming out of port 1 ($V1_r$) labeled S(1,1) or S_{11}, and a signal wave incident into port 1 ($V1_i$) coming out of port 2 ($V2_r$) termination into impedance Z_0, labeled S(2,1) or S_{21} (see Figure 4.3).

A second equation relates a microwave voltage signal (magnitude and phase) incident to port 2 ($V2_i$) to a signal coming out of port 2 ($V2_r$), labeled S(2,2) or S_{22}, and a signal wave incident into port 2 ($V2_i$) coming out of port 1 ($V1_r$) terminated into a impedance Z_0, labeled S(1,2) or S_{12} (see Figure 4.4).

All of the waves are complex, frequency-dependent transfer functions.

4.2.1 Passive Symmetrical Devices

When a passive device is symmetrical, that is, the device is a mirror image with respect to the center of the device, S(1,1) equals S(2,2) and S(1,2) equals S(2,1) [3].

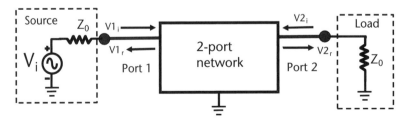

Figure 4.3 A two-port network excited by input voltage source V_i with a source impedance Z_0.

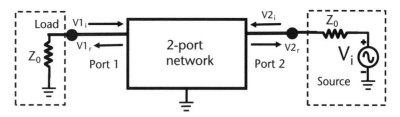

Figure 4.4   The same two-port network shown in Figure 4.3 with port 2 excited by input voltage source V_i and port 1 terminated with impedance Z_0.

4.2.2   Measuring the *S*-Parameter Matrix

An *S*-parameter function is described by matrix (4.1).

$$\begin{bmatrix} V1_r \\ V2_r \end{bmatrix} = \begin{bmatrix} S_{11} & S_{12} \\ S_{21} & S_{22} \end{bmatrix} \begin{bmatrix} V1_i \\ V2_i \end{bmatrix} \tag{4.1}$$

The algebraic matrix equation when expanded gives some insight into the functions of the microwave two-port network.

$$V1_r = S_{11} \cdot \left(V1_i \right) + S_{12} \cdot \left(V2_i \right) \tag{4.2}$$

$$V2_r = S_{21} \cdot \left(V1_i \right) + S_{22} \cdot \left(V2_i \right) \tag{4.3}$$

4.2.2.1   Measuring Parameter S_{11}

Refer to the diagram in Figure 4.3 where the two-port network is excited by a source into port 1 [4].

If $V2_r = 0$ or $S_{12} = 0$,

$$V1_r = S_{11} \cdot \left(V1_i \right) + S_{12} \cdot \left(V2_i \right) \rightarrow \frac{V1_r}{V1_i} = S_{11}\big|_{S_{12} \text{ or } V2_r = 0} \tag{4.4}$$

S_{11} under the conditions in (4.4) is the return loss at port 1 when the network is driven by a source into port 1 as shown in Figure 4.3. This reflects the mismatch of the input impedance looking into port 1 with respect to the characteristic impedance (Z_0). $V2_r \approx 0$, a condition of (4.4), occurs when the output impedance of the network is approximately matched to the characteristic impedance (Z_0). The reverse gain (S_{12}) of the device approximately zero ($S_{12} \approx 0$), an alternate condition of (4.4), occurs when the network unidirectional. Examples of unidirectional devices are an active device such as an amplifier or a passive device such as an isolator. It should be noted that S_{11} is the return loss when $S_{12} \cdot V2_r \approx 0$. The equivalent device impedance looking into port 1 (imaginary and real parts) can be found knowing S_{11}.

4.2.2.2   Measuring Parameter S_{21}

Referring to the diagram in Figure 4.3 where the two-port network is excited by a source into port 1.

If $V2_i = 0$ or $S_{22} = 0$,

$$V2_r = S_{21} \cdot \left(V1_i\right) + S_{22} \cdot \left(V2_i\right) \rightarrow \frac{V2_r}{V1_i} = S_{21}\big|_{V2_i=0} \qquad (4.5)$$

S_{21} under the conditions in (4.5) is the forward voltage gain at port 2 with respect to the voltage at port 1 when the network is driven by a source into port 1 as shown in Figure 4.3. The condition that $V2_i \approx 0$ is met when the output impedance of the network is approximately matched to the characteristic impedance (Z_0). The coefficient S_{22} is related to the output impedance of the device and not equal to zero ($S_{22} \neq 0$) and ideally approximately equal to Z_0.

4.2.2.3 Measuring Parameter S_{22}

Refer to the diagram in Figure 4.4 where the two-port network is excited by a source into port 2.
If $V1_i = 0$,

$$V2_r = S_{21} \cdot \left(V1_i\right) + S_{22} \cdot \left(V2_i\right) \rightarrow \frac{V2_r}{V2_i} = S_{22}\big|_{V1_i=0} \qquad (4.6)$$

S_{22} under the conditions in (4.6) is the return loss at port 2 when the network is driven by a source into port 2 as shown in Figure 4.4. This reflects the mismatch of the input impedance looking into port 2 with respect to the characteristic impedance (Z_0) under the condition that $V1_i = 0$. S_{21} is the forward gain of the device and in general cannot be considered approximately zero. $V1_i$ is the reflected signal at port 1 due to the voltage $V1_r$ coming through the reverse path of device. The reverse path usually has more loss through active devices and passive isolators than passive devices. $V1_i$ is multiplied by the forward gain of the device $\approx S_{21}$ and added to the signal reflected back from port 2. The forward gain of an active device increasing the signal level returning to port 2 is mitigated by the high loss in the reverse path.

4.2.2.4 Measuring Parameter S_{12}

Refer to the diagram in Figure 4.4 where the two-port network is excited by a source into port 2.
If $V1_i = 0$,

$$V1_r = S_{11} \cdot \left(V1_i\right) + S_{12} \cdot \left(V2_i\right) \rightarrow \frac{V1_r}{V2_i} = S_{12}\big|_{V1_i=0} \qquad (4.7)$$

S_{12} under the conditions in (4.7) is the reverse voltage gain at port 1 with respect to the voltage at port 2 when the network is driven by a voltage source into port 2 as shown in Figure 4.4. $V1_i \approx 0$ when the when the input impedance of the network is approximately matched to the characteristic impedance (Z_0). The coefficient S_{11} is related to the input impedance of the device and not equal to zero ($S_{11} \neq 0$) and ideally approximately equal to Z_0.

4.2.3   Notes on *S*-Parameters

S-parameter magnitudes can be presented in linear units or decibels and phase in degrees or radians at the same respective frequency as the magnitude. Since the incident signal is a voltage wave, the conversion from linear units to decibels is 20·Log [magnitude of *S(m,n)*]:

$$S_{n,m}(dB) = 20 \cdot Log\big[\|S(m,n)\|\big] \qquad (4.8)$$

where *n* is the input port and *m* is the output port.

$$\text{Magnitude of } S(m,n) = |S(m,n)| \qquad (4.9)$$

$$|S(m,n)| = \sqrt{\big(\text{Real part of } S(m,n)\big)^2 + \big(\text{Imag part of } S(m,n)\big)^2} \qquad (4.10)$$

The angle or phase of a complex *S*-parameter is usually presented in degrees (but can be given in radians). The angle of *S*-parameter *S(m,n)* is:

$$\text{angle}\ [S(m,n)] = \text{arc} \tan\left(\frac{\text{Imag Part of } S(m,n)}{\text{Real Part of } S(m,n)}\right) \qquad (4.11)$$

4.3   *S*-Parameters of Cascaded Devices: ABCD Parameters

S-parameters are measured and defined with the microwave network under test driven by a voltage source with a fixed source impedance (Z_0) and terminated at its output with the same output impedance (Z_0). Most microwave networks and test equipment are designed to operate with a nominal impedance, $Z_0 = 50\Omega$, and most *S*-parameters are given with respect to $Z_0 = 50\Omega$. Figure 4.5 shows two microwave networks with their respective *S*-parameter matrices S1 and S2. In Figure 4.6 the networks are cascaded. The source and load impedances between the microwave networks are uncertain and most likely not equal to Z_0 ($\neq Z_0$). This uncertainty makes it very difficult to calculate the composite *S*-parameters of the cascaded devices

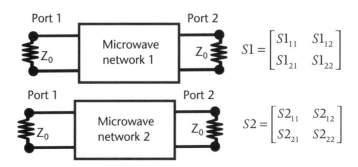

Figure 4.5 Two microwave networks with associated *S*-parameter matrices S1 and S2 measured with input and output impedance Z_0.

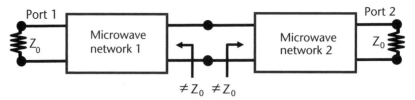

Figure 4.6 The two microwave networks in Figure 4.5 are cascaded. The cascaded network is driven by and terminated with impedance Z_0.

knowing the *S*-parameters of each individual device. This dilemma is resolved by calculating the input and output voltages and currents of each device and using this information to create another matrix called the ABCD matrix. The ABCD matrix characterizes the microwave network independent of impedance. The voltage and current out of the first network become the voltage and current into the second device enabling an analysis of the cascaded devices at any source and load imped-ance. The *S*-parameters of the cascaded microwave network can be found by using an inverse algorithm, ABCD Parameter → *S*-Parameters.

4.3.1 Defining ABCD Parameters

ABCD parameters define the microwave network in terms of its input and output voltage and current (see Figure 4.7). It is import to note that all the input and output voltages (V_1 and V_2) and currents (I_1 and I_2) are vector quantities, with magnitudes and phases a function of frequency [5].

The following equations relating to the microwave network describe the volt-ages and currents in terms of the A, B, C, and D coefficients.

$$V_1 = A \cdot V_2 + B \tag{4.12}$$

$$I_1 = C \cdot V_2 + D \tag{4.13}$$

Written as a matrix,

$$\begin{bmatrix} V_1 \\ I_1 \end{bmatrix} = \begin{bmatrix} A & B \\ C & D \end{bmatrix} \cdot \begin{bmatrix} V_2 \\ I_2 \end{bmatrix} \tag{4.14}$$

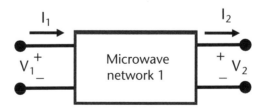

Figure 4.7 A microwave network described in terms of its input and output voltages and currents.

where:

$$A = \left.\frac{V_1}{V_2}\right|_{I_2=0} \qquad B = \left.\frac{V_1}{I_2}\right|_{V_2=0} \qquad C = \left.\frac{I_1}{V_2}\right|_{I_2=0} \qquad D = \left.\frac{I_1}{I_2}\right|_{V_2=0} \tag{4.15}$$

4.3.2 Cascading ABCD Networks

Figure 4.8 is a cascade of two networks $ABCD_1$ at the input and $ABCD_2$ at the output represented by their respective ABCD parameters A_n, B_n, C_n, and D_n where n is the network number.

As shown in Figure 4.8, the output currents and voltages of matrix 1 ($ABCD_1$) is equal to the input currents and voltages of a cascaded matrix 2 ($ABCD_2$). Since the matrix interface is compatible (i.e., $V_2 = V_2'$ and $I_2 = I_2'$), the resulting parameters of a cascaded matrix 3 ($ABCD_3$) can be found by multiplying two matrices $ABCD_1$ and $ABCD_2$ [6]. See (4.16) and (4.17).

$$\begin{bmatrix} V_1 \\ I_1 \end{bmatrix} = \begin{bmatrix} A_1 & B_1 \\ C_1 & D_1 \end{bmatrix} \cdot \begin{bmatrix} A_2 & B_2 \\ C_2 & D_2 \end{bmatrix} \tag{4.16}$$

Combining $ABCD_1$ and $ABCD_2$ results in matrix $ABCD_3$.

$$\begin{bmatrix} A_3 & B_3 \\ C_3 & D_3 \end{bmatrix} = \begin{bmatrix} A_1 & B_1 \\ C_1 & D_1 \end{bmatrix} \tag{4.17}$$

The resultant combined matrix ($ABCD_3$) in terms of input voltage V_1 and input current I_1 and output parameters V_3 and I_3 is given in (4.18).

$$\begin{bmatrix} V_1 \\ I_1 \end{bmatrix} = \begin{bmatrix} A_3 & B_3 \\ C_3 & D_3 \end{bmatrix} \tag{4.18}$$

4.3.3 Converting S-Parameters to ABCD Parameters and ABCD Parameters to S-Parameters

ABCD parameters are a convenient method of cascading devices, but they do not provide the insight necessary to evaluate the performance of a microwave device embedded in a transmission line. The solution to this problem is to convert transmission line S-parameters of each microwave device to ABCD parameters, calculate

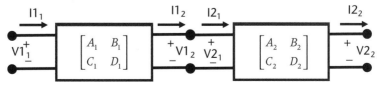

Figure 4.8 Two ABCD matrices in cascade.

the cascaded the ABCD parameters, and then convert the combined ABCD matrix back to *S*-parameters. The resultant *S*-parameters of the cascaded assembly can then be analyzed to show the expected performance.

Some key points to note about this process are:

1. Every device individually must have an associated set of *S*-parameters.
2. The interface transmission line between devices is a microwave circuit that must have its own set of *S*-parameters.
3. Bends and other perturbations in the transmission lines must have their own set of *S*-parameters and respective ABCD parameters.
4. ABCD parameters are voltage and current vectors, with an associated magnitude and phase. Transmission lines as an example can be longer than one wavelength (1λ), so the conversion process from *S*-parameters to ABCD parameters and back to *S*-parameters must not lose the signal delay factor (i.e., multiple wavelengths).

4.3.3.1   Conversion from *S*-Parameters to ABCD Parameters

The equations for conversion of *S*-parameters to ABCD parameters are:

$$A = \frac{(1 + S_{11}) \cdot (1 - S_{22}) + S_{12} \cdot S_{21}}{2 \cdot S_{21}} \tag{4.19}$$

$$B = Z_0 \cdot \frac{(1 + S_{11}) \cdot (1 + S_{22}) - S_{12} \cdot S_{21}}{2 \cdot S_{21}} \tag{4.20}$$

$$C = \frac{1}{Z_0} \cdot \frac{(1 - S_{11}) \cdot (1 - S_{22}) - S_{12} \cdot S_{21}}{2 \cdot S_{21}} \tag{4.21}$$

$$D = \frac{(1 - S_{11}) \cdot (1 + S_{22}) + S_{12} \cdot S_{21}}{2 \cdot S_{21}} \tag{4.22}$$

The *S*-parameters are frequency-dependent; therefore, the conversion must be done for each frequency of interest. Finer data resolution requires more frequency steps, which could have a significant impact on the computational time and available computer memory.

4.3.3.2   Conversion from ABCD Parameters to *S*-Parameters

The equations for conversion of ABCD parameters to *S*-parameters are:

$$S_{11} = \frac{A + \dfrac{B}{Z_0} - (C \cdot Z_0) - D}{A + \dfrac{B}{Z_0} + (C \cdot Z_0) + D} \tag{4.23}$$

$$S_{12} = \frac{2 \cdot (A \cdot D - B \cdot C)}{A + \dfrac{B}{Z_0} + (C \cdot Z_0) + D} \qquad (4.24)$$

$$S_{21} = \frac{2}{A + \dfrac{B}{Z_0} + (C \cdot Z_0) + D} \qquad (4.25)$$

$$S_{22} = \frac{-A + \dfrac{B}{Z_0} - (C \cdot Z_0) + D}{A + \dfrac{B}{Z_0} + (C \cdot Z_0) + D} \qquad (4.26)$$

4.3.3.3 Summary

Solving for the combined S-parameters of a multiple device network as shown in Figure 4.9 requires a computationally intensive procedure:

1. ABCD parameters are very difficult to measure and are just used as a mathematical convenience to calculate the combined S-parameters of multiple cascaded microwave devices.
2. The S-parameters of each device has to be known or measured over the frequency of interest at every frequency resolution element.
3. The S-parameters of each device at each frequency have to be converted to ABCD parameters. Each S-parameter at each frequency of interest may be a complex number. Interface transmission lines, including transmission line discontinuities such as bends are separate microwave devices with its own S-parameters and respective ABCD parameters.
4. The S-parameters and the ABCD parameters are complex numbers applicable to a single frequency.
5. At each frequency of interest, the S-parameters of each element are converted to ABCD parameters.
6. The ABCD parameters of each device is combined by multiplying the respective matrices to a single matrix.
7. The final single matrix, one for each frequency of interest, is converted back to an S-parameter matrix representing all of the devices in series as a single device with equivalent S-parameters.

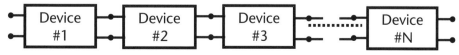

Figure 4.9 A microwave network consisting of N devices.

4.4   *S*-Parameters of Multiport Networks

The *S*-parameter matrix for an N-port device [7] (see Figure 4.10) contains N^2 coefficients (N^2 *S*-parameters), each one representing a possible input-output path. Each *S*-parameter refers to what happens at one port when it is excited by a signal incident at another port with all of the other ports terminated with impedance Z_0 (where Z_0 is usually but not necessarily 50Ω).

The *S*-parameter matrix for the device in Figure 4.10 is:

$$
\begin{bmatrix} V1_r \\ V2_r \\ V3_r \\ V4_r \\ V5_r \\ \vdots \\ VN_r \end{bmatrix} = \begin{bmatrix} S_{11} & S_{12} & S_{13} & S_{14} & \cdots & S_{1N} \\ S_{21} & S_{22} & S_{23} & S_{24} & \cdots & S_{2N} \\ S_{31} & S_{32} & S_{33} & S_{34} & \cdots & S_{3N} \\ S_{41} & S_{42} & S_{43} & S_{44} & \cdots & S_{4N} \\ S_{51} & S_{52} & S_{53} & S_{54} & \cdots & S_{5N} \\ \vdots & \vdots & \vdots & \vdots & \vdots & \vdots \\ S_{N1} & S_{N2} & S_{N3} & S_{N4} & \cdots & S_{NN} \end{bmatrix} \cdot \begin{bmatrix} V1_i \\ V2_i \\ V3_i \\ V4_i \\ V5_i \\ \vdots \\ VN_i \end{bmatrix}
\tag{4.27}
$$

All of the voltage waves entering the device have a subscript *i* and all of the voltage waves leaving the device have a subscript *r*.

4.4.1   Example of a Multiport Device: Branch Line Coupler

A common four-port network known as a branch line coupler is shown in Figure 4.11. The branch line coupler [6] is a four-port device with typically one port designated as the input (port 1) and two ports designated as the outputs (ports 1 and 3) in this diagram. Port 4 is an isolated port. The output ports 1 and 2, labeled A and B, are 90° out of phase and ideally half the power of the input signal. If the A and B outputs are identical and terminated in the same impedance (not necessarily Z_0), the reflected signals from ports 2 and 3 are directed to the isolated port 4. Under these conditions, the input port 1 appears to have no reflected signal and therefore looks like an ideal load to the source signal.

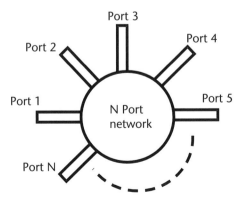

Figure 4.10   A multiport microwave device consisting of *N* ports.

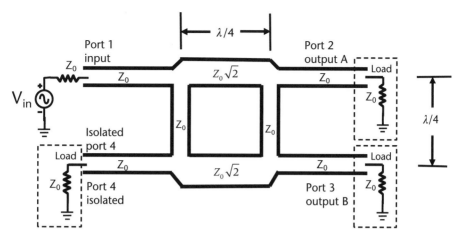

Figure 4.11 A branch line quadrature coupler.

The branch line coupler is a symmetrical device (i.e., if port 2 is the input, the outputs will appear on ports 1 and 4 with port 3 as the isolated port). The branch line coupler can also be used as a combiner. If two identical signals are present at ports 2 and 3, the signals will ideally sum at port 4. Port 1 will be isolated and only see signals if the input signals are not identical and do not perfectly cancel.

4.5 *S*-Parameter Summary

S-parameters are specified at a predefined impedance. The source and load impedances are generally equal over the microwave frequency range of interest. The characteristic impedance of most microwave transmission lines is 50Ω. Microwave devices are designed to match the transmission lines and therefore are also designed with input and output impedances equal to 50Ω.

S-parameter coefficients can be complex numbers (real and imaginary parts) and frequency dependent. *S*-parameter coefficients are voltage vectors with a magnitude and phase. The power related to an *S*-parameter is equal to the magnitude squared of the voltage waveform (e.g., power related to parameter $S_{21} = |S_{21}|^2$). The phase of the *S*-parameter is the arc tangent of the imaginary part over the real part.

Some networks are symmetrical (i.e., $S_{21} = S_{12}$ and $S_{11} = S_{22}$). Interchanging the input and output ports does not change the transmission or reflection properties. The branch coupler is an example of a symmetrical four-port network.

References

[1]   HP Application Note 95-1, *S-Parameters Techniques*, 1997.
[2]   Dunleavy, L., *S-Parameters*, University of South Florida.
[3]   Adam, S., *Microwave Theory and Applications*, Agilent Technologies.

[4]   Hassan, S., *Network Parameters*, Engineering Campus USM.

[5]   Mellitz, R., *Introduction to Frequency Domain Analysis*, Intel.

[6]   Pozar, D., *Microwave Engineering*, New York: John Wiley & Sons, 2012.

[7]   Ahmadi, F., "RF Communication Circuits," Sharif University of Technology.

CHAPTER 5

Microstrip Transmission Lines

5.1 Microstrip Transmission Lines

Microstrip transmission lines are widely used to interconnect microwave devices because of their open structure. This topology permits access to the top of the substrate, making it convenient for assembly and tuning. The theory and concepts presented with respect to this topology are applicable to many other transmission line structures such as stripline and coplanar waveguides. The versatility of microstrip makes it possible to integrate all three of these topologies on the same substrate as well as integrating a wide variety of MMIC and hybrid devices.

A typical microstrip transmission line is shown in Figure 5.1. It is fabricated like most printed circuit boards by typically etching the unwanted copper off the top side of a low-loss dielectric material commonly called the substrate. The backside of the substrate is usually a solid copper ground plane. Critical to the design are the dielectric constant of the substrate material (ε_r), the height of the material (h) between the copper planes, the thickness of the copper (t), the width of the microstrip conductor (w), and the length of the microstrip conductor (l).

Microstrip transmission lines radiate fields from all surfaces of the conductor. The primary energy is concentrated in the conductor and in the dielectric material between the conductor and the ground plane, but there exists some energy radiate into the air above the transmission line (see Figure 5.2). Leakage from the top of the conductor must be addressed in the design phase of the project to protect the emissions from interfering with other functions on the same substrate or in the same housing [1].

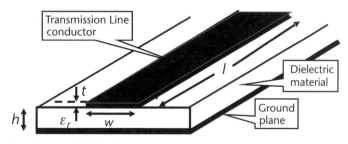

Figure 5.1   Microstrip transmission line.

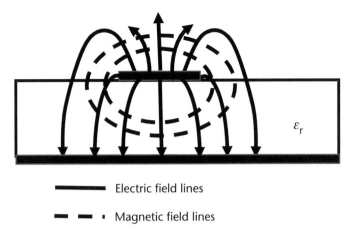

Electric field lines

Magnetic field lines

Figure 5.2 Microstrip radiated fields.

5.2 Dielectric Material

Dielectric material is the substance between the inner conductor and the outer conductor in the coaxial cable and the material between the conductor and the ground plane in the microstrip transmission line. The electromagnetic signal travels through the conductor and through the dielectric material each having a significant effect on signal velocity and insertion loss. The dielectric materials effect on electromagnetic waves is characterized by its dielectric constant or permittivity (ε).

The dielectric constant relates the extent a material concentrates electric flux. This characteristic is used to describe the amount of charge that can be held by a set of parallel metal plates separated by the dielectric material. Materials with high dielectric constants (units of farads per meter) are useful in the manufacture of high-value, low-density capacitors.

The dielectric constant (ε) is typically separated into two factors:

$$\varepsilon = \varepsilon_0 \cdot \varepsilon_r \tag{5.1}$$

where ε_0 is the dielectric constant of free space ($\varepsilon_0 = 8.85418782 \cdot 10^{-12}$ farad/meter) and ε_r (no units) is the relative dielectric constant of the material. A table in Appendix B shows the relative dielectric constant (ε_r) for some commonly used dielectrics materials.

The relative dielectric constant (ε_r) is classified by its ability to store energy when an external electric field is applied. When a DC voltage source is placed across a parallel plate capacitor, more charge is stored when a dielectric material is between the plates than if there was a vacuum between the plates. The capacitance (C) is directly proportional to the relative dielectric constant of the material between the plates ($\varepsilon_r = 1$ for a vacuum).

$$C = \frac{\left(\varepsilon_r \cdot \varepsilon_0 \cdot A\right)}{d} \tag{5.2}$$

where C is the capacitance (F), A is the area of the parallel plates, d is the distance between the plates, ε_0 is the dielectric constant of free space ($8.85418782 \cdot 10^{-12}$ farad/meter), and ε_r is the relative dielectric constant of a material.

The relative dielectric constant for a given material can be obtained by measuring the capacitance of two parallel plates in vacuum (or in air which has a relative dielectric constant close to a vacuum) and comparing the capacitance of the same parallel plates when a dielectric material is placed between the plates [2].

The material used in a particular design can have dielectric constant variations that can vary across manufacturing lots, over temperature and frequency [3]. In addition, most dielectric materials are anisotropic (i.e., the value of the dielectric constant could be different along a different axis of the material). The variation of dielectric constant plus the mechanical dimensional tolerances could add up significantly and should be considered in the product design.

5.3   Effective Conductor Width of a Microstrip Transmission Line

The effective conductor width is wider (Δw) than the actual conductor width (w) resulting from the fields emanating from the edge of the conductor (see Figure 5.2). Equations (5.3) through (5.5) calculate an effective conductor width (w_{eff}) as a function of the conductor width (w), substrate height (h), and conductor thickness (t) [1].

$$w_{\text{eff}} = w + \frac{t}{\pi} \cdot \left[1 + \ln\left(\frac{4 \cdot \pi \cdot w}{t} \right) \right] \quad \text{for} \quad \frac{w}{h} \leq \frac{1}{2 \cdot \pi} \tag{5.3}$$

$$w_{\text{eff}} = w + \frac{t}{\pi} \cdot \left[1 + \ln\left(\frac{2 \cdot h}{t} \right) \right] \quad \text{for} \quad \frac{w}{h} > \frac{1}{2 \cdot \pi} \tag{5.4}$$

$$\Delta w = w_{\text{eff}} - w \tag{5.5}$$

Note that ln ( ) is $\log_e$( ). The effective conductor width is used in all calculations relating to microstrip transmission lines.

5.4   Effective Dielectric Constant in a Microstrip Transmission Line

An electromagnetic wave traveling through a microstrip transmission line partially exists in the conductor, partially exists in the dielectric substrate, and partially exists in the air above the conductor and substrate. The combined propagation velocity and the respective dielectric constant called the effective dielectric constant (ε_{eff}) will be somewhere between the dielectric constant of free space (ε_0) and the dielectric constant of the substrate material (ε_r) [4]. The effective dielectric constant is a function of the of the dielectric constant (ε_r) of the material between the conductor and ground plane, the width of the conductor (w_{eff}) and the conductor's proximity to the ground plane (h). The value of the effective dielectric constant is given by the relationship in (5.6) and (5.7).

For smaller conductor widths (i.e., $w/h \leq 1$), the relationship is given in (5.6)

$$\varepsilon_{\text{eff}} = \frac{(\varepsilon_r + 1)}{2} + \frac{(\varepsilon_r - 1)}{2} \cdot \left[\frac{1}{\sqrt{1 + 12 \cdot \dfrac{h}{w_{\text{eff}}}}} + 0.04 \cdot \left(1 - \frac{w_{\text{eff}}}{h}\right)^2 \right] \qquad \frac{w_{\text{eff}}}{h} \leq 1 \quad (5.6)$$

For larger conductor widths (i.e., $w/h \geq 1$), the relationship is given in (5.3)

$$\varepsilon_{\text{eff}} = \frac{(\varepsilon_r + 1)}{2} + \frac{(\varepsilon_r - 1)}{2 \cdot \sqrt{1 + 12 \cdot \dfrac{h}{ww_{\text{eff}}}}} \qquad \frac{w_{\text{eff}}}{h} \geq 1 \qquad (5.7)$$

5.5 Wave Velocity and Wavelength of a Signal Traveling Through a Dielectric Material

Frequency (f) and wavelength (λ) are inversely proportional where the constant of constant of proportionality in a vacuum is the speed of light (c), as shown in (5.8).

$$f \cdot \lambda = c \qquad (5.8)$$

where c is the speed of light in distance (d) per second in centimeters (cm) per second ($c = 2.99792458 \cdot 10^{10}$ cm/second), f is the signal frequency in hertz, and λ is the wavelength in the same units as distance (d).

The speed of light (c) in a vacuum is related to the permeability in a vacuum (μ_0 the magnetic field constant) and the permittivity in a vacuum (ε_0 the electric field constant):

$$c = \frac{1}{\sqrt{\varepsilon_0 \cdot \mu_0}} \qquad (5.9)$$

In a material other than a vacuum, the velocity of the wave (v) is the related to the permeability of the material ($\mu_0 \cdot \mu_r$) and the permittivity of the material ($\varepsilon_0 \cdot \varepsilon_r$):

$$v = \frac{1}{\sqrt{\varepsilon_0 \cdot \varepsilon_r \cdot \mu_0 \cdot \mu_r}} \qquad (5.10)$$

where μ_r and ε_r are the relativity permeability and the permittivity respectively in a material where $\varepsilon_r \geq 1$ and $\mu_r \geq 1$.

Combining (5.9) and (5.10),

$$v = \frac{c}{\sqrt{\varepsilon_r \cdot \mu_r}} \qquad (5.11)$$

From (5.11), it is obvious that the velocity of an electromagnetic wave in a material is less than the velocity of the wave in a vacuum: $v < c$.

The relativity permeability (μ_r) of many typically used dielectric materials is approximately equal to unity ($\mu_r \approx 1$). This simplifies the velocity of an electromagnetic wave in a material (v) to:

$$v \approx \frac{1}{\sqrt{\varepsilon_0 \varepsilon_r \mu_0}} = \frac{c}{\sqrt{\varepsilon_r}} = f \cdot \lambda \qquad (5.12)$$

Solving (5.12) for wavelength (λ) gives the result shown in (5.13):

$$\lambda = \frac{c}{f \cdot \sqrt{\varepsilon_r}} \qquad (5.13)$$

Equation (5.13) is the wavelength of a signal traveling through a material with a relative dielectric constant (ε_r) assuming the relativity permeability (μ_r) is equal to unity ($\mu_r \approx 1$). The velocity of the signal traveling through the material is v always less than the velocity of a signal traveling in a vacuum c. The relative dielectric constant of some commonly used materials is shown in Appendix B (the subject of loss tangents will be discussed later).

5.6   The Effective Wave Velocity and Wavelength in a Microstrip Transmission Line

The calculations of wave velocity and wavelength in a microstrip transmission line are almost the same as (5.11) and (5.13), respectively, except that the relative dielectric constant (ε_r) is substituted with the effective dielectric constant (ε_{eff}). In both equations, the relativity permeability (μ_r) is assumed equal to unity ($\mu_r \approx 1$) [5].

Equation (5.11) is rewritten as:

$$v = \frac{c}{\sqrt{\varepsilon_{\text{eff}}}} = f \cdot \lambda \qquad (5.14)$$

and the rewritten calculation of the wavelength becomes:

$$\lambda = \frac{c}{f \cdot \sqrt{\varepsilon_{\text{eff}}}} \qquad (5.15)$$

where c is the speed of light in a vacuum, ε_r is the relative dielectric constant of the material between the conductor and the ground plane, v is the velocity of the signal in the microstrip transmission line with conductor width, w, and the conductor height above the ground plane, h, f is the signal frequency in hertz, and ε_{eff} is the effective dielectric constant.

5.7 Calculating the Impedance of a Microstrip Transmission Line

The impedance of a microstrip transmission line is critical to matching devices and creating microstrip functions. The closed-form equations for calculating the impedance was developed in the 1960s and has been incrementally improved upon by considering higher-order effects. In most applications, the manufacturing variations of the conductor and substrate and the chassis enclosure effects on the microstrip line make it unnecessary to theoretically fine-tune the line impedance calculation to an error less than a few percent. The equations below take into account some of the higher-order effects considering the effective line width, the effective dielectric constant (ε_{eff}), the conductor line thickness (t), and the ratio of line effective width (w_{eff}) to substrate height (h) as shown in Figure 5.3. The impedance results obtained is sufficiently accurate for most applications. The calculation of the impedance of a microstrip line is separated into two equations: (1) (5.16) for $w/h \le 1$, and (2) (5.17) for $w/h \ge 1$. For small conductor widths, $w_{eff}/h \le 1$. Using the effective width (w_{eff}) and the effective relative dielectric constant (ε_{eff}) found in (5.6), the characteristic impedance (Z_0) is:

$$Z_0 = \frac{60}{\sqrt{\varepsilon_{eff}}} \cdot \ln\left(\frac{8 \cdot h}{w_{eff}} + \frac{w_{eff}}{4 \cdot h} \right) \quad \text{for } \frac{w_{eff}}{h} \le 1 \tag{5.16}$$

For large conductor widths, $w_{eff}/h \ge 1$. Using the effective width (w_{eff}) found in (5.5) and the effective relative dielectric constant (ε_{eff}) found using (5.7), the characteristic impedance (Z_0) is:

$$Z_0 = \frac{1}{\sqrt{\varepsilon_{eff}}} \cdot \frac{120 \cdot \pi}{\dfrac{w_{eff}}{h} + 1.393 + \dfrac{2}{3} \cdot \ln\left(\dfrac{w_{eff}}{h} + 1.444 \right)} \quad \text{for } \frac{w_{eff}}{h} \ge 1 \tag{5.17}$$

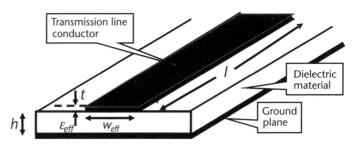

Figure 5.3 Microstrip transmission line parameters showing effective width and effective dielectric constant.

Table 5.1 Standard Thickness for Copper on Printed Circuit Boards

PCB Copper Thickness	
1/2 oz.	0.689 mil
1 oz.	1.378 mils
2 oz.	2.756 mils

For example, calculating microstrip impedance knowing the substrate height (h), conductor width (w), substrate relative dielectric constant (ε_r), and conductor thickness (t), assume the following parameters relating to Figure 5.3 given in mils (1 mil = 0.001 inch): substrate height (h) = 31 mils, conductor width (w) = 94 mils, substrate relative dielectric constant (ε_r) = 2.2, and conductor thickness (t) = 1.378 mils (this is a standard thickness for 1-ounce copper from Table 5.1).

Using (5.4), the effective width (w_{eff}) = 96.10828 mils. Next, determine the width over height ratio (w_{eff}/h) = 3.10027. Since $w_{eff}/h \geq 1$, use (5.7) and (5.17). Using (5.7), the relative effective dielectric constant: ε_{eff} = 1.97774. Appling the effective dielectric constant to (5.17), the characteristic impedance (Z_0) is Z_0 = 50.07633Ω.

5.8   Calculating the Line Width for a Desired Impedance

Equations (5.16) through (5.17) contain a constant factor width (w_{eff}) over height (h) factor (w_{eff}/h) suggesting that the equations can be reconfigured to solve for w_{eff}/h in terms of Z_0. Given a known substrate with a relative dielectric constant (ε_r), a substrate height (h), and the desired characteristic impedance (Z_0) in ohms, the effective conductor width (w_{eff}) over height (h), that is, (w_{eff}/h) is solved using two sets of equations, one equation for ($w_{eff}/h \leq 2$) and another equation for ($w_{eff}/h \geq 2$). The procedure for finding the correct w_{eff}/h is to solve the equations for $w_{eff}/h \leq 2$ [(5.18) and (5.19)] and solve the equations for $w_{eff}/h \geq 2$ [(5.16) and (5.17)] and use the result that best fits the w_{eff}/h criteria.

For small conductor widths, $w_{eff}/h \leq 2$. The interim equation (5.17) solves for a factor ($F1$) in terms of the desired impedance Z_0

$$ F1 = \frac{Z_0}{60} \cdot \sqrt{\frac{\varepsilon_r + 1}{2}} + \frac{\varepsilon_r - 1}{\varepsilon_r + 1} \cdot \left(0.23 + \frac{0.11}{\varepsilon_r} \right) \quad \text{for } \frac{w_{eff}}{h} \leq 2 \quad (5.18) $$

Note that (5.18) uses the relative dielectric constant (ε_r) and not the effective dielectric constant (ε_{eff}).

Using the interim factor ($F1$), (5.19) solves for (w_{eff}/h).

$$ \frac{w_{eff}}{h} = \frac{8 \cdot e^{F1}}{e^{2 \cdot F1} - 2} \quad \text{for } \frac{w_{eff}}{h} \leq 2 \quad (5.19) $$

For larger conductor widths, $w_{eff}/h \geq 2$.

The interim equation (5.20) solves for a factor ($F2$) in terms of the desired impedance Z_0

$$F2 = \frac{377 \cdot \pi}{2 \cdot Z_0 \cdot \sqrt{\varepsilon_r}} \qquad \text{for } \frac{w_{\text{eff}}}{h} \geq 2 \tag{5.20}$$

Note that (5.20) uses the relative dielectric constant (ε_r) and not the effective dielectric constant (ε_{eff}).

Using the interim factor ($F2$) and the relative dielectric constant (ε_r), (5.21) solves for (w_{eff}/h) when $w_{\text{eff}}/h \geq 2$.

$$\frac{w_{\text{eff}}}{h} = \frac{2}{\pi} \cdot \left[F2 - 1 - \ln(2 \cdot F2 - 1) + \frac{\varepsilon_r - 1}{2 \cdot \varepsilon_r} \cdot \left(\ln(F2 - 1) + 0.39 - \frac{.61}{\varepsilon_r} \right) \right] \qquad \text{for } \frac{w_{\text{eff}}}{h} \geq 2 \tag{5.21}$$

Once it is determined which set of equations best fits the criteria $w_{\text{eff}}/h \leq 2$ or $w_{\text{eff}}/h \geq 2$, the effective width can be calculated by multiplying w_{eff}/h times h:

$$w_{\text{eff}} = \left(\frac{w_{\text{eff}}}{h} \right) \cdot (h) \tag{5.22}$$

The actual width (w) can be by calculated from the effective width (w_{eff}), the substrate height (h), and the conductor thickness (t) as shown in (5.23).

$$w = w_{\text{eff}} - \frac{t}{\pi} \cdot \left(\ln\left(\frac{2 \cdot h}{t} \right) + 1 \right) \tag{5.23}$$

For example, calculate the microstrip conductor width (w) knowing the substrate height (h), the desired impedance (Z_0 in ohms), the substrate relative dielectric constant (ε_r), and the conductor thickness (t).

Assume the following parameters relating to Figure 5.3 are given in mils (1 mil = 0.001 inch): substrate height (h) = 31 mils, impedance (Z_0) = 70.7Ω, substrate relative dielectric constant (ε_r) = 2.2, and conductor thickness (t) = 1.378 mils (this is a standard thickness for 1-ounce copper from Table 5.1).

- *Step 1:* Use (5.18) and (5.19) to calculate w_{eff}/h for $w_{\text{eff}}/h \leq 2$, F1 = 1.59549, and w_{eff}/h = 1.76791.
- *Step 2:* Use (5.16) and (5.17) to calculate w_{eff}/h for $w_{\text{eff}}/h \geq 2$, F2 = 5.64716, and w_{eff}/h = 1.76044.
- *Step 3:* Use the result in Step 1 and (5.18) to calculate the effective width w_{eff}: w_{eff} = 54.8053 mils.
- *Step 4:* Use (5.23) to calculate the actual width (w): w = 52.69702 mils.

With regard to microstrip transmission-line impedances, note that the calculated results are not exact, but should be accurate enough to satisfy most requirements. Around the crossover point $w_{\text{eff}}/h = 2$, the calculated conductor width is almost continuous, the results of either set of equations (for $w_{\mathrm{eff}}/h \leq 2$ and for $w_{\mathrm{eff}}/h \geq 2$) converge to approximately the same result. The typical impedance range of a microstrip line is 20Ω to 125Ω. The line widths as the impedance gets higher (>125 Ω) become very thin and it is tough to maintain manufacturing tolerances better than 10%. The lower limit (20Ω) is due to the possible appearance of higher-order modes because the conductor widths become a significant percentage of a wavelength (λ) as the frequency increases. There are many quality microwave simulation programs that will perform these calculations.

5.9   Optimizing Bends in Microstrip Transmission Lines

The design of most microstrip circuits requires the microstrip transmission line to change directions (i.e., turn through a large angle or bend). The bending of the conductor causes a discontinuity as the wave travels down the transmission line resulting in some signal radiating off the transmission line or being reflected back to the source. In either case, only part of the signal is continuing through the bend, resulting in excess loses and possible interference with other functions in the circuit. There are many ways of eliminating this problem; one is to implement a curve in the line with an arc of radius at least three times the width of the microstrip line (a generally accepted approximation for a smooth transition, not a calculation). Another more commonly used space-saving technique is to chamber the bend. Figure 5.6 shows a 90° bend in a microstrip transmission line. Equation (5.24) is a calculation of the optimum chamber depth D as a function of the line width (w) and height of the substrate (h) for that bend.

$$D = \sqrt{2} \cdot w \cdot \left(0.52 + 0.65 \cdot e^{(-1.35) \cdot \frac{w}{h}} \right) \tag{5.24}$$

Other bend angles can be optimized, but it is wise to use a microwave simulation program to assess the performance of the selected configuration.

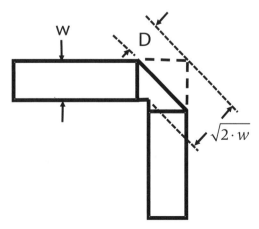

Figure 5.4   Microstrip mitered bend.

5.10   Transmission Line Losses

Signal is lost propagating through a microstrip transmission line due to losses in the conductor (α_c), losses in the dielectric medium (α_d), radiation losses (α_r), and surface-wave propagation losses (α_s). Radiation losses (α_r) and surface-wave propagation losses (α_s) are usually very small can be neglected. The total signal loss is the multiplication of the each of the losses given in linear units or the addition of each of the losses if they are given in decibels. The total loss (α_t) is given in (5.25) where the individual losses and total loss is in decibels:

$$\alpha_t(\text{dB}) = \alpha_c(\text{dB}) + \alpha_d(\text{dB}) + \alpha_r(\text{dB}) + \left(\alpha_s[\text{dB}]\right) \approx \alpha_c(\text{dB}) + \alpha_d(\text{dB}) \quad (5.25)$$

The complex nature of the electromagnetic signals, the physical structure forming the transmission line, and the surrounding mechanical structures preclude an exact closed form analysis of the expected signal loss. The discussion of transmission line losses is qualitative rather than quantitative with some closed-form equations included that give approximate results sufficient for many applications. The main components of loss are discussed to highlight and avoid issues that will lead to a nonoptimum design.

5.10.1   Transmission-Line Conductor Losses

Conductor losses (α_c) are mainly due to the resistivity of the conductor material; the typical resistivity (ρ) of copper is 1.68×10^{-8} ohms/meter. In addition to resistivity, the roughness of the conductor material can have a significant effect on insertion loss as the operational frequency increases. Cold rolled copper is smoother than electrodeposited copper and results in a lower insertion loss at higher frequencies.

5.10.1.1   Skin Depth

An additional frequency dependence on loss in the conductor material is due to the phenomenon of skin effect, which is the tendency of a signal to concentrate near the outer part of a conductor. DC has a uniform distribution over the conductor cross section, but AC signals tend to congregate closer to the conductor surface with less signal penetration into the conductor as frequency increases. The conductor's effective cross section is therefore reduced and the effective resistance of the material increases because of the effective decrease in the cross-sectional area. The skin depth or penetration depth δ is the depth below the conductor surface at which the current density has decreased to e^{-1} (approximately 37%) of its value at the surface. The skin effect depth δ is given in (5.26) [6]:

$$\delta_s = \sqrt{\frac{1}{\pi \cdot f \cdot \mu \cdot \sigma}} \quad (5.26)$$

where f = frequency (Hz), $\mu = \mu_0 \cdot \mu_r$ permeability of the material [$\mu_0 = 4 \cdot \pi \cdot 10^{-7}$ (H/m) and $\mu_r = 1$ for copper], and σ = conductivity [conductivity of copper is 5.96×10^7 (S/m) at 20°C].

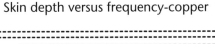

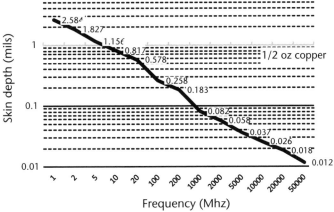

Figure 5.5   Skin in copper as a function frequency.

Figure 5.5 is a chart of skin depth versus frequency for copper. Indicated on the chart is the thickness of standard 0.5-ounce copper and 1-ounce copper traces as reference comparison to expected skin depth. Table 5.2 shows the percentage signal penetration as a function of the number of skin depths. It is suggested that the conductor thickness should be at least three ($N = 3$) skin depths ($3 \cdot \delta$) to ensure that most of the signal passes through the conductor.

5.10.1.2   Plating Microwave Conductors

Conductive surfaces are frequently plated with a high-conductivity material to reduce the signal loss. For this application to be successful, the plating thickness should also be at least three ($N = 3$) skin depths ($3 \cdot \delta$).

5.10.1.3   Microwave Coaxial Cable Line Conductors

This analysis suggests that conductors with larger surface areas have lower losses. Coaxial cables (see Figure 5.6) reduce the transmission loss for a fixed transmission

Table 5.2   Number of Skin Depths (δ) as a Function of Signal Penetration (Applicable to Any Material or Frequency)

$N = Number\ of\ Skin\ Depths\ (\sigma's)$	Percent Signal Penetration
1	63.21%
2	86.47%
3	95.02%
4	98.17%
5	99.33%
6	99.75%

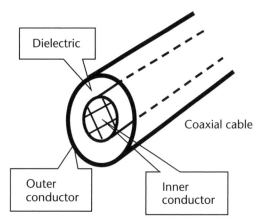

Figure 5.6 Coaxial cable with a dielectric between the inner conductor and the outer shield.

impedance by making the outer diameter of the cable thicker, requiring a larger diameter inner conductor and more effective signal surface area.

5.10.1.4   Microstrip Transmission-Line Conductors

In microstrip circuits (see Figure 5.7), the conductor width (w) is proportional to the height above the ground plane (h), therefore increasing the height of the conductor with respect to the ground plane for a fixed impedance requires an increase in the conductor width. Increasing the conductor width lowers the loss due to the increase in surface area along the conductor. The conductor thickness (t) should be at least three skin depths ($3 \cdot \delta$).

When using topologies such as microstrip, the fields are concentrated between the conducting surface and the dielectric material. Under these circumstances, plating the conductor surface that does not make contact with the dielectric material has less effect on the conductor loss.

5.10.1.5   Surface Roughness

A perfectly smooth surface will have a problem adhering to the dielectric material. A rough surface (see Figure 5.8) increases the effective resistivity of the material, increasing the insertion loss [7]. The increase in resistivity is a function of frequency increasing as frequency increases [8].

5.10.1.6   Microstrip Transmission-Line Conductor Loss Equation

An approximate equation for insertion loss in the conductor portion of a microstrip transmission line is:

$$\alpha_c = \frac{8.6859 \cdot R_s \cdot K_r}{Z_0 \cdot W} \qquad (5.27)$$

where W is the conductor width, the units of conductor width are the same as the length units for α_c [α_c is the insertion loss due to the conductor in dB per unit length

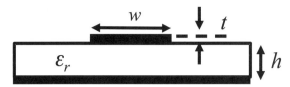

Figure 5.7 Cross-section of a microstrip transmission line.

(given by W)], R_s is the effective resistivity of the conductor, K_r is the roughness factor, and Z_0 is the characteristic impedance of the transmission line.

In (5.28), R_s is a function of the conductivity of the conductor material and the signal frequency

$$R_s = \sqrt{\frac{2 \cdot \pi \cdot f \cdot \mu}{2 \cdot \sigma}} \qquad (5.28)$$

where f is the signal frequency in hertz, μ is the permeability ($\mu_r \cdot \mu_0$), where ($\mu_r \approx 1$) and $\mu_0 = 1.25664 * 10^{-6} \cdot$ henry/m, and σ is the conductivity of the conductor material (copper $\approx 5.88 * 10^7 \cdot$ S/m).

K_r is the conductor roughness factor [9]

$$K_r = 1 + \frac{2}{\pi} \cdot \arctan \left[1.4 \cdot \left(\frac{\Delta r}{\delta_s} \right)^2 \right] \qquad (5.29)$$

where Δr is the roughness, δ_s is the skin depth [see (5.26)], and Δr and δ_s are in the same units.

For the conductor loss example, assume the following parameters: $f = 2,000$ MHz, Z_0 is 50Ω, $W = 93.9$ mils, $\Delta r = 0.4$ micron (1 micron = 10^{-6} meters), $K_r = 1.06597$, $R_s = 0.01159\Omega$, and $\alpha_c = 0.8997$ dB/meter.

5.10.2 Dielectric Material Losses

5.10.2.1 Loss Tangent

The dielectric constant (ε) is actually a complex quantity, having a real part (ε_{real}) and an imaginary part (ε_{imag}); see (5.30):

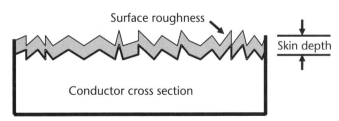

Figure 5.8 Cross section of a microstrip transmission line showing surface roughness and skin depth (not to scale).

$$\varepsilon = \varepsilon_{\text{real}} + j \cdot \varepsilon_{\text{imag}} \qquad j = \sqrt{-1} \qquad (5.30)$$

Equation (5.30) can be expressed as:

$$\varepsilon = \varepsilon_{\text{real}} \cdot \left[1 + j \cdot \frac{\varepsilon_{\text{imag}}}{\varepsilon_{\text{real}}} \right] \qquad (5.31)$$

The tangent of the complex quantity in (5.24) is called the loss tangent, Tan (δ).

$$\text{Tan}(\delta) = \left(\frac{\varepsilon_{\text{imag}}}{\varepsilon_{\text{real}}} \right) \text{ and } \delta \text{ is in radians} \qquad (5.32)$$

The Tan (δ) is usually a very small number; therefore:

$$\text{Tan}(\delta) = \delta \text{ (radians) using the small angle approximation} \qquad (5.33)$$

The real part of the loss tangent ($\varepsilon_{\text{real}}$) is given as a characteristic of the dielectric material between the conductor and the ground plane. The imaginary part of the loss tangent ($\varepsilon_{\text{imag}}$) is frequency-dependent, making the loss tangent a function of frequency.

5.10.2.2   Losses in the Dielectric Material

The loss in the dielectric material [10] is found from solving (5.34) [11].

$$\alpha_d = 27.3 \cdot \frac{\varepsilon_r \cdot (\varepsilon_{\text{eff}} - 1) \cdot \tan(\delta)}{\varepsilon_{\text{eff}} \cdot (\varepsilon_r - 1) \cdot \lambda_{\text{eff}}} \qquad (5.34)$$

where α_d is the insertion loss due to the dielectric material in dB per unit length (length is the same units as the effective wavelength λ_{eff}), tan (δ) is the loss tangent given by the dielectric material manufacturer, and λ_{eff} is the effective wave length in the microstrip transmission line

$$\lambda_{\text{eff}} = \frac{\lambda_0}{\sqrt{\varepsilon_{\text{eff}}}} \qquad (5.35)$$

where λ_0 is the wavelength in free space ($\lambda_0 = c/f$), c is the speed of light, f is the signal frequency (Hz), ε_r is the relative dielectric constant of the material between the conductor and the ground plane, and ε_{eff} is the effective dielectric constant found in (5.6) and (5.7).

Dielectric Material Loss Example
Assume the following parameters: f = 2,000 MHz, Z_0 is 50Ω, W = 93.9 mils, H = 31 mils, t = 1.378 mils, ε_r = 2.2, Δr = 0.4 micron (1 micron = 10^{-6} meters), tan

$(\delta) = 0.0009$, $\varepsilon_{\text{eff}} = 1.87176$, $\lambda_0 = 0.1499$m, $\lambda_{\text{eff}} = 0.10956$m, and $\alpha_d = 0.19148$ dB/meter. The total loss in the microstrip transmission line is $\alpha_t = \alpha_c + \alpha_d$: $\alpha_t = \alpha_c + \alpha_d = 1.09118$ dB/meter.

5.10.3 Summary of Transmission-Line Losses

There are many conducting surfaces. The primary surfaces are the bottom of the strip and the top of the ground plane. The conduction on the top surface is the plating the upper edge with a thin layer of a loss metal such as nickel can have a noticeable increase in insertion loss.

With the skin effect, at low frequencies, the current flowing in a conductor will spread out as much as possible: skin effect constricts that flow (less signal penetration as frequency increases). Conductor thickness should be three to five skin depths

Another conducting surface is conduction at the edges of the conductors: more conduction as the conducting strip gets thicker and less copper is better as long as it is greater than five skin depths $(5 \cdot \delta_s)$. Wider conductor widths result in lower losses. Return current in the ground plane contributes to the frequency-dependent losses. The ground plane is very wide, reducing the negative effects. Attenuation is mainly due to conductor and dielectric losses. Losses are dependent on frequency.

References

[1]   Bahl, I. J., and D. K. Trivedi, "A Designer's Guide to Microstrip Line," *Microwaves*, May 1977.

[2]   Keysight Technologies, "Basics of Measuring the Dielectric Properties of Materials," Application Note, 2017.

[3]   Rogers Corp., "RO4000 Series High Frequency Circuit Materials," 2015.

[4]   Rogers Corp., "Width and Effective Dielectric Constant Data for Design of Microstrip Transmission Lines on Various Thicknesses, Types and Claddings," 2015.

[5]   Pathak, R., "Characterizing Losses in Microstrip Transmission Lines," University of Wisconsin-Madison, 2005.

[6]   Intel, "Interconnect I—Class 21," 2002.

[7]   Bogatin, E., et al., "Which One Is Better? Comparing Options to Describe Frequency Dependent Losses," *DesignCon*, 2013.

[8]   Rogers Corp., "Copper Foils for High Frequency Materials," 2015.

[9]   Horn, A., et al., "Effect of Conductor Profile on the Insertion Loss, Phase Constant, and Dispersion in Thin High Frequency Transmission Lines," *DesignCon*, 2010.

[10]   Che, W., et al., "Formulas of Dielectric and Total Attenuations of a Microstrip Line," *Radio Science*, Vol. 45, 2010.

[11]   Hornung, R., "Insertion Loss and Loss Tangent," Arlon, Inc.

CHAPTER 6
Circuit Matching and VSWR

6.1 Introduction

An electromagnetic field is a combination of electric and magnetic force vectors containing a defined amount of energy. In a microwave circuit, this energy is guided from device to device in a transmission line. A key aspect in designing microwave circuits is moving this energy with maximum power transfer to the destination load and minimum energy reflected from the destination load.

6.2 Maximum Power Transfer

The key to maximum power transfer is matching the source impedance with the load impedance. The circuit in Figure 6.1 is analyzed to find the maximum power transferred to load resistor RL as a function of the ratio of load resistor to source resistor (RL/Rs). Figure 6.2 is a graph of the result of the analysis with the maximum power is normalized to 1W. As shown in Figure 6.2, maximum power is delivered to the load when the source resistance (Rs) is equal to the load resistance (RL). Load resistances higher and lower than the source impedance result in less power delivered to the load.

$$\text{Maximum Power Transfer} \quad Rs = RL \qquad (6.1)$$

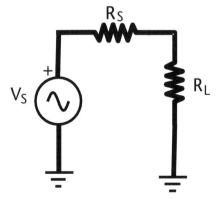

Figure 6.1 Lumped-element circuit used to calculate normalized power.

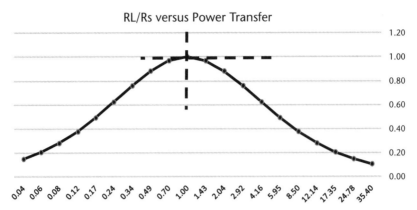

Figure 6.2 Graph of normalized power transfer as a function of *RL/Rs*.

A similar analysis can be performed for complex load and source impedances. Under this condition, maximum power transfer occurs when the load impedance is equal to the complex conjugate of the source impedance [1].

6.3 Electromagnetic Waves Traveling Through an Infinite Transmission Line

The effect on an electromagnetic microwave traveling down an infinite transmission line (no reflections) shown in Figure 6.3 is a finite distributed group delay and distributed attenuation as a function of distance x, the distance that the wave has traveled [2].

6.3.1 Distributed Attenuation

The distributed attenuation in decibels per unit length is due to the resistivity of the conductor (a series resistance) and the conductivity (a parallel resistance) of the dielectric material between the conductor and the ground plane. Figure 6.4 is an illustration of a microstrip transmission line showing the source circuit, the distributed conductor losses, the distributed dielectric losses, and the direction of signal travel. The illustrations show a microstrip transmission line, but the same issues of distributed loss can be related to other transmission-line topologies [3].

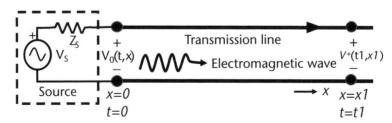

Figure 6.3 A signal moving down a transmission line in the *x* direction.

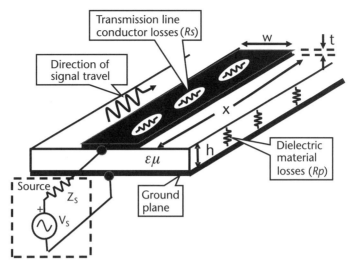

Figure 6.4 A microstrip transmission line with its respective series and parallel losses.

The signals' distributed conductor loss (Rs) and dielectric constant loss (Rp) are calculated independently. The results in decibels (dB) are summed together, giving a composite loss for the transmission line in decibels per unit length [dB/ (unit value of x)].

6.3.2   Distributed Delay

Signal delays become significant when the phase of a signal at the source is different than the phase of the signal at the destination. Phase differences are relative terms that require further definition that is application-related. In the absence of a specific application, there are many rules of thumb that are used in the industry. For this chapter, a signal transitioning from a source to a destination will have a significant delay if the phase difference for a sinusoidal signal results in an amplitude change greater than 5% at the signal's peak. A 5% amplitude change translates to a phase change of 18° for a cosine wave around 0°. Knowing the frequency of the signal and the signal's velocity through the medium that it is traveling, the phase change can be calculated along with defining the parameter of the significant distance (D_{SL}). The procedure is outlined below:

1. The signal traveling through a transmission line with an effective dielectric constant (ε_{eff}) has an effective group velocity (Vs) given by

$$Vs = \frac{c}{\sqrt{\varepsilon_{\text{eff}}}} \qquad (6.2)$$

where c is the speed of light in a vacuum and ε_{eff} is the effective dielectric constant of the material.

2. The wavelength (λ) of the signal in a dielectric media is found from the signal velocity (Vs) and signal frequency (f) in hertz:

$$\lambda = \frac{Vg}{f} \tag{6.3}$$

3. A 5% change in peak amplitude of a sinusoidal signal is approximate 18° or one-twentieth of a wavelength. The distance between source and load (D_{SL}) is the significant distance defined in this chapter as ≥5% of a wavelength ($D_{SL} \geq \lambda/20$):

$$\text{Significant delay: } D_{SL} \geq \frac{\lambda}{20} = \frac{Vg}{20 \cdot f} = \frac{c}{20 \cdot f \cdot \sqrt{\varepsilon_{\text{eff}}}} \tag{6.4}$$

where c is the speed of light in a vacuum, ε_{eff} is the effective dielectric constant of the material, f is the frequency of the signal in Hertz, and c and D_{SL} are in the same units of length.

6.4   Reflected Waves in a Transmission Line

When signals travel down a transmission line with impedance (Z_0) hit a change in impedance (Z_L), part of the signal is transmitted through the impedance change and part of the signal is reflected to the source [4]; see Figure 6.5.

The ratio of the reflected signal (V_r) with respect to the incident signal (V_i) is the reflection coefficient (Γ) [5]:

$$\Gamma = \frac{V_r}{V_i} \tag{6.5}$$

The incident voltage wave (V_i) and the reflected voltage wave (V_r) are vector quantities with an associated amplitude and phase. The reflection coefficient (Γ) is therefore a complex number with a respective amplitude (ρ) and phase (Φ).

$$|\Gamma| = \rho \tag{6.6}$$

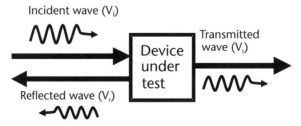

Figure 6.5 Incident electromagnetic wave in a transmission line interfacing with a device having an input impedance different than the characteristic impedance of the transmission line.

$$\arg \Gamma = \Phi \tag{6.7}$$

The reflection coefficient (Γ) can be expressed as a vector quantity

$$\Gamma \rightarrow \rho \angle \Phi \tag{6.8}$$

The complex reflection coefficient is related to the source (Z_S) and load (Z_L) impedances (see Figure 6.6), both of which could be complex quantities [6].

$$\Gamma = \frac{V_r}{V_i} = \frac{Z_L - Z_S}{Z_L + Z_S} \tag{6.9}$$

It must also be noted that the source and load impedances are may be a function of frequency (f):

Source impedance: $Z_S(f)$

Load impedance: $Z_L(f)$

The reflection coefficient is therefore also a complex function that is frequency-dependent [$\Gamma(f)$]:

Figure 6.6 shows a source consisting of an ideal voltage generator (V_S) and an impedance Z_S. The source creates an incident signal (V_i) driving into a transmission line with an associated impedance Z_S connected to a load (device under test) with an input impedance Z_L. There is no reflected signal from the interface between the source and transmission line because the source and transmission line impedance are the same ($\Gamma = 0$). The incident wave (V_i) travels down the transmission line until it reaches the load (device under test) with an input impedance Z_L [7]. If Z_L is not equal to Z_S ($Z_L \neq Z_S$), a portion of the incident signal is reflected into the transmission line (V_r). The reflected signal (V_r) characteristic is determined by:

$$V_r = \Gamma \cdot V_i = V_i \cdot \frac{Z_L - Z_S}{Z_L + Z_S} \tag{6.10}$$

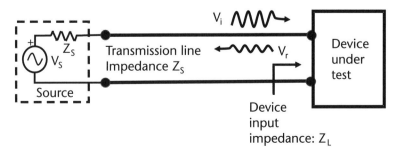

Figure 6.6 A source with output impedance Z_S drives a transmission line with impedance Z_S into a device under test with input impedance Z_L, when $Z_L \neq Z_S$.

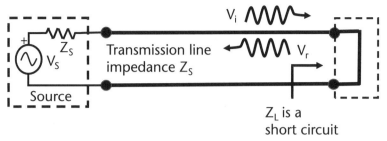

Figure 6.7 Signal reflected from a short circuit.

6.4.1 Reflections from a Short-Circuit Load Impedance

If the load impedance (Z_L) is a short circuit $(Z_L = 0)$ (see Figure 6.7), the entire incident signal (V_i) is reflected to the source 180° out of phase as shown in (6.11).

$$V_r = \Gamma \cdot V_i = V_i \cdot \frac{0 - Z_S}{0 + Z_S} = -1 \cdot V_i \qquad (6.11)$$

6.4.2 Reflections from an Open-Circuit Load Impedance

If the load impedance (Z_L) is an open circuit $(Z_L = \infty)$ (see Figure 6.8), the entire incident signal (V_i) is reflected to the source in phase as shown in (6.12).

$$V_r = \Gamma \cdot V_i = V_i \cdot \frac{Z_L - Z_S}{Z_L + Z_S} = V_i \cdot \frac{\infty - Z_S}{\infty + Z_S} \approx +1 \cdot V_i \qquad (6.12)$$

An open circuit at microwave frequencies is difficult to realize because signals at the end of an open-circuited transmission line tend to radiate into the air.

6.5 Voltage Standing Wave Ratio

The reflection coefficient (Γ) is the ratio of the reflected wave to incident wave. Superposition of these two waves creates a standing wave pattern [1] as shown in Figure 6.9.

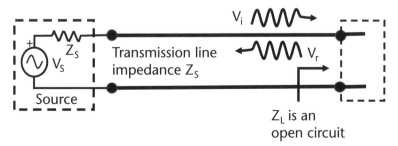

Figure 6.8 Signal reflected from an open circuit.

Figure 6.9   Standing-wave pattern showing the minimum and maximum peaks

The voltage standing wave ratio (VSWR) is a ratio of the maximum to minimum peak value of the standing wave.

$$VSWR = \frac{Vp_{max}}{Vp_{min}} \qquad VSWR \geq 1 \qquad (6.13)$$

When the minimum peak equals the maximum peak, there is no reflected wave and the VSWR is an ideal value of 1:1. If the entire incident signal is reflected to the source, the standing wave exhibits periodic complete cancellation ($Vp_{min} = 0$) and the VSWR goes to infinity (∞).

6.5.1   VSWR as a Function of the Reflection Coefficient

The maximum and minimum peaks of the standing wave created from the incident and reflected waves suggest a relationship between VSWR and the magnitude of the reflection coefficient ($|\rho|$). Equation (6.14) shows the relationship between these parameters.

$$VSWR = \frac{1+\rho}{1-\rho} \qquad (6.14)$$

For example, assuming that the magnitude of the reflection coefficient is 0.5 ($|\Gamma| = 0.5$), the VSWR = 3:1.

6.5.2   Reflection Coefficient as a Function of the VSWR

Solving (6.14) for reflection coefficient (ρ) results in (6.15), which shows reflection coefficient as a function of VSWR.

$$|\Gamma| = \rho = \frac{|V_r|}{|V_i|} = \frac{VSWR - 1}{VSWR + 1} \qquad (6.15)$$

6.5.3   VSWR as a Function of the Source and Load Impedance

As shown in (6.9), the reflection coefficient (Γ) can represented as a function of the source impedance (Z_S) and load impedance (Z_L): $\Gamma = (Z_L - Z_S)/(Z_L + Z_S)$. The source and load impedances are complex numbers resulting in a complex reflection coefficient (Γ).

Dividing the numerator and denominator in (6.9) by Z_S:

$$\Gamma = \frac{Z_L/Z_S - 1}{Z_L/Z_S + 1} \qquad (6.16)$$

The magnitude of the reflection coefficient is:

$$|\Gamma| = \left| \frac{Z_L/Z_S - 1}{Z_L/Z_S + 1} \right| = \rho \qquad (6.17)$$

Dividing the numerator and denominator in (6.9) by Z_L:

$$\Gamma = \frac{1 - Z_S/Z_L}{1 + Z_S/Z_L} = -\left(\frac{Z_S/Z_L - 1}{Z_S/Z_L + 1} \right) \qquad (6.18)$$

The magnitude of the reflection coefficient is:

$$|\Gamma| = \left| \frac{Z_S/Z_L - 1}{Z_S/Z_L + 1} \right| = \rho \qquad (6.19)$$

From (6.17) and (6.19), it is obvious that if Z_S and Z_L are real, (Z_L/Z_S) or (Z_S/Z_L), whichever is greater than unity, can be substituted in (6.15) to find the reflection coefficient (ρ). This leads to the relationships shown in (6.20) and (6.21), which express VSWR in terms of the source and load impedance.

If Z_S and Z_L are real,

$$VSWR = \frac{Z_L}{Z_S} \text{ if } Z_L > Z_S \qquad (6.20)$$

or if Z_S and Z_L are real,

$$VSWR = \frac{Z_S}{Z_L} \text{ if } Z_S > Z_L \qquad (6.21)$$

The requirement that the source impedance (Z_S) and load impedance (Z_L) be real holds for an exact solution of VSWR. If the real parts of Z_S and Z_L are greater than the imaginary parts such that $|Z_S| \approx \text{Real}(Z_S)$ and $|Z_L| \approx \text{Real}(Z_L)$, (6.22) and (6.23) are valid:

$$\text{If } |Z_S| \approx \text{Real}(Z_S) \text{ and } |Z_L| \approx \text{Real}(Z_L): VSWR \approx \frac{|Z_L|}{|Z_S|} \text{if } |Z_L| > |Z_S| \qquad (6.22)$$

or

$$\text{If } |Z_S| \approx \text{Real}(Z_S) \text{ and } |Z_L| \approx \text{Real}(Z_L): \; VSWR \approx \frac{|Z_S|}{|Z_L|} \text{ if } |Z_S| > |Z_L| \qquad (6.23)$$

The conditions mentioned in (6.22) and (6.23) are not uncommon since power is only transferred into the real part of the impedance. Device impedances must be designed with a significant real part for efficient operation.

6.6   Mismatch Loss

In a lossless system, the total signal power is conserved. Figure 6.10 describes the signal path showing the incident power wave (P_i) coming from a source impedance Z_S, the reflected power wave (P_r) due to the impedance change ($Z_S \neq Z_L$) and the transmitted power wave (P_t) [8].

Since the impedance discontinuity is assumed lossless, the total power into the discontinuity (Z_L) equals the power emanating from the discontinuity.

$$P_i = P_r + P_t \qquad (6.24)$$

where P_i is the incident power from source impedance Z_S, P_r is the reflected power from load impedance Z_L, and P_t is the transmitted power through load impedance Z_L.

The power transmitted through the mismatch is:

$$P_t = P_i - P_r \qquad (6.25)$$

The power loss or mismatch loss in decibels (MLdB) is:

$$MLdB = -10 \cdot \text{Log}\left(\frac{P_i}{P_t}\right) = -10 \cdot \text{Log}\left(\frac{P_i}{P_i - P_r}\right) \qquad (6.26)$$

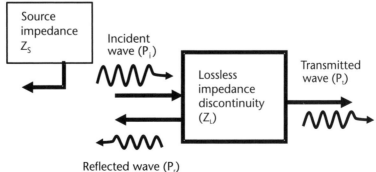

Figure 6.10   Incident wave through an impedance discontinuity.

The square of the reflection coefficient (ρ^2) relates the reflected signal power (P_r) to the incident signal power (P_i)

$$\rho^2 = \frac{P_r}{P_i} \tag{6.27}$$

Substituting (6.27) into (6.25), the transmitted power (P_t) is:

$$P_t = P_i \cdot \left(1 - \rho^2\right) \tag{6.28}$$

The mismatch loss in decibels (MLdB) can therefore be expressed in term of the reflection coefficient (ρ):

$$MLdB = -10 \cdot \log\left(\frac{P_i}{P_t}\right) = -10 \cdot \log\left(1 - \rho^2\right) \tag{6.29}$$

Note that MLdB is mismatch loss and therefore a positive decibel. If the discussion was in terms of gain (GaindB) through the system, the transmitted power (P_t) and the incident power (P_i) ratio would be inverted and the factor GaindB would be negative.

The magnitude of the reflection coefficient (ρ) can expressed in terms of VSWR (6.15). Substituting (6.15) into (6.29) gives an expression for mismatch loss in terms of VSWR:

$$MLdB = -10 \cdot \log\left(1 - \rho^2\right) = -10 \cdot \log\left(1 - \left(\frac{VSWR - 1}{VSWR + 1}\right)^2\right) \tag{6.30}$$

Table 6.1 lists some sample VSWR values and the corresponding results for the reflection coefficient (ρ), the return loss in decibels, and mismatch loss in decibels.

With regard to mismatch loss, any nonideal component of a microwave system that has an input and output will have a mismatch loss at that interface, the mismatch losses are usually added to the loss or subtracted from the gain of the device at the interface, and mismatch losses are typically not accumulated as a separated item in the system chain analysis.

6.7   Mismatch Uncertainty

Consider the interface of two devices as shown in Figure 6.11. The output of device number one (Device 1) has a VSWR, VSWR1 greater than 1 and the input of device number two (Device 2) has a VSWR, VSWR2 greater than 1. The source and load impedance not being perfectly matched indicates the presence of reflected signals (i.e., an expected mismatch loss). The mismatch loss calculated in Section 6.5 assumed that the signal reflected from the load is absorbed by a perfectly matched source

Table 6.1 Reflection Coefficient (ρ), Return Loss, and Mismatch Loss as a Function of Typical Values of VSWR

(1) VSWR:1	(2) Reflection Coefficient Rho (ρ)	(3) Return Loss dB	(4) Mismatch Loss dB
1.10	0.048	26.44	0.010
1.15	0.070	23.13	0.021
1.25	0.111	19.08	0.054
1.35	0.149	16.54	0.097
1.50	0.200	13.98	0.177
1.60	0.231	12.74	0.238
1.70	0.259	11.73	0.302
1.80	0.286	10.88	0.370
1.90	0.310	10.16	0.440
2.00	0.333	9.54	0.512
2.50	0.429	7.36	0.881
3.00	0.500	6.02	1.249
3.50	0.556	5.11	1.603
4.00	0.600	4.44	1.938
5.00	0.667	3.52	2.553

impedance. Most of the time as shown in Figure 6.11, the source impedance is not ideal (perfectly matched). Under these conditions the signal reflected from the input of Device 2 going back to the source is again reflected again at the output of Device 1 creating another incident signal (V_2) [9]. This original incident signal (V_1) and the twice reflected signal (V_2) both travel to Device 2 coherent in frequency but with an uncertain relative phase [8].

Note that the phase difference between the original incident signal (V_1) and the twice reflected signal (V_2) is a function of the phase of the reflection coefficient at the input to Device 2 (Γ_2), the reflection coefficient at the output of Device 1 (Γ_1) and the distance between Device 1 and Device 2.

Figure 6.12 shows the two signals V_1 and V_2 as vectors added together, creating a composite vector V_3. The angle (θ) between the original vector (V_1) and the composite vector (V_3) is uncertain and can be any value from 0° to 360°.

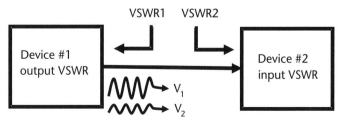

Figure 6.11 Interface of the output of Device 1 with the input of Devices 2. V_1 is the original incident signal and V_2 is the signal from Device 1 and reflected from Device 2.

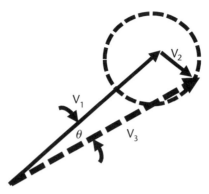

Figure 6.12 The incident signal is voltage vector V_1. V_2 is reflected from the input of Device 2 and the output of Device 1. Vector V_3 is the sum vector.

6.7.1 Amplitude Uncertainty

The amplitude of the vector V_2 is the reflection coefficient of VSWR2 (ρ_2) times the reflection coefficient of VSWR1 (ρ_1) added to the incident vector V_1 to create the sum vector V_3. The extreme values of amplitude uncertainty of sum vector V_3 is:

$$V_3 \max = 1 + \rho_1 \cdot \rho_2 \tag{6.31}$$

$$V_3 \min = 1 - \rho_1 \cdot \rho_2 \tag{6.32}$$

The maximum level with respect to the input signal in decibels is $V_1 \mathrm{dB}_{max}$.

$$V_3 \mathrm{dB}_{max} = 20 \cdot \log\left(1 + \rho_1 \cdot \rho_2\right) \tag{6.33}$$

The minimum level with respect to the input signal in decibels is $V_1 \mathrm{dB}_{min}$.

$$V_3 \mathrm{dB}_{min} = 20 \cdot \log\left(1 - \rho_1 \cdot \rho_2\right) \tag{6.34}$$

6.7.2 Phase Uncertainty

The phase uncertainty is the phase change (θ) of the composite vector V_3 as the interference vector V_2 rotates around the original incident vector V_1; see Figure 6.12. The maximum phase uncertainty will occur when the vector V_2 is perpendicular to vector V_1. Under this condition, the peak phase change ($\Delta\theta$) is:

$$\Delta\theta \ (\text{radians}) = \pm\sin^{-1}\left(\rho_1 \cdot \rho_2\right) \tag{6.35}$$

Using the small angle approximation:

$$\Delta\theta \ (\text{radians}) \approx \pm\left(\rho_1 \cdot \rho_2\right) \tag{6.36}$$

Converting radians to degrees, the uncertainty in phase is:

$$\Delta\theta \,(\text{degrees}) = \pm(\rho_1 \cdot \rho_2) \cdot \frac{180}{\pi} \tag{6.37}$$

6.7.3 VSWR Uncertainty

VSWR is typically given with respect to the nominal characteristic impedance of the system which most of the time is 50Ω. When two devices are connected, the VSWR between the two devices is uncertain, assuming that the source and load complex impedances were not actually measured at each frequency over the range of interest.

From (6.22) and (6.23), (VSWR = Z_L/Z_S, if $Z_L > Z_S$ or VSWR = Z_S/Z_L if $Z_S > Z_L$), it is shown that VSWR is a function of the magnitude of the source impedance (Z_S) and load impedance (Z_L) if the source and load impedance is real or approximately real. The VSWR uncertainty factor is an approximation also assuming the Z_S and Z_L are real or approximately real. The uncertainty factor is bound by the minimum and maximum VSWR given that the source impedance (Z_S) and load impedance (Z_L) both assume real quantities.

$$VSWR_{\text{MAX}} = VSWR_S \cdot VSWR_L \tag{6.38}$$

$$VSWR_{\text{MIN}} = \frac{VSWRZ_S}{VSWR_L} \qquad \text{for } VSWR_S > VSWR_L \tag{6.39}$$

or

$$VSWR_{\text{MIN}} = \frac{VSWR_L}{VSWRZ_S} \qquad \text{for } VSWR_L > VSWR_S \tag{6.40}$$

where $VSWR_S$ is the VSWR of the source and $VSWR_L$ is the VSWR of the load.

For example, for the characteristic impedance (Z_0) equal to 50Ω

$$VSWR_S = 3:1 \qquad \left(Z_S = 3 \cdot Z_0 \text{ or } Z_S = \frac{Z_0}{3} \right) \tag{6.41}$$

Therefore, $Z_S = 150\Omega$ or $Z_S = 16.7\Omega$

$$VSWR_L = 1.5:1 \qquad \left(Z_L = 1.5 \cdot Z_0 \text{ or } Z_L = \frac{Z_0}{1.5} \right) \tag{6.42}$$

Therefore, $Z_L = 75\Omega$ or $Z_L = 33.3\Omega$

$$VSWR_{\text{MAX}} = 4.5:1 \qquad \left(\frac{150\Omega}{33.3\Omega} \text{ or } \frac{75\Omega}{16.7\Omega} \right) \tag{6.43}$$

$$VSWR_{\text{MIN}} = 2.0:1 \qquad \left(\frac{150\Omega}{75\Omega} \text{ or } \frac{33.3\Omega}{16.7\Omega} \right) \tag{6.44}$$

6.8   Matching Impedances

Matching impedance within or as an add-on to a device is important because it reduces the signal transmission uncertainties in amplitude and phase. Some common techniques used to improve the impedance match between devices are: (1) use a quarter-wave matching transformer, and (2) use a matched resistive attenuator. Both techniques can be used on the input or output of the device [10].

6.8.1   Matching with a Quarter-Wave Transmission Line

A popular technique used to match devices in a microstrip circuit is to insert a quarter-wave transmission line at a calculated impedance geometrically between the load impedance and the desired impedance [11] (see Figure 6.13). This design technique in its simplest form is typically effective over a ±15% range around the designed center frequency depending upon the acceptable VSWR. Multiple sections of quarter-wave transmission lines can be used to increase the effective bandwidth.

Equation (6.45) is a generalized equation that calculates the impedance looking into a transmission line of characteristic impedance Z_T terminated in a real impedance R_L.

$$Z_{IN} = Z_T \cdot \frac{R_L + j \cdot Z_T \cdot \tan(\beta_S \cdot l)}{Z_T + j \cdot R_L \cdot \tan(\beta_S \cdot l)} \tag{6.45}$$

where Z_{IN} is the impedance looking into the transmission line, Z_T is the impedance of the transmission line, β_S is in radians per unit wavelength [$\beta_S = (2 \cdot \pi)/\lambda$], in which λ is the wavelength of the actual signal, and l is the length of the transmission-line transformer.

If the transformer length (l) is a quarter-wave at the designed frequency of interest (F_0) (i.e., the frequency of optimization). The wavelength (λ_0) is the respective wavelength at F_0 found by solving (6.46).

$$Vg = F_0 \cdot \lambda_0 \rightarrow \lambda_0 = \frac{Vg}{F_0} \tag{6.46}$$

where Vg is the propagation velocity of the signal through the transmission line.

The length of the transmission line is given by

$$l = \frac{\lambda_0}{4} = \frac{Vg}{4 \cdot F_0} \tag{6.47}$$

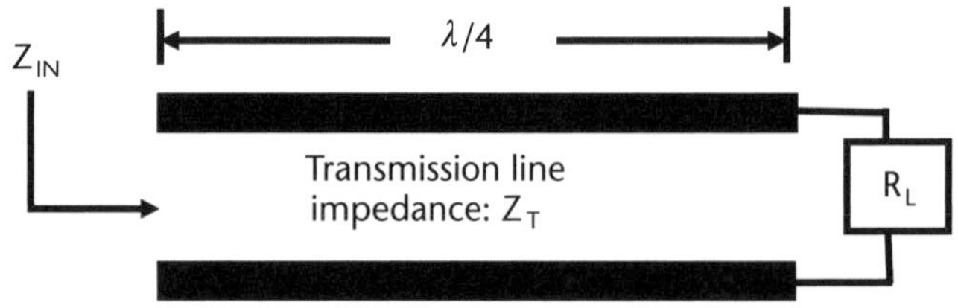

Figure 6.13   A quarter-wave transmission line matching circuit.

if the actual signal is at the optimum design frequency ($\lambda = \lambda_0$)
From (6.45),

$$\beta_S \cdot l = \frac{2 \cdot \pi}{\lambda} \cdot \frac{\lambda_0}{4} = \frac{\pi}{2} \tag{6.48}$$

$$\tan(\beta_S \cdot l) = \tan\left(\frac{\pi}{2}\right) \to \infty \tag{6.49}$$

and (6.45) becomes

$$Z_{\mathrm{IN}} = Z_T \cdot \frac{R_L + j \cdot Z_T \cdot \tan\left(\dfrac{\pi}{2}\right)}{Z_T + j \cdot R_L \cdot \tan\left(\dfrac{\pi}{2}\right)} = \frac{(Z_T)^2}{R_L} \tag{6.50}$$

Equation (6.50) can be solved for the impedance of a quarter-wave transmission line (Z_T) required to transform a real load impedance (R_L) to a desired characteristic impedance (Z_{IN}).

$$Z_T = \sqrt{(Z_{\mathrm{IN}} \cdot R_L)} \tag{6.51}$$

Equation (6.45) can be used to solve for the resultant input impedance at frequencies other than the optimal design frequency (F_0). The usable bandwidth (ΔF) of the design (a single-section quarter-wave transformer) is approximately 30% of F_0 for a theoretical VSWR less than 1.2:1.

6.8.2   Using a Matched Resistive Attenuator to Improve VSWR

A common technique used to improve the VSWR by reducing the reflected signal from the input of a device is to place an attenuator at the device input as shown in Figure 6.14. The attenuator reduces the reflected signal two times the value of the attenuation, while the transmitted signal is reduced by the attenuation value. The same technique could be used to improve the output VSWR of a device by placing the attenuator at the device output.

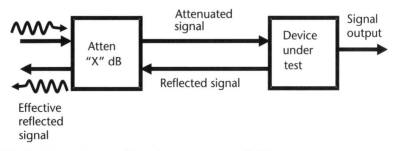

Figure 6.14   Device under test with an input attenuator (X dB).

Using (6.14) and inserting $\rho = |\Gamma|$ = return loss,

$$VSWR = \frac{1+\rho}{1-\rho} \text{ where } \rho = \frac{\left|\text{Reflected Voltage } (Vr)\right|}{\left|\text{Incident Voltage } (Vi)\right|} \tag{6.52}$$

VSWR in terms of ρ is:

$$\rho = \frac{VSWR - 1}{VSWR + 1} \tag{6.53}$$

The return loss in decibels [RL(dB)] is:

$$RL(\text{dB}) = 20 \cdot \log(\rho) = 20 \cdot \log\left(\frac{VSWR - 1}{VSWR + 1}\right) \tag{6.54}$$

Solving for VSWR in terms of [RL(dB)]:

$$VSWR = \frac{10^{\frac{RL(\text{dB})}{20}} + 1}{10^{\frac{RL(\text{dB})}{20}} - 1} \quad \text{Note: } VSWR : 1 \tag{6.55}$$

Table 6.2 shows the device VSWR (1), return loss (2) with respect to the VSWR in column (1), and the final VSWR when attenuation is added in the respective columns (3) through (7).

Table 6.2 Improved VSWR as a Function of Added Attenuation

		Added Attenuation				
(1) Device VSWR	(2) Return Loss dB Improved	(3) dB 1 VSWR	(4) dB 3 VSWR	(5) dB 5 VSWR	(6) dB 7 VSWR	(7) dB 9 VSWR
1.20	−20.83	−1.16	−1.10	−1.06	−1.04	−1.02
1.30	−17.69	−1.23	−1.14	−1.09	−1.05	−1.03
1.50	−13.98	−1.38	−1.22	−1.14	−1.08	−1.05
2.00	−9.54	−1.72	−1.40	−1.24	−1.14	−1.09
2.50	−7.36	−2.03	−1.55	−1.31	−1.19	−1.11
3.00	−6.02	−2.32	−1.67	−1.38	−1.22	−1.13
3.50	−5.11	−2.58	−1.77	−1.43	−1.25	−1.15
4.00	−4.44	−2.82	−1.86	−1.47	−1.27	−1.16
4.50	−3.93	−3.04	−1.94	−1.50	−1.29	−1.17
5.00	−3.52	−3.25	−2.00	−1.53	−1.31	−1.18
5.50	−3.19	−3.44	−2.06	−1.56	−1.32	−1.19

The attenuators used to improve VSWR are assumed ideal. The attenuator has a perfectly matched VSWR input and output. The attenuator reflection coefficient and throughput has only real parts and no imaginary parts.

References

[1] "Agilent Fundamentals of RF and Microwave Power Measurements (Part 3)," Agilent Technologies, 2011.

[2] Wu, T-L., "Microwave Filter Design," Lecture slides, National Taiwan University.

[3] Cruz-Pol, S., "Transmission Lines," Lecture slides, University of Puerto Rico Mayaguez.

[4] Rosu, I., "Microstrip, Stripline, and CPW Design," Whitepaper, 2014.

[5] Adam, S., "Microwave Theory and Applications," Hewlett Packard.

[6] Jackson, D., "Transmission Line Theory," Dept. of ECE, 2015.

[7] Leddige, M., "Transmission Line Basics II," Intel Corp.

[8] Dobbert, M., and J. Gorin, "Revisiting Mismatch Uncertainty with the Rayleigh Distribution," Agilent Technologies, 2011.

[9] Castro, A., "RF/uW Measurement Uncertainty: Calculate, Characterize, Minimize," Aerospace & Defense Symposium, 2013.

[10] Amanogawa, "Transmission Lines," Digital Maestro Series, 2006.

[11] Pozar, D., *Microwave Engineering*, 4th ed., New York: John Wiley & Sons, 2011.

CHAPTER 7

Noise in Microwave Circuits

7.1 Introduction

Noise is a very broad term with a lot of meanings and definitions depending on the application. With respect to microwave amplifier circuits, the study of noise can be limited to thermal noise, flicker noise, and phase noise. Thermal noise of a device under test (DUT) can be modeled as a voltage (or power) generator in series or linearly adding to the input signal [1]. Flicker noise relates to thermal noise close to the carrier and is usually added to the thermal noise resulting in an increase in the noise spectral density proportional to the carrier offset [2]. Phase noise modulates the carrier unlike thermal noise which adds to the carrier. As a carrier modulation phase noise is usually expressed in dBc/Hz (i.e., level with respect to the carrier in a normalized 1-Hz bandwidth). Phase noise decreases as the carrier offset frequency increases and becomes negligible when its level goes below the thermal noise level.

7.2 Properties of Noise

Thermal noise is usually the dominant noise issue in an amplifier design and is therefore discussed in depth. The mechanisms causing flicker noise and phase noise are explained along with their effect on the carrier and under what conditions these issues can become significant to the system performance.

7.2.1 Thermal Noise

Thermal noise, sometimes called Johnson-Nyquist noise, Johnson noise, or Nyquist noise, is the electronic noise generated by the random motion of electrons inside resistive devices or the real part of the impedance in reactive devices with no externally applied potential [3]. The electron motion exists at all temperatures greater than absolute zero (0K) and increases as the temperature increases. Noise power increases proportional to the increase in temperature in degrees kelvin (K). The random motion of electronics produces no net current flow, but the motion can be measured as a mean square noise value across a resistance [4].

From quantum mechanics the noise voltage ($V_n(f)$) spectrum is:

$$V_n(f) = \sqrt{\frac{4 \cdot R \cdot B \cdot h \cdot f}{e^{\left(\frac{h \cdot f}{k \cdot T}\right)} - 1}} \tag{7.1}$$

where h is Planck's constant, k is Boltzmann's constant, T is temperature in degrees kelvin, R is the resistance in ohms, f is the frequency in hertz, and B is the bandwidth in hertz.

For frequencies in the microwave region (frequencies < 100 GHz), $h \cdot f << k \cdot T$

$$e^{\left(\frac{h \cdot f}{k \cdot T}\right)} - 1 \approx \frac{h \cdot f}{k \cdot T} \tag{7.2}$$

Equation (7.1) simplifies to:

$$V_n(f) = \sqrt{4 \cdot R \cdot B \cdot k \cdot T} \tag{7.3}$$

The equivalent noise circuit is shown in Figure 7.1.

V_n is a voltage noise generator with output resistance R. The load for maximum power transfer is also R, the same as the source resistance [5]. The noise power (P_n) into the load is:

$$P_n = \left(\frac{Vn}{2R}\right)^2 R = \frac{V_n^2}{4 \cdot R} = \frac{4 \cdot R \cdot B \cdot k \cdot T}{4 \cdot R} = k \cdot T \cdot B \tag{7.4}$$

Noise power (P_n) at 298K is 4.114×10^{-9} picowatts or -203.857 dBW $\rightarrow$ -173.875 dBm.

Thermal noise is a random process with its instantaneous amplitude being unpredictable and its average value over a long period of time being very predictable. Typically, averaging times should be much greater than the inverse of the noise bandwidth (B_n) (i.e., averaging time >> $1/B_n$). The noise amplitude exhibits a Gaussian distribution with a mean (μ) and a standard deviation (σ) where μ is zero for unbiased noise and σ is the measured noise power [root mean square (RMS) level]. Figure 7.2 depicts noise as a function of time with an approximation of a Gaussian distribution curve vertically on the right. The curve shows the measured noise power of one standard deviation (1σ) is contained in a Gaussian distribution of noise with a 68.2% probability and three standard deviations ($3 \cdot \sigma$) with a 99.7% probability.

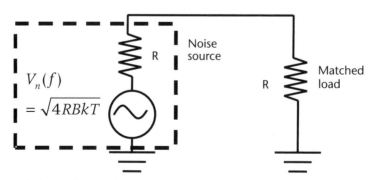

Figure 7.1 Equivalent noise circuit.

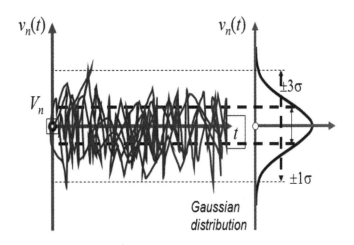

Figure 7.2   Thermal noise.

7.2.2   Flicker Noise

Flicker noise is the dominant noise at low frequencies in semiconductors, occurring only when electric current is flowing. It usually originates from carriers of DC that are held for some time and then released. This is assumed to be due to impurities in the semiconductors' material.

The spectral density function of flicker noise is approximately proportional to $1/f^\alpha$ where alpha (α) is in the range of 0.5 to 1.5 and typically assumed to be 1 ($\alpha = 1$) when actual measured data is not available [6]. The power spectral density function starts at a very low frequency and decreases at a rate of 10 dB per decade until it reaches the level of thermal noise and then it becomes negligible. The frequency at the intersection of flicker noise and thermal noise is called the corner frequency (fc) [7]; see Figure 7.3. Many times, the corner frequency is assumed to be between 1 kHz and 10 kHz, but it can be much lower or significantly higher. The best determination for corner frequency is empirically through measured data.

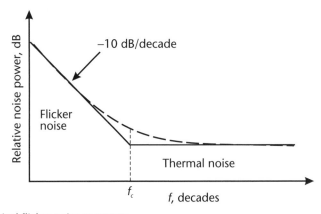

Figure 7.3   Typical flicker noise spectrum.

7.2.3   Phase Noise

Phase noise is due to the thermal noise phase modulating a carrier. It is primarily associated with an oscillator, but can also be an important parameter in amplifier designs. The resultant phase modulated carrier with a narrowband modulation index β in a normalized 1-Hz bandwidth has a single-sided spectral component equal to $20 \cdot \text{Log}(\beta/2)$. The noise with an assumed constant modulation level decreases at a rate of 20 dB per decade as the frequency offset from the carrier increases. Flicker noise increases the noise closer to the carrier by 10 dB per decade making the total noise increasing as the offset from the carrier decreases by 30 dB per decade below the flicker noise corner frequency (*fc*) [8]. Offset frequencies above the point where phase noise and thermal noise intersect is usually dominated by thermal noise and therefore has a flat spectral density function assuming other external effects are not included in the result [9]. The notation dBc/Hz is commonly used to indicate the phase noise level (i.e., noise power with respect to the carrier). This not an algebraic notation, dBc refers to the level of noise with respect to the carrier and the notation "/Hz" refers to the measurement resolution bandwidth of 1 Hz. Figure 7.4 shows a typical phase noise response. This spectral density curve does not include the effects of other external frequency-dependent networks.

7.3   Noise Figure

The concept of noise figure (NF), and by extension noise factor (F), was introduced in the 1940s by Harold Friis. He defined the noise factor as the ratio of the signal-to-noise power ratio at the input to the signal-to-noise power ratio at the output, which is a measure of the ability of a demodulator to resolve the information contained in the signal. Noise figure (NF in dB), used and misused interchangeably with noise factor (F), is the decibel equivalent of noise factor (F in linear units).

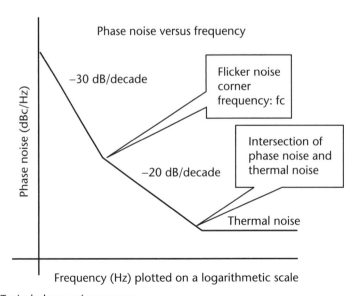

Figure 7.4   Typical phase noise response.

The noise figure is the decrease (or degradation) in the signal-to-noise ratio as the signal goes through a network. To analyze the system effects, each component in the network chain, characterized by its gain (G) and noise figure (NF) (see Figure 7.5), degrades the signal-to-noise ratio and therefore increases the network's noise figure. As an example, an ideal amplifier would amplify the signal and noise at its input maintaining the same signal-to-noise ratio at its input and at its output. A real amplifier adds its own internal noise to the input signal-to-noise ratio degrading the output signal-to-noise of the device.

The signal-to-noise ratio [10] is a very important parameter in most microwave systems and minimizing the degradation is a primary design criterion. NF is a measure of degradation in signal to noise in decibels as the signal progresses through a system. Noise factor (F) is the linear degradation in the signal-to-noise ratio [11]. The relationship between these parameters is shown in (7.5) and (7.6).

$$F = \frac{S_{in}/N_{in}}{S_{out}/N_{out}} \tag{7.5}$$

$$NF = 10\text{Log}(F) \tag{7.6}$$

where S_{in} is the input signal level, N_{in} is the input noise level, S_{out} is the output signal level, N_{out} is the output noise level, F is the noise factor, and NF is the noise figure.

In any real amplifier or component in a system, noise factor (F) is always greater than 1 $(F > 1)$, and the noise figure of that amplifier or component is always greater than 0 $(NF > 0$ dB). Designing a system is always a trade-off involving minimizing noise figure to maintain the highest possible signal-to-noise ratio, minimizing distortion due to high signal levels and maximizing the systems dynamic range (i.e., operating range of the microwave system).

Signal sensitivity in communication, radar, or electronic warfare (EW) receiving systems is primarily limited by thermal noise and the final output signal-to-noise ratio through the chain of microwave and RF components. Signal to noise at the input of a microwave system is the best it will be (i.e., each component in the receiver cascade, while performing its intended function, degrades the output signal to noise).

7.4   Noise Temperature

Many low noise systems use the concept of noise temperature (T_e) instead of noise factor and noise figure. These parameters are interrelated and interchangeable and are used as a convenience when adding the effects of an external noise source in a very low noise receiving system. Thermal noise as previously stated is $k \cdot T \cdot B$,

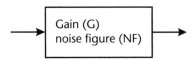

Figure 7.5   Typical device characterized by its gain (G) and noise figure (NF).

where k is Boltzmann's constant, T is the temperature in degrees kelvin, and B is the system bandwidth in hertz. When noise is added to a system (ΔN), internally or externally, it can be represented as an increase in the noise temperature (T). The total increase in added noise is represented as an added term with a noise equivalent temperature T_e [12]. This can be derived mathematically as follows. The gain (G) of the device or system is given in (7.7) where S_{out} is the output signal and S_{in} is the input signal:

$$G = \frac{S_{\text{out}}}{S_{\text{in}}} \tag{7.7}$$

The noise factor is the ratio of the input signal-to-noise ratio ($S_{\text{in}}/N_{\text{in}}$) to the output signal-to-noise ratio ($S_{\text{out}}/N_{\text{out}}$). The inverse of the gain (G^{-1}) = $S_{\text{in}}/S_{\text{out}}$ and the input noise (N_{in}) = $k \cdot T_0 \cdot B$.

$$F = \frac{\left(S_{\text{in}}/N_{\text{in}}\right)}{\left(S_{\text{out}}/N_{\text{out}}\right)} = \frac{N_{\text{out}}}{k \cdot T_0 \cdot B \cdot G} \tag{7.8}$$

Equation (7.9) separates the output noise into two factors, the input thermal noise ($N_{\text{in}} = k \cdot T_0 \cdot B$) and the added noise ($\Delta N$):

$$N_{\text{out}} = N_{\text{in}} + \Delta N = k \cdot T_0 \cdot B + \Delta N \tag{7.9}$$

Substituting $k \cdot T_0 \cdot B + \Delta N$ for N_{out} in the noise factor (F_n) in (7.8) and substituting $k \cdot T_0 \cdot B$ for N_{in} results in F as a function of ΔN and $k \cdot T_0 \cdot B$:

$$F = \frac{G\left(k \cdot T_0 \cdot B + \Delta N\right)}{k \cdot T_0 \cdot B \cdot G} = 1 + \frac{\Delta N}{k \cdot T_0 \cdot B} \tag{7.10}$$

The added noise (ΔN) in terms of noise temperature is given in (7.11).

$$\Delta N = kT_e B \tag{7.11}$$

Substituting for $\Delta N = k \cdot T_e \cdot B$ in the noise factor in (7.10) and noting that $k \cdot B$ cancels out give the result shown in (7.12).

$$F = 1 + \frac{k \cdot T_e \cdot B}{k \cdot T_0 \cdot B} = 1 + \frac{T_e}{T_0} \tag{7.12}$$

The result is noise factor as a function of the standard noise temperature (T_0 = 290K) [13] and the equivalent noise temperature (T_e) where T_e represents the temperature that will produce the same noise as the added noise ΔN. The effective noise temperature can be determined from the noise factor F_n as shown in (7.13).

$$T_e = (F - 1) \cdot T_0 \tag{7.13}$$

NF is 10 times the log of the noise factor (F_n) and is shown in (7.14) as a function of the noise temperature (T_e) and the standard noise temperature (T_0) [11].

$$NF = 10 * \text{Log}_{10}\left[1 + \left(\frac{T_e}{T_0}\right)\right] \tag{7.14}$$

The following tables are samples of the calculations relating noise temperature (T_e), noise factor (F), and noise figure (NF). Table 7.1 translates noise temperature (T_e) into noise factor (F) and noise figure (NF), and Table 7.2 translates noise figure (NF) to noise factor (F) and noise temperature (T_e).

7.5   Modeling the Noise Figure of a Single Amplifier

Every nonideal amplifier adds internal noise to the input signal and amplifies the input signal plus the internally added noise. A practical model of this effect is shown in Figure 7.6 where S1 is the input signal and N1 is the noise internally generated by the amplifier. Once the input noise (N1) is modeled as a separate input, the amplifier with gain G1 can be considered an ideal gain block with respect to the calculation of noise figure. It should also be noted that the input signal S1 may be contain noise and have an associated input signal to noise ratio. If the input noise and the noise

Table 7.1   Noise Conversion: Degrees Kelvin to Noise Factor and Noise Figure (dB)

T_0		290K
T_e	F	NF
Degrees Kelvin		dB
10	1.034	0.147
20	1.069	0.290
40	1.138	0.561
70	1.241	0.939
100	1.345	1.287
150	1.517	1.811
200	1.690	2.278
250	1.862	2.700
300	2.034	3.085
400	2.379	3.765
500	2.724	4.352
700	3.414	5.332

Table 7.2   Noise Conversion: Noise Figure (dB) to Noise Factor and Degrees Kelvin

T_0		290K
NF	F	T_e
dB		Degrees Kelvin
0.100	1.023	6.755
0.200	1.047	13.667
0.300	1.072	20.741
0.400	1.096	27.979
0.500	1.122	35.385
0.600	1.148	42.965
0.700	1.175	50.720
0.800	1.202	58.657
0.900	1.230	66.778
1.000	1.259	75.088
1.100	1.288	83.592
1.200	1.318	92.294

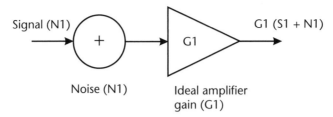

Figure 7.6 Noise figure model of an amplifier stage.

generated by the amplifier are the same characteristic (i.e., random thermal noise), the noise power will add arithmetically.

Figure 7.7 is a model of a single-stage amplifier with gain of G1dB and noise figure of F1dB. Both gain and noise figure are typically expressed in decibels with the respective linear values of gain (G1) and noise factor (F1):

$$G1 = 10^{\left(\frac{G1dB}{10}\right)} \tag{7.15}$$

$$F1 = 10^{\left(\frac{F1dB}{10}\right)} \tag{7.16}$$

The noise reflected to the amplifier input (N_{in}) is $k \cdot T \cdot B \cdot F1$, and the noise at the output of the amplifier (N_{out}) is $k \cdot T \cdot B \cdot F1 \cdot G1$.

For example, an amplifier has 12 dB of gain and a 4-dB noise figure. The input signal is −55 dBm (dBm is dB with respect to 1 mW), the bandwidth is 300 MHz, and the temperature is 298K.

Find the input and output noise level: (a) effective input noise level, (b) output noise level, (c) the effective input noise temperature (T_e) in degrees kelvin, and (d) output signal-to-noise ratio. The answers are as follows:

a. Effective input noise level:
    $N_{in} = k \cdot T \cdot B \cdot F1$
    $F1 = 10^{(F1db/10)} = 10^{(4/10)} = 2.512$
    $N_{in} = 1.033$ picowatts
    $N_{in}(dB) = 10 \cdot Log(N_{in}) = -89.857$ dBm

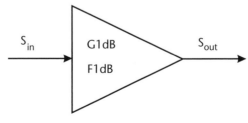

Figure 7.7 Model of a single-stage amplifier.

b. Output noise level:

$N_{out} = k \cdot T \cdot B \cdot F1 \cdot G1$

$N_{out} = 16.379$ picowatts

$N_{out}(dB) = 10 \cdot Log(N_{out}) = -77.857$ dBm

c. The effective input noise temperature (T_e) in degrees kelvin:

$$F = \frac{SNR_{in}}{SNR_{out}} = \frac{\dfrac{S}{N_0}}{\dfrac{[G1 \cdot S]}{\left[G1 \cdot (N_0 + N1)\right]}} = \frac{N_0 + N1}{N_0} = 1 + \frac{T_e}{T_0} \tag{7.17}$$

$$T_0 = (F1 - 1) = T_e = (290°K) \cdot (2.512 - 1) = 438.447°K \tag{7.18}$$

d. Output signal-to-noise ratio:

Input signal is -55 dBm

Output signal S_{out} (dBm) $= -55$ dBm $+ 12$ dB $= -43$ dBm

Output noise $= N_{out}(dB) = -77.857$ dBm

Output signal to noise (dB) $= 34.857$ dB

Output signal to noise (linear) $= 3.06 \cdot 10^3$

7.6   Noise Figure of Passive Devices

Devices composed totally of resistive and reactive components generally called passive devices have a gain less than unity and characteristically have the same thermal noise $(N_0 = k \cdot T \cdot B)$ at its input and output. The noise figure or noise factor (F) of these devices is calculated similarly to that of an amplifier using the parameters of gain and signal-to-noise ratio.

$$F = \frac{\left(S_{in}/N_{in}\right)}{\left(S_{out}/N_{out}\right)} \tag{7.19}$$

where, in the case of passive devices:

$$N_{in} = N_{out} = N_0 = k \cdot T \cdot B \tag{7.20}$$

$$\frac{S_{in}}{S_{out}} = \text{Loss (Gain} < 1) \tag{7.21}$$

Using (7.19) and (7.20), the noise components cancel out resulting in the noise factor (F) being equal to S_{in}/S_{out}, the inverse of the gain of the device [15]. For example, Figure 7.8 shows a 3-dB attenuator with a gain of 0.5 (-3 dB). The noise factor is $F = 2$ and the noise figure $(NF) = 3$ dB.

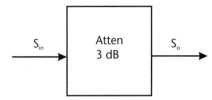

Figure 7.8 Noise figure of a 3-dB attenuator.

7.7 Noise Figure of Cascaded Devices

Noise is a random process. The noncoherent noise power from each element in a cascade of elements adds algebraically [16].

7.7.1 Cascading Two Amplifiers

The noise model for a single amplifier (Figure 7.7) can be extended to two amplifiers in cascade, shown in Figure 7.9.

As shown the output of the first amplifier stage ($S1_{out}$) is:

$$S1_{out} = S2_{in} = G1 \cdot \left(S1_{in} + N_0 + N1\right) \qquad (7.22)$$

where $S1_{in}$ is the input signal, N_0 is thermal noise ($k \cdot T \cdot B$) [T is the ambient temperature (K)], $N1$ is the added noise from the first amplifier: ($N1 = k \cdot T_e \cdot B$) (T_e is the equivalent noise temperature of noise factor $F1$), and $S1_{out}$ is the output of the first stage.

Figure 7.10 is a schematic representation of the cascaded amplifiers. The added noise (noise figure) of each amplifier is summed into the input signal and analyzed similar to the single-stage amplifier.

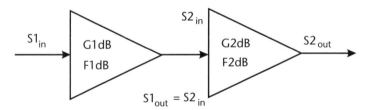

Figure 7.9 Two amplifiers in cascade.

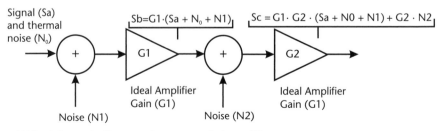

Figure 7.10 Schematic diagram of two cascaded amplifier stages.

The input to the first amplifier is signal S_a plus thermal noise N_0 ($N_0 = k \cdot T \cdot B$). Added to the input is noise $N1$, the added amplifier noise reflected to the input. Similar to the analysis of a single-stage amplifier following the input summer circuit is an ideal gain stage with gain $G1$. The output of the first amplifier is signal S_b where:

$$S_b = G1 \cdot (S_a + N_0 + N1) \tag{7.23}$$

Added to S_b is the noise at the input of the second amplifier ($N2$). The output of second amplifier is:

$$S_c = G1 \cdot G2(S_a + N_0 + N1) + G2 \cdot N2 \tag{7.24}$$

where N_0 is thermal noise ($N_0 = k \cdot T \cdot B$), $N1$ is the internally generated noise from the input amplifier, $N2$ is the internally generated noise from the second amplifier, $G1$ and $G2$ are the first and second amplifier gains, respectively, and S_c is the composite signal out of the second amplifier.

The signal-to-noise ratio at the output (SNR_{out}) is [17]:

$$SNR_{out} = \frac{G1 \cdot G2 \cdot S_a}{G1 \cdot G2 \cdot (N_0 + N1) + G2 \cdot N2} \tag{7.25}$$

The signal-to-noise ratio entering the two amplifier stages is:

$$SNR_{in} = \frac{S_a}{N_0} \tag{7.26}$$

The total noise factor (F_t) for the two-stage amplifier cascade is:

$$F_t = \frac{SNR_{in}}{SNR_{out}} \tag{7.27}$$

Substituting (7.25) and (7.26) into (7.27),

$$F_t = \frac{\dfrac{S_a}{N_0}}{\left[\dfrac{G1 \cdot G2 \cdot S_a}{G1 \cdot G2 \cdot (N_0 + N1) + G2 \cdot N2} \right]} \tag{7.28}$$

Cancelling parameters,

$$F_t = \frac{N_0 + N1 + \left(\dfrac{N2}{G1} \right)}{N_0} \tag{7.29}$$

Simplifying (7.29),

$$F_t = 1 + \frac{N1}{N_0} + \frac{N2}{N_0 \cdot G1} \tag{7.30}$$

F_1 is the noise factor of the first stage.

$$F_1 = 1 + \frac{N1}{N_0} \tag{7.31}$$

The second-stage noise factor F_2 is:

$$F_2 = 1 + \frac{N2}{N_0} \tag{7.32}$$

Solving for $N2$,

$$N_2 = \left(F_2 - 1\right) \cdot N_0 \tag{7.33}$$

Substituting into (7.30),

$$F_t = 1 + \frac{N1}{N_0} + \frac{N2}{N_0 \cdot G1} = 1 + \frac{N1}{N_0} + \frac{\left(F_2 - 1\right) \cdot N_0}{N_0 \cdot G1} \tag{7.34}$$

Simplifying the equation in terms of noise factors F_1 and F_2, the total noise factor F_t is:

$$F_t = F_1 + \frac{\left(F_2 - 1\right)}{G1} \tag{7.35}$$

where F_t is the total noise factor of the two cascaded amplifiers, F_1 is the noise factor of the first amplifier, F_2 is the noise factor of the second amplifier, and $G1$ is the gain of the first amplifier. Note that the gain of the second amplifier ($G2$) does not enter into the calculation of total noise factor.

The total noise figure (NF_t) of the two stages is:

$$NF_t = 10 \cdot \mathrm{Log}_{10}\left(F_t\right) \tag{7.36}$$

7.7.2   Noise Figure of a Cascade of Multiple Devices

Calculating the noise figure of multiple devices is simply applying and reapplying the equation determined for two devices, (7.35). Figure 7.11 shows three devices that are cascade.

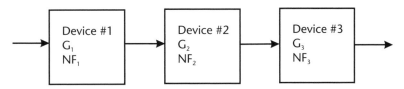

Figure 7.11   Determining the noise figure of three devices in series.

Since noise figure accumulates and degrades as a signal progresses through a system, the analysis of a system's noise figure should naturally start from the input device (in Figure 7.11, it is labeled as Device 1) and progress through the cascade ultimately determining the gain and noise figure of the three devices in series (Devices 1 through 3). Procedurally, the noise figure of Device 1 combined with Device 2 can be determined using (7.35), that is, $F_t = F_1 + (F_2 - 1)/G1$. For F_t, the subscript 12 is substituted signifying the noise factor is the combination of Devices 1 and 2.

$$F_{12} = F_1 + \frac{(F_2 - 1)}{G1} \tag{7.37}$$

The new model of the combined devices (1 and 2) is shown in Figure 7.12. Applying (7.34) to the combined Devices 1 and 2,

$$F_t = F_{12} + \frac{(F_3 - 1)}{G1 \cdot G2} \tag{7.38}$$

F_t is the combined noise factor of all three devices. Equation (7.39) is the result of substituting the results of (7.37) (F_{12}) into (7.38).

$$F_t = F_1 + \frac{(F_2 - 1)}{G1} + \frac{(F_3 - 1)}{G1 \cdot G2} \tag{7.39}$$

Extending this analysis to N devices shown in Figure 7.13 results in the general noise factor and, by extension, noise figure analysis (7.40).

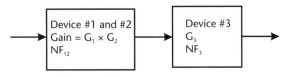

Figure 7.12   The characteristics of Devices 1 and 2 are combined.

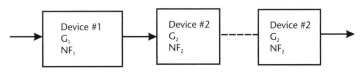

Figure 7.13   Typical chain of N devices.

$$F_t = F_1 + \frac{(F_2 - 1)}{G1} + \frac{(F_3 - 1)}{G1 \cdot G2} + \cdots + \frac{(F_N - 1)}{G1 \cdot G2 \cdot G3 \cdots G_N} \qquad (7.40)$$

7.8   Noise Figure Calculation Example

An example of a typical system microwave system is shown in Figure 7.14. The components with the pertinent parameters for a noise figure calculation are listed Table 7.3 column 2, "component name," column 3, "Gain (dB)," and column 4, "Noise Figure (dB)." The table is an Excel spreadsheet that calculates the accumulative gain (dB), noise figure (dB), carrier power (dBm) (if the input carrier power is entered), the carrier-to-noise ratio (C/N) in decibels, the noise output in a 1-Hz bandwidth, and the noise output in a selected bandwidth (in the illustrated table, the selected bandwidth was 10 MHz).

Note that the terminology carrier-to-noise ratio is typically used to describe predetection signals, whereas the signal-to-noise ratio is commonly used to describe predetection and postdetection signals. Noise figure is always used for predetection signals, so in this context, signal-to-noise ratio (S/N) and carrier-to-noise ratio (C/N) are used interchangeably.

The calculated information in Table 7.3 is contained in columns 5 through 10. The cumulative effect at the output of each component is shown in the row associated with the respective component. The total cumulative result is shown on the line associated with the last component and repeated on line 2.

In this example, the total system (Devices 1 through 5) results are as follows:

Column 5, Total gain: 28 dB;

Column 6, Total system NF: 2.49 dB;

Column 7, Carrier output power: −12 dBm (assuming an input power of −40 dBm);

Column 8, C/N ratio: 61.4 dB (assuming an input power of −40 dBm and a 10-MHz bandwidth);

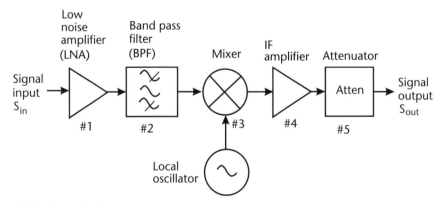

Figure 7.14   Example of a microwave system.

Table 7.3 Calculated Gain, Noise Figure, Signal Level, C/N, and Noise Level

Line	Temperature (°C)	25		BW	10	MHz			
1	2	3	4	5	6	7	8	9	10
2	System total			28.0	2.49	−12.0	61.4	−143.4	−73.4
3		Gain	NF	Total gain	Total NF	Carrier power	C/N Out	Noise Out	Noise Out
4		dB	dB	dB	dB	dBm	dB	dBm/Hz	dBc/10 MHz
5	Input level			0.0	0.00	−40.0		−−173.9	−103.9
6		0	0	0.0	0.00	−40.0	63.9	−173.9	−103.9
7	1. Low noise amplifier	20	2	20.0	2.00	−20.0	61.9	−151.9	−81.9
8	2. Bandpass filter	−1	1	19.0	2.01	−21.0	61.8	−152.8	−82.8
9	3. Mixer	−8	8	11.0	2.19	−29.0	61.7	−160.7	−90.7
10	4. IF amplifier	20	4	31.0	2.49	−9.0	61.4	−140.4	−70.4
11	5. Attenuator	−3	3	28.0	2.49	−12.0	61.4	−143.4	−73.4

Column 9, Output noise power in a 1-Hz bandwidth: −143.4 dBm/Hz;

Column 10, Output noise power in a 10-MHz bandwidth: −73.4 dBm/Hz.

Note that the noise figure of the mixer was assigned the same magnitude as the mixer loss. This assumption is very common and usually correct, but it is possible that since the mixer is constructed using semiconductor devices, the noise may on occasion be slightly higher than the loss through the mixer. In addition, a poorly designed local oscillator can add noise seen at the mixer output, effectively increasing the mixer noise figure.

References

[1] Tektronix, "Noise Figure, Overview of Noise Measurement Methods," 2014.

[2] Hansen, M., "Achieving Accurate On-Wafer Flicker Noise Measurements Through 30 MHz," Cascade Microtech, Inc., May 2009.

[3] Carrasco, R., "Noise in Communication Systems," Lecture slides, University of Newcastle-upon-Tyne, 2009.

[4] Vaseghi, S., "Noise and Distortion," Chapter 2, in *Advanced Digital Signal Processing and Noise Reduction*, 2nd ed., New York: John Wiley & Sons, 2000.

[5] Rosu, I., "Understanding Noise Figure," Whitepaper.

[6] Strauch, M., "1/f Noise," Lecture slides, 2011.

[7] Konczakowska, A., and B. Wilamowski, "Noise in Semiconductor Devices," (Chapter 11), in *Fundamentals of Industrial Electronics*, B. M. Wilamowski and J. D. Irwin (eds.), Boca Raton, FL: CRC Press, 2010.

[8] AN1026 Application Note, "1/f Noise Characteristics Influencing Phase Noise," California Eastern Laboratories, 2003.

[9]   Ham, D., "Statistical Electronics, Noise Processes in RF Integrated Circuits (Oscillators and Mixers)," Harvard RF and High-Speed IC & Quantum Circuits Laboratory, 2003.

[10]   Schwartz, M., "Improving the Noise Performance of Communication Systems," Technical paper, Columbia University.

[11]   Agrawal, V., and F. Dai, "RFIC Design and Testing for Wireless Communications, Lecture 4: Testing for Noise," TI India Technical University, July 2008.

[12]   Application Note, 57-1 "Fundamentals of RF and Microwave Noise Figure Measurements," Agilent Technologies, Inc. 2010.

[13]   Purnachandar, P., and R. Gupta, "Signal Chain Noise Figure Analysis," Texas Instruments, Application Report, October 2014.

[14]   Mohr, R., "Mohr on Receiver Noise Characterization, Insights & Surprises," *Microwave Theory & Techniques Society of the IEEE Long Island Section*, 2010.

[15]   Razavi, B., "Basic Concepts in RF Design," RF Microelectronics.

[16]   Niknejad, A., "Lecture 12: Noise in Communication Systems," University of California, Berkeley, 2005.

[17]   Lankford, D., "A Derivation of Friis' Noise Factor Cascade Formula," Technical paper, July 2011.

CHAPTER 8

Nonlinear Signal Distortion

8.1 Introduction

Signal distortion occurs when the output signal does not look exactly like the input signal, with the exception of amplitude scaling and the signal delayed in time. Distortion can manifest itself in many ways broken down into two main categories [1]: distortion originating in the frequency domain (i.e., nonideal frequency response), and distortion originating in the time domain (i.e., nonlinearity). This chapter focuses mainly on nonlinearity in the time domain and its manifestation in the frequency domain.

8.2 Distortion Originating in the Frequency Domain

An ideal communications channel has a flat amplitude response [2] and a linear phase over the frequency of interest (see Figure 8.1).

Since group delay is the derivative of phase (Φ) with respect to frequency (f) [see (8.1)], if phase is a linear function of frequency, the result of the group delay (GD) calculation is a constant over the frequency of interest.

$$\text{Group Delay:} \quad GD = \frac{d\Phi(f)}{df} \tag{8.1}$$

Constant group delay as well as a flat amplitude response is a necessary but not a complete criterion to successfully reproduce a modulated signal [3]. All the sidebands associated with the modulated carrier must always have the same relative

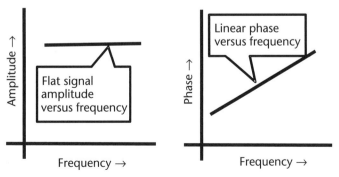

Figure 8.1 Ideal transmission path in the frequency domain.

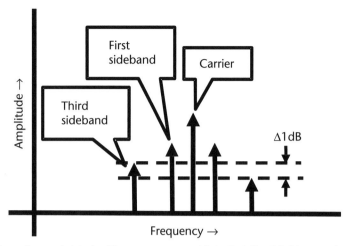

Figure 8.2 A carrier modulated with a square wave. Note that the third harmonic of the square wave is 1 dB lower on one side of the carrier.

amplitude and time delay with respect to the carrier and each other's sideband to avoid frequency-domain distortion.

As an example, if a third-order sideband on one side of a 10-MHz pulse-modulated waveform was lowered in a simulation program to be 1 dB below the other third-order sideband (see Figure 8.2), the resultant demodulated pulse wave form would go from a flat response on the top of the pulse to a ripple of approximately 15% (see Figure 8.3).

When the same sideband was delayed approximately 1 ns with respect to the carrier and the other sideband, the distortion significantly increased.

8.3   Distortion of a Single Sinusoidal Signal Due to Device Nonlinearity

A device is considered linear in the time domain when the output signal is the same as the input signal, with the exceptions that the output may be scaled in amplitude and delayed in time. In the real world, nothing is linear, and all devices have

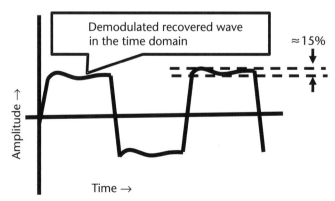

Figure 8.3 The resultant demodulated square wave when the third sideband is 1 dB lower than one side of the carrier.

nonlinear transfer functions. The output signal is never an exact scaled version of the input signal; linearity is an approximation of the real world.

Distortion is the result of a nonlinear transfer function. Nonlinearity is usually associated with active devices, but passive components are also nonlinear and introduce distortion but usually to a much less extent.

The problem is understanding the causal effects of nonlinear transfer function (distortion), the expected level of the nonlinear signal, and the acceptable level of distortion that will satisfy the overall desired system characteristics.

Figure 8.4 shows a sinusoidal waveform passed through a nonlinear active device. The amount of distortion is a function of the signal level and the nonlinear characteristics of the device. Note that the positive peak of the output waveform is distorted and the negative peak is not. This is an exaggeration of the nonlinear effect meant to show that active semiconductor devices generally do not have the same nonlinear characteristic in the positive and negative direction.

8.3.1 Mathematical Representation of a Nonlinear Transfer Function

The transfer function of any nonlinear device can be represented by a Taylor series; see (8.2) [4].

$$v_0(t) = a_0 + a_1 \cdot v_{in}(t) + a_2 \cdot v_{in}(t)^2 + a_3 \cdot v_{in}(t)^3 + \cdots + a_n \cdot v_{in}(t)^n \quad (8.2)$$

where a_0 is the DC coefficient, a_1 is the linear coefficient, a_2 is the second-order coefficient, a_3 is the third-order coefficient, and a_n is the nth-order coefficient. In a well-behaved system (i.e., no abrupt discontinuities), the higher-order coefficients decrease and usually can be ignored for an initial approximate analysis.

When a device is assumed to be linear, mathematically all coefficients greater than a_1 are assumed to be equal to zero (a_2, a_3, ..., $a_n = 0$). When a device is AC-coupled, the a_0 coefficient is zero ($a_0 = 0$). The equation for a linear system (8.3) is:

$$v_0(t) = a_1 \cdot v_{in}(t) \quad \text{AC-coupled linear system} \quad (8.3)$$

where a_1 is the linear transfer function of the device.

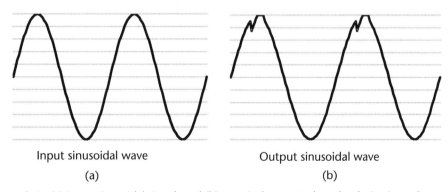

Input sinusoidal wave   Output sinusoidal wave

(a)   (b)

Figure 8.4   (a) Input sinusoidal signal, and (b) a typical output when the device is nonlinear.

8.3.2 Harmonic Distortion Due a Device Nonlinearity

Signal distortion through a device occurs when the device exhibits a nonlinear transfer function. A single sinusoidal signal through the device produces harmonics of the signal with levels calculated by applying the signal to the representative Taylor series [2] (8.2).

If the input signal is sinusoidal [e.g., $Vp \cdot \cos(\omega \cdot t)$] and it is applied to a Taylor series (8.2), the output signal will have an infinite number of harmonics of the input signal ($2 \cdot \omega$, $3 \cdot \omega$, $4 \cdot \omega$, ..., $n \cdot \omega$), where n is the order of the Taylor series coefficient, Vp is the peak voltage, and ω is the signal frequency in radians per second of the input signal.

Using the applicable trigonometric identities and the values of the higher-order coefficients of the Taylor series, the levels at each harmonic can be evaluated.

8.3.3 Gain Through a Nonlinear Device

If a signal going into a nonlinear device is sinusoidal [e.g., $Vp \cdot \cos(\omega \cdot t)$], the output as per (8.2) will contain an infinite number of harmonics of the input frequency ($n \cdot \omega$) in addition to the fundamental frequency [$Vo \cdot \cos(\omega \cdot t)$], where Vo is the peak output voltage at the frequency of the input signal (ω). The linear gain (G) of the device is a function of the output with respect to input at the same frequency (i.e., the output power contained in the harmonics is not considered as part of the linear gain function).

$$GdB = 20 \cdot \mathrm{Log}(G) = 20 \cdot \mathrm{Log}\left(\frac{Vo}{Vp}\right) \tag{8.4}$$

where ω is the radian frequency of the input and output signal, the output signal applied to the Taylor series in (8.2) has an infinite number of harmonics of the input signal ($2 \cdot \omega$, $3 \cdot \omega$, $4 \cdot \omega$, ..., $n \cdot \omega$), where n is the order of the Taylor series coefficient, GdB is the gain (G) in decibels, Vo is the peak output voltage at the fundamental frequency, and Vp is the peak input voltage.

With regard to signal gain, note that since gain (G) is measured at the fundamental frequency (ω) when calculating gain, the even terms (n is an even number) are neglected because only odd terms (n is an odd number) trigonometrically contain components of the fundamental frequency $\cos(\omega \cdot t)$.

8.3.4 Gain Compression

Any real device has a limited output power capability. Typically, the gain of a well-behaved device (one with no abrupt discontinuities) goes into compression and ultimately saturation as the input is increased [5]; see Figure 8.5.

The linear region of a device is usually assumed to be from a very low signal level (i.e., small signal) to when the gain of the device decreases by 1 dB stated as an output power (P1dB) in decibel units usually with respect to 1 mW (dBm) [6]. At the 1-dB compression point (P1dB), the second harmonic of the input signal is approximately 20 to 26 dB below the fundamental output signal (−20 dBc to −26 dBc).

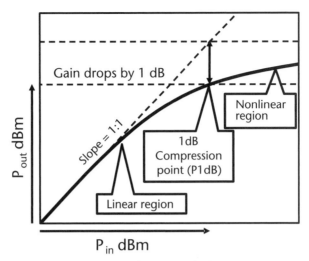

Figure 8.5 Typical transfer characteristics of an active device.

Note that dBc is decibels with respect to the carrier, dBm is decibels with respect to 1 mW, and dBW is decibels with respect to 1W. In many solid-state amplifiers, the output level usually goes into saturation (P_{sat}) in less than 2 dB after the 1-dB compression point ($P1dB$) [7]. Typically,

$$P_{sat} - P1dB \leq 2 \text{ dB} \tag{8.5}$$

where P_{sat} and $P1dB$ are in units of dBm or dBW.

8.4 Intermodulation Interference Frequencies

Intermodulation products are the results of interaction between the input signals caused by the nonlinearity of the processing device. The simplest example of inter-modulation distortion is when two signals are present at the input with equal amplitudes (Vp).

$$v_{in}(t) = Vp \cdot \cos(\omega_1 \cdot t) + Vp \cdot \cos(\omega_2 \cdot t) \tag{8.6}$$

where ω_1 and ω_2 are the respective carrier frequencies in radians per second, $\omega_1 = 2 \cdot \pi \cdot F_1$ and $\omega_2 = 2 \cdot \pi \cdot F_2$ where F_1 and F_2 are the respective frequency in hertz.

In addition to creating harmonics of each of the signals, intermodulation prod-ucts are created. Intermodulation products are theoretically an infinite number of combinations of the sum and difference of all of the harmonics of each of the input signals. The frequency of intermodulation (IM) products exists at frequencies are given by

$$\text{IM}_{n,m} = n \cdot F_1 \pm m \cdot F_2 \tag{8.7}$$

where n and m are integers from zero to infinity ($0 \leq n \leq \infty$ and $0 \leq m \leq \infty$) [4].

8.5   Second-Order Intermodulation Distortion

Intermodulation distortion is the result of two or more signals applied to the input of a device with a nonlinear characteristic given by a Taylor series as shown in (8.2). Second-order intermodulation distortion results from two or more signals interacting in the a_2 term (second-order coefficients) of the Taylor series. To simplify the analysis, the first-order approximation assumes $a_0 = 0$ and all coefficients higher than a_2 are equal to zero ($a_3, ..., a_n = 0$). The simplified nonlinear transfer function for the device [8] is given in (8.8).

$$v_0(t) = a_1 \cdot v_{in}(t) + a_2 \cdot v_{in}(t)^2 \tag{8.8}$$

Separating $v_0(t)$ into its first-order component ($v_0(t)_{a1}$) and second-order component ($v_0(t)_{a2}$) terms:

$$v_0(t) = v_0(t)_{a1} + v_0(t)_{a2} \tag{8.9}$$

where the $v_0(t)_{a1}$ is the linear term of the Taylor series,

$$v_0(t)_{a1} = a_1 \cdot v_{in}(t) = a_1 \cdot Vp \cdot \cos(\omega_1 \cdot t) + a_1 \cdot Vp \cdot \cos(\omega_2 \cdot t) \tag{8.10}$$

and the $v_0(t)_{a2}$ term is the second-order distortion term of the Taylor series.

$$v_0(t)_{a2} = a_2 \cdot v_{in}(t)^2 = a_2 \cdot Vp^2 \cdot \left[\cos(\omega_1 \cdot t) + \cos(\omega_2 \cdot t)\right]^2 \tag{8.11}$$

Equation (8.12) expands the second-order distortion term in (8.11).

$$\begin{aligned} v_0(t)_{a2} &= a_2 \cdot Vp^2 \cdot \left[\cos(\omega_1 \cdot t)\right]^2 + a_2 \cdot Vp^2 \cdot \left[\cos(\omega_2 \cdot t)\right]^2 \\ &\quad + 2 \cdot a_2 \cdot Vp^2 \cdot \cos(\omega_1 \cdot t) \cdot \cos(\omega_2 \cdot t) \end{aligned} \tag{8.12}$$

Assuming AC coupling eliminates the DC terms [4].

$$\begin{aligned} v_0(t)_{a2} &= \frac{a_2}{2} \cdot Vp^2 \cdot \cos(2 \cdot \omega_1 \cdot t) + \frac{a_2}{2} \cdot Vp^2 \cdot \cos(2 \cdot \omega_2 \cdot t) \\ &\quad + a_2 \cdot Vp^2 \cdot \cos([\omega_1 - \omega_1] \cdot t) + a_2 \cdot Vp^2 \cdot \cos([\omega_1 - \omega_1] \cdot t) \end{aligned} \tag{8.13}$$

8.5.1   Levels and Spectrum of Second-Order Intermodulation Products

Equations (8.10) and (8.13) and Figure 8.6 shows the frequencies ω_1 and ω_2 in radians per second and the respective frequency F_1 and F_2 in hertz derived from ω_1 and ω_2 where $\omega = 2 \cdot \pi \cdot F$. The amplitude of the fundamental frequencies is $a_1 \cdot Vp$, the linear term of the output signal. Second harmonics of the fundamental inputs are at frequencies $2 \cdot \omega_1$ and $2 \cdot \omega_2$ and their respective amplitudes are equal to $(a_2)/2 \cdot Vp^2$. Also shown are the intermodulation products, the sum frequency $(\omega_1 + \omega_2)$ and the difference frequency $(\omega_1 - \omega_2)$, with their respective amplitudes equal to $a_2 \cdot Vp^2$. It should be noted that the harmonics $(2 \cdot \omega)$ are half the level (−6 dB) of the intermodulation products $(\omega_1 + \omega_2$ and $\omega_1 + \omega_2)$ [9].

8.5.2 Frequency Spectrum of Second-Order Intermodulation Products When the Carriers Are Closely Spaced

If the two carriers $F1$ and $F2$ are closely spaced, all the second-order intermodulation products will be almost an octave away from the original carriers.

$$\text{Closely spaced carriers: } F2 - F1 \ll F1 \text{ and } F2 - F1 \ll F2 \qquad (8.14)$$

Then

$$F2 + F1 \approx 2 \cdot F1 \text{ and } F2 + F1 \approx 2 \cdot F2 \qquad (8.15)$$

and

$$F2 - F1 \ll F1 \text{ or } F2 \qquad (8.16)$$

If the input signals are closely space in frequency, which is typical in communication systems, all the second-order intermodulation products, including harmonics, are almost an octave away from the original carriers. Under these conditions, system filters can reduce the levels of these interference products such that they do not have to be considered when evaluating the system performance.

8.5.3 Frequency Spectrum of Even-Order Intermodulation Products When the Carriers Are Closely Spaced

The results obtained calculating the second-order intermodulation effects of two closely spaced carriers can be expanded to any even order coefficients of the Taylor series (a_2, a_4, a_6, ...) and any number of closely space carriers. Expanding the

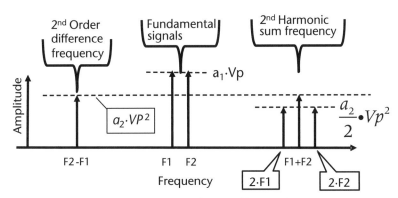

Figure 8.6 Second-order intermodulation products and harmonics.

trigonometric identities shows that all the even-order intermodulation products are at least approximately an octave away from the fundamental carriers when the bandwidth (BW) is relatively narrow (BW << Carrier frequency). Under these conditions, the system design can reduce the level of all the even-order intermodulation products and harmonics such that the resultant levels can have an insignificant effect on the system operation.

8.5.4 Second-Order Intercept Point and Second-Order Intermodulation Levels

Assume that the transfer function has only first-order and second-order terms as presented in (8.8) and restated in (8.17).

$$v_0(t) = a_1 \cdot v_{in}(t) + a_2 \cdot v_{in}(t)^2 \tag{8.17}$$

If the peak voltage levels of the two input signals are equal and have a value Vp, the input signal $V_{in}(t)$ would form as shown in (8.18).

$$v_{in}(t) = Vp \cdot \cos(\omega_1 \cdot t) + Vp \cdot \cos(\omega_2 \cdot t) = Vp \cdot \left[\cos(\omega_1 \cdot t) + \cos(\omega_2 \cdot t) \right] \tag{8.18}$$

The output signal $[V_0(t)]$ would be:

$$\begin{aligned} v_0(t) = &\ a_1 \cdot Vp \cdot \left[\cos(\omega_1 \cdot t) + \cos(\omega_2 \cdot t) \right] \\ &+ a_2 \cdot Vp^2 \cdot \left[\cos(\omega_1 \cdot t) + \cos(\omega_2 \cdot t) \right]^2 \end{aligned} \tag{8.19}$$

From (8.19), the normalized power in the first-order term in decibels (P_1dB) is:

$$P_1 dB \sim 20 \cdot \log(Vp) \tag{8.20}$$

and the power in the second-order term in decibels (P_2dB) is:

$$P_2 dB \sim 20 \cdot \log(Vp^2) = 40 \cdot \log(Vp) \tag{8.21}$$

Figure 8.7 is a plot of a linear output and a second-order intermodulation product relative to the input power of one of two equal power input signals applied to a nonlinear device.

The input powers are P_{F1} (dBm) $= P_{F2} = P_{in}$ (in dBm) for each of the respective signals at frequencies F_1 and F_2. From (8.20) and (8.21), the power in the second-order intermodulation interference signals increase at twice the rate (2:1) of the fundamental signals (1:1). The linear response and the second-order intermodulation response intersect at a point called the second-order intercept point (IP2) [10]. The second-order intercept point with respect to the output is designated as OIP2 and the second-order intercept point with respect to the input is designated as IIP2.

The output second-order intercept point in dBm is usually approximated to be 20 dB higher than the output 1-dB compression point (P1dB) of the device.

If the second harmonic was plotted against the fundament signals, the second harmonic slope (2:1) would also be twice the fundamental slope (1:1), but typically the intercept point of the second harmonic line would be 6 dB higher than the intercept point of the second-order intermodulation intercept point reflecting the fact that the second harmonic is 6 dB below the second-order intermodulation products as shown in (8.13).

Calculation of the output level of each second-order intermodulation interference product (IM_2 in dBc) can be determined using (8.22).

$$IM_2 = -\left(OIP2 - P_{out}\right) \qquad (8.22)$$

where IM_2 is a single second-order intermodulation product in dBc, which is the level of a single intermodulation product with respect to a single output carrier in decibels, OIP2 is the output second-order intermodulation intercept point (OIP2) in dBm, and P_{out} is output level of one of two equal power carriers in dBm.

Example 8.1

Two input carriers one at a frequency of 1.0 GHz (F_1) and the other at a frequency of 1.1 GHz (F_2) have an equal output level of −10 dBm from a device with an output second-order intermodulation intercept point of +30 dBm.

Find the output frequency spectrum of each of the signals and its anticipated second-order intermodulation interference level [$IM_2 = -(OIP2 - P_{in})$] in dBc and dBm.

The tabulated results are shown in Table 8.1.

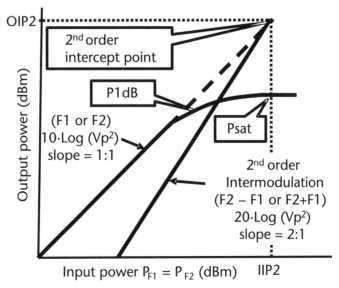

Figure 8.7   Input power and output power for each of two equal carriers and the power of a single second-order intermodulation interference signal produced from the two input signals.

Table 8.1   Second-Order Intermodulation Interference Results for OIP2 = +30 dBm

Frequency Designation	Output Signal Frequency	Output Signal Level (dBc)	Output Signal Level (dBm)
F_1	1.0 GHz	0 dBc	−10 dBm
F_2	1.1 GHz	0 dBc	−10 dBm
$F_2 - F_1$	0.1 GHz	−40dBc	−50 dBm
$F_2 + F_1$	2.1 GHz	−40 dBc	−50 dBm
$2 \cdot F_1$	2.0 GHz	−46 dBc	−56 dBm
$2 \cdot F_2$	2.2 GHz	−46 dBc	−56 dBm

8.5.5   Notes on Even-Order Intermodulation Products and Intercept Points:

In a narrow bandwidth (BW) system (BW << Carrier frequency), all of the even-order intermodulation interference products can be filtered to a level where this interference can be ignored. Usually, second harmonics and second-order intermodulation interference are important when the system bandwidth is close to an octave wide. There are circuit configurations that can be employed to significant raise the even order intermodulation intercept points. The process involves the use of parallel circuits and harmonic cancellation techniques. The device saturation power (P_{sat}) is typically within 1 dB of the 1-dB compression point and significantly below the output second-order output interception point.

8.6   Third-Order Intermodulation Distortion

Intermodulation distortion results when two or more signals are applied to the input of a device that has a nonlinear characteristic given by a Taylor series as shown in (8.2). Third-order intermodulation distortion results from two or more signals interacting in the a_3 term (third order) in the Taylor series. To simplify the analysis of third-order intermodulation distortion without losing substance to the conclusions, the following assumptions are made: (1) the coefficients $a_0 = a_2 = 0$ and (2) all coefficients higher than a_3 are equal to zero ($a_4, \ldots, a_n = 0$). The nonlinear transfer function for the device is then given by (8.23).

$$v_0(t) = a_1 \cdot v_{in}(t) + a_3 \cdot v_{in}(t)^3 \tag{8.23}$$

This simplification of the analysis is valid for AC-coupled narrow bandwidth systems where the frequencies of the input signal ($F_1, F_2, \ldots, F_N$) are much greater than the bandwidth containing all of the signals. Under these conditions, the even-order terms of the Taylor series can easily be filtered to a low enough level such that intermodulation interference from the even terms can be ignored.

To further simplify the analysis the initial input will assume only two signals with equal peak amplitudes (Vp).

$$v_{\text{in}}(t) = Vp \cdot \sin(\omega_1 \cdot t) + Vp \cdot \sin(\omega_2 \cdot t) \tag{8.24}$$

Plugging (8.24) into (8.23),

$$v_0(t) = a_1 \cdot Vp \cdot \sin(\omega_1 \cdot t) + a_1 \cdot Vp \cdot \sin(\omega_2 \cdot t)$$
$$+ a_3 \cdot Vp^3 \cdot \left[\sin(\omega_1 \cdot t) + \sin(\omega_2 \cdot t)\right]^3 \tag{8.25}$$

Evaluating (8.25) and eliminating all the DC terms, the $3 \cdot \omega$ terms, and out-of-band sum frequencies and assuming that the signals at frequencies ω_1 and ω_2 multiplied by the a_1 coefficient are much greater than the same signals generated in the third-order terms multiplied by the a_3 coefficient,

$$v_0(t) = a_1 \cdot Vp \cdot \sin(\omega_1 \cdot t) + a_1 \cdot Vp \cdot \sin(\omega_2 \cdot t) - \frac{3}{4} \cdot a_3 \cdot Vp^3$$
$$\cdot \sin\left(\left[2 \cdot \omega_2 - \omega_1\right] \cdot t\right) - \frac{3}{4} \cdot a_3 \cdot Vp^3 \cdot \sin\left(\left[2 \cdot \omega_1 - \omega_2\right] \cdot t\right) \tag{8.26}$$

8.6.1   The Frequency Spectrum of Third-Order Intermodulation Products

The resultant (8.26) has four frequency components, shown in Figure 8.8. The original sinusoidal signals with equal amplitudes (Vp) at ω_1 (F_1 in hertz) and ω_2 (F_2 in hertz) are amplified by the linear coefficient a_1. The two third-order intermodulation interference components at frequencies $2 \cdot \omega_1 - \omega_2$ ($2 \cdot F_1 - F_2$) and $2 \cdot \omega_2 + \omega_1$ ($2 \cdot F_2 - F_1$) each have a magnitude equal to $\left|(3/4) \cdot a_3 \cdot Vp^3\right|$.

The separation between the two carriers (F_1 and $F2$) is $\Delta F = F_2 - F_1$ and the intermodulation products are on either side of the carriers. The lower third-order intermodulation product is ΔF below F_1 and the higher third-order intermodulation product is ΔF above F_2. If the carriers F_1 and F_2 are close together, the intermodulation products are close together. Under these conditions, it is not practical to reduce the level of the intermodulation products through the process of filtering.

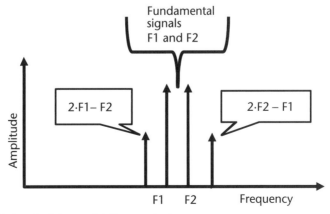

Figure 8.8   Third-order intermodulation products.

8.6.2 Calculating the Level of Third-Order Intermodulation Products

Assume that the transfer function is approximated to have only first-order and third-order terms. The justification of this assumption is that the system is narrowband such that the even-order and harmonic terms are eliminated by filtering and the higher-order odd terms are lower in amplitude than the first-order and third-order terms [11].

From (8.26), the normalized power of each equal carrier in decibels (P_1dB) is:

$$P_1 \text{dB} = 20 \cdot \log(|a_1|) + 20 \cdot \log(Vp) \tag{8.27}$$

and the normalized power in the third-order term in decibels (P_3dB) is:

$$P_3 \text{dB} = 20 \cdot \log\left(\left|\frac{3}{4} \cdot a_3\right|\right) + 20 \cdot \log(Vp^3) = 20 \cdot \log\left(\left|\frac{3}{4} \cdot a_3\right|\right) + 60 \cdot \log(Vp) \tag{8.28}$$

In (8.27) the $20 \cdot \log(|a_1|)$ term and in (8.28) the $20 \cdot \log(|(3/4) \cdot a_3|)$ term are constants.

In Figure 8.9, (8.27) and (8.28) are plotted as a function of normalized input power ($10 \cdot \log[Vp^2]$). The third-order term has a slope of 3:1 with respect to the linear term slope of 1:1. The input power [P_{IN} (dBm)] and output power [P_{OUT} (dBm)] are typically expressed in decibels with respect to a milliwatt (dBm). The two lines are linear functions for power levels significantly below the output 1-dB compression point (P_1dB) and the output saturation level (P_{sat}). The linear functions for P_1dB (1) and P_3dB (2) are extrapolated beyond P_1dB with dashed lines until they intersect at an imaginary point call the third-order intermodulation intercept point (IP3).

The third-order intercept point with respect to the output is designated OIP3 and third-order intercept point with respect to the input is designated IIP3. The output

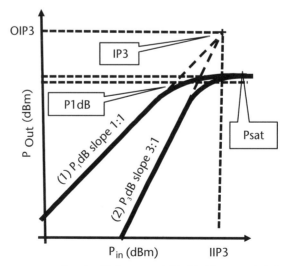

Figure 8.9 Plot of two output signals: (1) linear and (2) third intermodulation products as a function of input power.

third-order intercept point in dBm is usually approximated to be 10 dB higher than the output 1-dB compression point (P_1dB) of the device but can be significantly lower (6 dB to 7 dB above P_1dB) depending on the level of the higher-order nonlinearities.

The level of the third-order intermodulation interference product (IM_3 in dBc) can be calculated using (8.29).

$$IM_3 = -2 \cdot \left(OIP3 - P_{out} \right) \tag{8.29}$$

where IM_3 is a single third-order intermodulation product in dBc, OIP3 is the third-order intermodulation intercept point in dBm, P_{out} is the device output level of one of two equal power carriers in dBm, and dBc is the level of a single intermodulation product with respect to a single input carrier in decibels (dBc).

Example 8.2

Two input carriers one at a frequency of 1.0 GHz (F_1) and the other at a frequency of 1.1 GHz (F_2) have an equal output level of -10 dBm from a device with an output third-order intermodulation intercept point (OIP3) equal to $+20$ dBm.

Find the output frequency spectrum of each of the third-order intermodulation signals and its anticipated level [$IM_3 = -2 \cdot (OIP3 - P_{in})$] in dBc and dBm.

The tabulated results are shown in Table 8.2.

8.6.3   Determining the Relative Level of IP3 and P_1dB

The 1-dB compression point (P_1dB) is easy to measure because it can be determined using a single carrier while the third-order intercept point requires two carriers with equal signal levels at the output of the device. The determination of the output third-order intercept point is usually thought to be more accurate and therefore more useful when the nonlinearity of an entire system is to be determined [12].

The relative level of IP3 and P_1dB is important for a quick estimate of the third order intermodulation interference signal levels given P_1dB. The commonly used factor is that IP3 is 10 dB above P_1dB both in decibels with respect a normalized power, 1 mW (dBm) or 1W (dBW). This approximation changes significantly under different circumstances and is seen to be between 6 dB and 15 dB, with exceptions beyond these figures on either side [13].

Table 8.2 Results of Example 8.2 for an Output Third-Order Intermodulation Intercept Point Equal to +20 dBm

Frequency Designation	Output Signal Frequency	Output Signal Level (dBc)	Output Signal Level (dBm)
F_1	1.0 GHz	0 dBc	−10 dBm
F_2	1.1 GHz	0 dBc	−10 dBm
$2 \cdot F_2 - F_1$	1.2 GHz	−60 dBc	−70 dBm
$2 \cdot F_1 - F_2$	0.9 GHz	−60 dBc	−70 dBm

The derivation of the delta between P_1dB and IP3 is shown below for reference purposes only to show some but not all the factors involved in this commonly used approximation.

8.6.4 Third-Order Intercept Point

The input third-order intercept point (IIP3) occurs at an imaginary level where the third order intermodulation product $((3/4) \cdot a_3 \cdot Vp^3_{IIP3})$ equals the fundamental signal $(a_1 \cdot Vp^3_{IIP3})$, where the input level is Vp^3_{IIP3}.

$$a_1 \cdot Vp_{IIP3} = \frac{3}{4} \cdot a_3 \cdot Vp^3_{IIP3} \tag{8.30}$$

Solving for the normalized input power level (IIP3) at the third-order intercept point:

$$Vp^2_{IIP3} = \frac{4}{3} \cdot \frac{a_1}{a_3} = IIP3 \tag{8.31}$$

The relationship between the input third-order intercept point (IIP3) and the output third-order intercept point (OIP3) is the linear power gain of the device (a_1^2).

The normalized output third-order intercept point (OIP3) is:

$$OIP3 = Vp^2_{OIP3} = a_1^2 \cdot Vp^2_{IIP3} = a_1^2 \cdot IIP3 = a_1^2 \cdot \frac{4}{3} \cdot \frac{a_1}{a_3} \tag{8.32}$$

8.6.5 Relative Level of P_1dB with Respect to IP3

The 1-dB gain compression point is normally expressed in terms of output power $(P_1$dB$)$. The input voltage that produces the 1-dB gain compression $(Vp1$dB$)$ is shown in (8.33). The equation expresses the value of $Vp1$dB that when applied to the first order and third order terms of the Taylor series reduces the linear gain by 1 dB $(a_1 \cdot 0.891 \cdot Vp1dB)$ [14].

$$a_1 \cdot 0.891 \cdot Vp1dB = a_1 \cdot Vp1dB - \frac{3}{4} \cdot a_3 \cdot Vp1dB^3 \tag{8.33}$$

Reconfiguring (8.33),

$$0.109 = \frac{3}{4} \cdot \frac{a_3}{a_1} \cdot Vp1dB^2 \tag{8.34}$$

Substituting IIP3 from (8.31) and noting that $Vp1dB^2$ is the normalized input power $P1dB_{in}$,

$$0.109 = \frac{Vp1dB^2}{IIP3} = \frac{P1dB_{in}}{IIP3} \tag{8.35}$$

Converting the ratio of both sides of (8.35) to dBm $P1dB_{in} \rightarrow P1dB_{in}(dBm)$ and $IIP3 \rightarrow IIP3(dBm)$:

$$10 \cdot \log(0.109) = 10 \cdot \log\left(P1dB_{in}\right) - 10 \cdot \log(IIP3) = -9.636 \text{ dB} \tag{8.36}$$

Therefore, the differential power between the input intercept point [IIP3(dBm)] and the input power [$P1dB_{in}(dBm)$] is 9.636 dB shown in (8.37).

$$IIP3 \text{ (dBm)} - P1dB_{in} \text{ (dBm)} = 9.636 \text{ dB} \tag{8.37}$$

It should be noted that the output third-order intermodulation intercept point (OIP3) is related to the input third-order intermodulation intercept point (IIP3) by the gain factor a_1, while the output 1-dB compression point ($P1dB_{out}$) and the input level that produces 1-dB compression on the output is related by $0.891 \cdot a_1$ (1 dB below linear gain factor a_1). Therefore, the differential power between the output intercept point [OIP3(dBm)] and the output power [$P1dB_{out}(dBm)$] is 10.636 dB, as shown in (8.38).

$$OIP3 \text{ (dBm)} - P1dB_{out} \text{ (dBm)} = 10.636 \text{ dB} \tag{8.38}$$

The difference between (8.37) and (8.38) is that the third-order intercept point is only valid in a quasilinear region below the 1-dB compression point and therefore related by gain constant a_1. The signal level by definition is compressed 1 dB at the 1-dB compression point increasing the differential between the intercept point and compression point by 1 dB when referenced to the output.

Note that on $P1dB$ and IP3 that the 1-dB compression point is approximately 10 dB below the third-order intercept point when all the higher-order nonlinear coefficients are neglected and the signal is significantly below $P1dB$. At higher frequencies, the higher-order coefficients are not insignificant. Under these conditions the difference between the IP3 and $P1dB$ can be much lower than the nominal 10 dB.

8.7   Spectrum of Higher-Order Intermodulation Products

Communication systems are typically narrowband and with the band of interest passed through a bandpass filter such that all intermodulation products produced in a nonlinear system not close to the carrier are attenuated to a level that does not affect the desired signal recovery. The resultant frequency spectrum of the simplified case of two input signals (F_1 and F_2) close in frequency ($|F_1 - F_2| << F_1$ and ($|F_1 - F_2| << F_2$) is shown in Figure 8.10 for third-order, fifth-order, and seventh-order intermodulation products derived from the Taylor series (8.2) when two equal level input signals are applied [15].

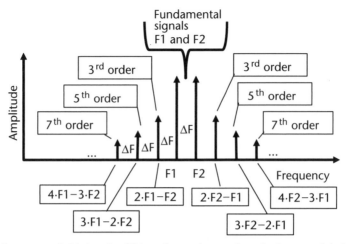

Figure 8.10 Spectrum of third-order, fifth-order, and seventh-order intermodulation products from two input signals at frequencies F_1 and F_2.

Note that all the intermodulation products are equally spaced and decreasing in amplitude. The spacing of adjacent intermodulation products are ΔF, the same as the spacing between the primary carriers F_1 and F_2.

8.8   Intermodulation Analysis of Cascaded Devices

Intermodulation interference products degrade a systems performance by generating unwanted spurious signals. In the case of odd-order interference products, the unwanted signals are close to the carrier and impossible to filter without degrading the desired signals. Since third-order intermodulation signals are usually the highest-level odd-order interference, they are typically the focus when analyzing the expected performance of a microwave system [16].

8.8.1   Calculating the Third-Order Intercept Point of Two Devices in Cascade

Every device in a chain generates intermodulation products, and when the devices are cascaded as shown in Figure 8.11, the interference signals accumulate in the worst case coherently in phase. Finding the worst-case accumulative effects of these interference signals is necessary to determine the viability of the system design.

The total level of the intermodulation interference can be found by calculating the interference products in each device and adding the interference signals

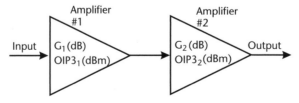

Figure 8.11   Two amplifiers in cascade.

coherently to find the worst-case interference level. Alternately, the effective third-order intercept point of the two devices in cascade can be determined and that intercept point can be used to calculate the interference in a single equation [(8.29): $IM_3 = -2 \cdot (OIP3 - P_{out})$].

Equation (8.39) calculates the effective output third-order intercept point of the cascade of two devices [17] shown in Figure 8.11.

$$OIP3_{eff} = \frac{1}{\dfrac{1}{OIP3_1 \cdot G_2} + \dfrac{1}{OIP3_2}} \tag{8.39}$$

$$G_{eff} = G_1 \cdot G_2 \tag{8.40}$$

where $OIP3_{eff}$ is the effective output intercept point of Devices 1 and 2 combined, $OIP3_1$ is the output third-order intercept point of the input device (Device 1), $OIP3_2$ is the output third-order intercept point of the output device (Device 2), G_1 is the gain of the output device (Device 1), G_2 is the gain of the output device (Device 2), and G_{eff} is the gain of the combined devices.

The units of $OIP3_{eff}$, $OIP3_1$, and $OIP3_2$ are power, usually in milliwatts or watts. Gain of individual and combined devices (G_1, G_2, and G_{eff}) are in linear units. Typically the device intercept point is given in dBm and the device gain is given in decibels. Under these conditions, the units have to be converted from decibels to linear units before being used in (8.39) and (8.40). The final results are converted back into decibels because (8.29) used to determine the interference levels is in decibels (dBm and dBc).

Note that with regard to the effective intercept point ($OIP3_{eff}$) in (8.39) the effective intercept point $OIP3_{eff}$ is less than output intercept point of the last stage ($OIP3_2$). The effective intercept point $OIP3_{eff}$ is lower than the output intercept point of the previous stage ($OIP3_1$) times the gain of the output stage (G_2). The interecpt point always degrades because every stage in a chain is nonlinear and therefore adds intermodulation products to that produced in the previous stages.

Example 8.3

See Figures 8.12 and 8.13. Find the combined intercept point of the two cascaded devices in Figure 8.11. The device parameters are shown in Table 8.3.

$OIP3_{eff}$ is 200.76 mW $\rightarrow$ 23.027 dBm.

Table 8.3   Device Parameters for Example

	Gain	OIP3	Gain	OIP3
Description	dB	dBm	Linear	mW
Device 1	15	17	31.623	50.119
Device 2	13	24	19.953	251.189

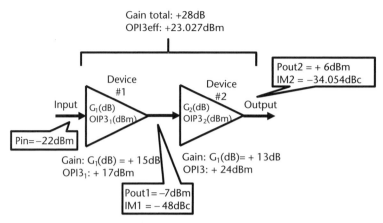

Figure 8.12 Example 8.3: two amplifiers in cascade.

Assume that the input power of each carrier is −22 dBm. What is the level of each fundamental signal and each third intermodulation product at the output of the first stage?

P_{in} = −22 dBm and P_{out1} = −7 dBm
IM_1 = −48 dBc (−55 dBm)

What is the level of each fundamental signal and each third intermodulation product at the output of the second stage?

P_{out2} = +6 dBm
IM_2 = −34.054 dBc (−28.054 dBm)

8.8.2 Calculating the Third-Order Intercept Point of Multiple Devices in Cascade

The intermodulation products of the chain contains more than two devices is calculated by successively applying the techniques used for two devices starting from

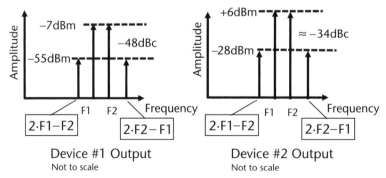

Figure 8.13 Example 8.3: fundamental and third-order intermodulation products at the output of Devices 1 and 2.

the input two devices and working forward. The two input devices are calculated to have an effective third-order intermodulation intercept point [OIP3$_{\text{eff}}$ and a total gain (G_{tot})]. The first devices are modeled as a single device. This single device is now the input to the third stage and the combined total gain and output intermodulation intercept point is calculated for the output of the third device. This procedure is repeated for any number of N devices in series.

Example 8.4

See Figure 8.14 and Table 8.4. Figure 8.14 is a typical receiving system of six devices each labeled with their respective gains (G_N in dB) and output intercept point (OIP3$_N$ in dBm) where N is the device number.

In any microwave system, each of the amplifiers is selected for their gain and intercept characteristics consistent with the design requirements. The mixer intercept point is the local oscillator power feeding the mixer. Passive devices, that is, filters and attenuators, are typically very linear when not operating at very high levels and are usually assigned a very high intercept point. In this example, the intercept point for the passive devices has been assigned an arbitrary level of +90 dBm. Table 8.4 lists all the six devices and their respective parameters in columns 1 to 4.

Columns 5 through 9 show the cumulative results from the input to the output of the devices the respective numbered row. Row 6 and columns 5 through 9 show the cumulative results of the entire system.

Columns 5 and 6 are the cumulative gain and output third order intercept point from the input row 0 to the output of the respective device number (row number). Column 7 is the third-order intercept point reflected to the system input (IIP3). This number is calculated by subtracting the cumulative system gain (column 5) from the output intercept point (OIP3, column 6).

Column 8 is the output power for the respective device and row number of each of the two equal power carriers entering the system. The level of the input power is on line 0, column 8. Column 9 is the level of one of the two equal third-order intermodulation products at the output of the device in the respective row. The level is in dBc, the single intermodulation product in decibels below the level of a single carrier.

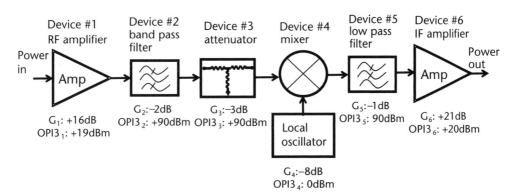

Figure 8.14   Example of a typical receiver, RF input to IF output.

Table 8.4 The Effects of Successive Series Devices in a Typical Receiver

1 Device Number	2 Description	3 Device Gain dB	4 Device OIP3 dBm	5 System Gain dB	6 System OIP3 dBm	7 System IIP3 dBm	8 System Each Carrier dBm	9 System Each IM dBc
0							−30.0	
1	RF amplifier	16	19	16.0	19.00	3.00	−14.0	−66.0
2	Bandpass filter	−2	90	14.0	17.00	3.00	−16.0	−66.0
3	Attenuator	−3	90	11.0	14.00	3.00	−19.0	−66.0
4	Mixer	−8	0	3.0	−0.97	−3.97	−27.0	−52.1
5	Lowpass filter	−1	90	2.0	−1.97	−3.97	−28.0	−52.1
6	IF amplifier	21	20	23.0	16.48	−6.52	−7.0	−47.0

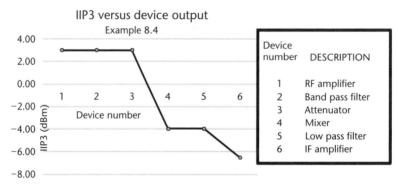

Figure 8.15 Plot of input IP3 as it degrades through each device in the system, Example 8.4.

Note that the input intermodulation intercept point (IIP3), column 7 and the intermodulation product column 9 are a good indicator of the how each device is degrading the system intermodulation performance. Figure 8.15 shows the input third-order intercept point at the output of each device reflected to the input. The graphs indicate the negative impact that each device has on the total system third-order intermodulation intercept point. In this graph device, number 4 (Mixer) has the greatest negative effect on the third-order intermodulation performance. If the system OIP3 specification is not met, the system designer would look at improving the intercept point of Device 4 and replot the results to determine if the performance of any other device needs to be enhanced.

References

[1] Pedro, J., and N. Carvalho, "A Novel Nonlinear Distortion Characterization Standard for RF and Microwave Communication Systems," *Engineering Science and Education Journal*, June 2001.

[2] National Instruments, "Understanding Frequency Performance Specifications," 2017.

[3] Kerczewski, R., "A Study of the Effect of Group Delay Distortion on an SMSK Satellite Communications Channel," National Aeronautics and Space Administration, Lewis Research Center, 1987.

[4]   Hausman, H., "High Dynamic Range Low Noise Amplifiers for Communications Surveillance," *43rd Annual AOC International Symposium and Convention*, November 2006.

[5]   Razavi, B., "Basic Concepts in RF Design," *RF Microelectronics*.

[6]   Lixin, Z, J. Zhi, and L. Xinyu, "Microwave Dynamic Large Signal Waveform Characterization of Advanced InGaP HBT for Power Amplifiers," *Journal of Semiconductors*, Vol. 30, No. 12, December 2009.

[7]   Niknejad, A., "Lecture 9: Intercept Point, Gain Compression and Blocking," University of California, Berkeley, 2005.

[8]   Agrawal, V., and F. Dai, "RFIC Design and Testing for Wireless Communications," TI India Technical University, 2008.

[9]   Hall, D., "Understanding Intermodulation Distortion Measurements," *Electronics Design*, October 2013.

[10] Agahi, D., W. Domino, and N. Vakilian, "Two-Tone vs. Single-Tone Measurement of Second-Order Nonlinearity," *Microwave Journal*, March 2002.

[11] Simon, M., "Interaction of Intermodulation Products Between DUT and Spectrum Analyzer," Rohde & Schwarz, December 2012.

[12] Frenzel, L., "What's the Difference Between the Third-Order Intercept and the 1-dB Compression Points?" *Electronic Design*, October 24, 2013.

[13] Cho, C. W. Eisenstadt, "A Simple Technique for IIP3 Prediction from the Gain Compression Curve," *Wamicon*, 2005.

[14] Troyanovsky, B. "Frequency Domain Algorithms for Simulating Large Signal Distortion in Semiconductor Devices," Stanford University, Ph.D. dissertation, November 1997.

[15] Agilent Technologies, "Generation and Conditioning of Multitone Test Signals," *Microwave Design and Measurement Seminar*, December 2002.

[16] Van Dyke, R., "Effective Cascaded Intermodulation Analysis," *RF Design*, September 2004.

[17] Van Dyke, R., "Practical Intercept Measurements and Cascaded Intermod Equations," Agilent Technologies, May 2007.

CHAPTER 9

System Cascade and Dynamic Range Analysis

9.1 Introduction to Cascade Analysis and Dynamic Range

The dynamic range of a device or a system is the input level between a minimum signal and maximum signal that will meet all the respective specifications at the output. In a linear or quasilinear system, the relative dynamic range (the delta between the maximum signal level and the minimum signal level) at input is equal to the relative dynamic range at the output.

Minimum signals are usually limited by the minimum acceptable signal to noise ratio required to recover the signal. The maximum signal is at level that produces an acceptable level of signal compression, third-order intermodulation interference, harmonic signals, and spectral regrowth [1]. Signal compression occurs when a single output signal distorts. Third-order intermodulation interference is signals related to multiple desired signals. Harmonics signals are signals that are integral multiples of the desired signal. Spectral regrowth is an increase in the desired modulated signals sidebands causing signal distortion or adjacent carrier interference.

The maximum signal is a related to the number of carriers simultaneously entering the system, the relative amplitude of the carriers, and the spacing between carriers. Dynamic range is usually specified with a single carrier or two equal amplitude carriers. The effect of three or more carriers at equal levels, two carriers at unequal levels, and many random carriers that can be modeled as a flat noise spectral density function can be approximated from the two equal carrier data.

A cascade analysis calculates the accumulative effect of the gain, noise figure, and third-order intercept point of every series component in the system to determine the total noise and expected third-order intermodulation interference at the system output. This information, combined with other system specifications such as minimum signal-to-noise ratio (S/N) and maximum acceptable interference, can be used to determine the system's dynamic range.

9.2 Minimum Signal Level Limitations

Dynamic range on the low signal side is the minimum S/N that will recover the transmitted signal with an acceptable error [2]. Possibly added to the noise is interference from other carriers present in the system chain. If unwanted carriers are at a higher level than the desired carrier, they can produce third intermodulation products that are close to the desired carrier such that a filter will not alleviate the interference issue. Under these conditions, noise is not the only issue in the recovery of small signals; the actual issue might be noise plus interference and the equation for minimum S/N becomes the minimum signal-to-noise plus interference ratio [S/(N + I)]. In the digital communications the minimum S/(N + I) is converted to an expected maximum bit error rate (BER) [3].

Figure 9.1 shows a microwave signal as a vector pointing to one sector of one quadrant of a constellation of 16 points. Each sector represents a 4-bit digital word. The figure shows a small vector at the tip of the signal vector that represents an uncorrelated interference vector rotating around the signal vector. The amplitude of the interference vector with respect to the signal vector can be described in terms of dBc (dB below the carrier). Added to the interference vector is a noise vector randomly varying in amplitude and phase. The root mean square (RMS) amplitude of the noise vector at any phase represents one standard deviation (1σ) of a Gaussian probability function. The probability that the noise power will be within or exceed one sigma (1σ), two sigma (2σ), and up to five sigma (5σ) is given in Table 9.1.

The probability that the composite (signal to noise plus interference) reaches an error threshold is determined by the level of the interference vector plus how many standard deviations the noise is away from a threshold crossover point (shown as a dashed line in Figure 9.1). When the signal crosses the threshold, the digital word changes from the desired word to an error word.

In addition to thermal noise and interference signals the signal vector will have associated errors due to: (1) phase noise, (2) signal amplitude compression, (3) signal phase change due to signal amplitude compression, and (4) group delay distortion. Signal distortions (1) through (4) plus signal interference (I) plus the thermal noise (N) determine the expected BER [4]. If the interference plus distortion is small compare to the RMS noise level, the S/N is the number of standard deviations from the tip of the signal vector to the threshold level and the BER is determined from Table 9.1 where number of standard deviations is the S/N.

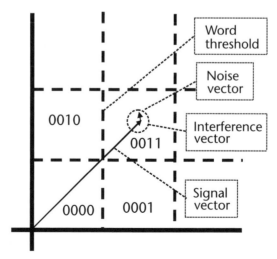

Figure 9.1 Signal vector with interference and noise.

Table 9.1 Probability of Being Within or Exceeding the Number of Standard Deviations (1σ is the RMS Value of Noise)

Number of Standard Deviations	Probability Within the Number of Standard Deviations (σ)	Probability of Exceeding the Number of Standard Deviations in Parts per Million
1	68.268949%	317,310.51
2	95.449974%	45,500.26
3	99.730020%	2,699.80
4	99.993666%	63.34
5	99.999943%	0.57

Many times, the allowable distortion is less than 1σ and the maximum third-order interference is equal to one standard deviation (1σ) (i.e., the maximum third-order interference is no greater than the RMS noise level). Under those conditions, the BER is the number of standard deviation minus one. The maximum difference in signal level, the maximum signal level with respect to the minimum signal level, is determined by the number of carriers and the third-order intermodulation intercept point.

9.3   Maximum Signal Level Limitations

The maximum signal into and out of a system could have multiple definitions depending on the type of signals, the modulation on those signals, the quantity of signals present at the same time, and the tolerance for signal recovery in the presences distortion. Assuming that all signals are modulated with the intent of the system to recover the modulation, the maximum acceptable signal criteria fall into two major categories, single signal distortion and the tolerance for intermodulation distortion caused by multiple signals at the same or different levels [5]. The generalized worst-case condition of all signals equal in amplitude at the maximum allowable signal level (meeting the respective system criteria) will be assumed.

9.3.1   Continuous-Wave Single Signal Maximum Levels

The specified maximum allowable continuous wave (CW) signal level that produces an acceptable level of distortion varies significantly over a range of system applications. Typically, the maximum signal level is specified with respect to the 1-dB gain compression point (P1dB).

The specified allowable distortion for CW single signal linear system levels could vary significantly from system to system but is typically not higher than 3 dB below the P1dB. This approximation is related to the peak power of a sinusoidal signal, which is 3 dB above the average power. Sinusoidal signal levels above P1dB begin to significantly compress and distort.

With regard to CW single-signal saturated system levels, the saturation of a CW signal is typically less than 3 dB above the P1dB.

9.3.2 Maximum Level for a Single-Signal Modulated Carrier

Single signals that are modulated have an issue beyond the distortion of the carrier due to the intermodulation distortion between the signal sidebands. Generally, the issue is related to the third-order and fifth-order intermodulation products of the carrier sidebands, but sometimes higher-order intermodulation distortion can be an issue. These products cause intermodulation distortion within the carrier bandwidth distorting the recovered information.

Single CW modulated signals [e.g., frequency modulation (FM) and phase modulation (PM)] may approach the P1dB and still be recoverable. Other types of amplitude modulated (AM) or amplitude- and phase-modulated signals such as higher-order quadrature amplitude modulated (QAM) signals may require that the maximum average signals level be significantly lower than the P1dB for adequate modulation recovery. A 16QAM (16-point quadrature amplitude modulation) constellation is shown in Figure 9.2. In QAM, signals are transmitted at a point and received in an area bounded by threshold levels, shown as dashed lines in Figure 9.2. The peak-to-average ratio of any type of amplitude modulation is greater than 3 dB.

9.4   A Typical Spurious-Free Dynamic Range Calculation

The term spurious-free dynamic range is often used by device manufacturers to compare their devices against other manufacturers' devices. Since spurious-free is an ideal unattainable concept, manufacturers often set the spurious level to the noise level in a normalized 1-Hz bandwidth when they define a spurious-free dynamic range. This assumption, along with the assumption that third-order intermodulation interference is the dominant form of spurious signals, leads to method of calculating third-order intermodulation interference in a device based on the device's

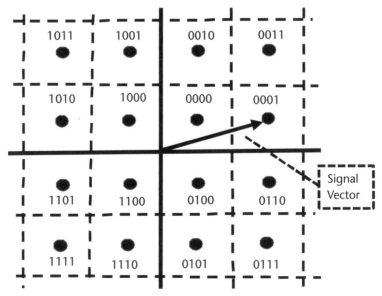

Figure 9.2 Typical 16QAM constellation.

third-order intermodulation intercept point and equating that level to the devices thermal noise level based on the device's noise figure and a normalized 1-Hz bandwidth both reflected to the input of the device [6]. Figure 9.3 graphically depicts the concept of a spurious-free dynamic range.

This calculation does not consider the actual bandwidth of the system and therefore is only useful as a relative measure of the differential between a device third-order intercept point and normalized noise level which is related to the device's noise figure. The spurious-free dynamic range is typically a large differential between minimum and maximum signals because of the 1-Hz noise bandwidth and therefore may or may not be useful in any application.

9.4.1   Calculating the Normalized Thermal Noise

Equation (9.1) calculates the input thermal noise power (P_N) normalized in 1-Hz bandwidth [7].

$$P_N = 10 \cdot \log(k \cdot T \cdot BW) + NF \tag{9.1}$$

where P_N is the input noise power (dBW), dBW is decibels with respect to 1W, dBm is decibels with respect to 1 mW (dBW = dBm − 30), NF is the device noise figure (dB), k is Boltzmann's constant ($k = 1.38064852 \cdot 10^{-23}$ J/K), T is the temperature in degrees kelvin, BW is the bandwidth normalized to 1 Hz, and

$$10 \cdot \log(k \cdot T \cdot BW) \approx -174 \text{ dBm in a 1-Hz bandwidth at of 298K} \tag{9.2}$$

Example 9.1: Noise Power

The device noise figure is 7 dB

$$P_N \approx -167 \text{ dBm (in a 1-Hz bandwidth at } T = 298\text{K)}$$

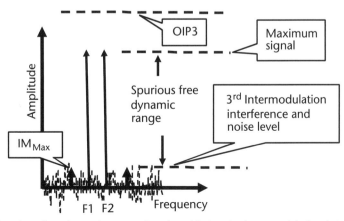

Figure 9.3 Spurious-free dynamic range showing third-order intermodulation interference level equal to a normalized thermal noise level.

9.4.2 Third-Order Intermodulation Interference

The maximum input level (S_{max}) defined in this section only is the level of one and two equal carriers at frequencies F_1 and F_2 that produce third-order intermodulation interference signals at frequencies $2 \cdot F_1 - F_2$ and $2 \cdot F_2 - F_1$ such that each is at the thermal noise level in a 1-Hz bandwidth all reflected to the input of the device.

The procedure for determining the input maximum signal level in dBm (S_{max}) involves calculating the third-order intermodulation interference of the device in dBc (IM3) knowing the input third-order intermodulation intercept point (IIP3).

The third-order intermodulation interference (IM3) in dBc is:

$$IM3 = -2 \cdot \left(IIP3 - S_{max} \right) \tag{9.3}$$

$$IM3 = -2 \cdot \left(IIP3 + 2 \cdot S_{max} \right) \tag{9.4}$$

The absolute value of the interference level equals $S_{max} + IM3$.

9.4.3 Calculating Spurious-Free Dynamic Range

The maximum signal level is determined by equating that interference level created at the output to the device, reflected to the device input ($S_{max} + IM3$) to the noise level in dBm (P_N) calculated in (9.1).

The dynamic range in dB (DRdB) is the maximum signal in dBm (S_{max}) minus the noise level (P_N) in dBm in a 1-Hz bandwidth.

Computing dynamic range is accomplished by setting the noise level [(P_N) equal to $S_{max} + IM3$], where IM3 is a negative number in dBc.

$$P_N = S_{max} + IM3 \tag{9.5}$$

Substituting (9.4) into (9.5),

$$P_N = S_{max} - 2 \cdot IIP3 + 2 \cdot S_{max} = -2 \cdot IIP3 + 3 \cdot S_{max} \tag{9.6}$$

The maximum signal in dBm (S_{max}) is:

$$S_{max} = \frac{1}{3} \cdot P_N + \frac{2}{3} \cdot IIP3 \tag{9.7}$$

Calculating the dynamic range in dB ($DRdB$) is accomplished by subtracting P_N from S_{max}.

$$DRdB = S_{max} - P_N = \frac{1}{3} \cdot P_N + \frac{2}{3} \cdot IIP3 - P_N = -\frac{2}{3} \cdot P_N + \frac{2}{3} \cdot IIP3 \tag{9.8}$$

Normally the output third-order intermodulation intercept point (OIP3) is specified for a device. The input third-order intermodulation intercept point (IIP3) both in dBm is

$$IIP3 = OIP3 - G \qquad (9.9)$$

where G is the device gain in decibels

9.5   Dynamic Range Analysis: Example

A typical RF-to-IF microwave downconverter is shown in Figure 9.5. Each of the six devices is shown with their respective noise figure (NF), gain (G), and output third-order intermodulation intercept point (OIP3).

The passive devices are assumed to have an output IP3 (OIP3) of +90 dBm. The noise figure of a passive device is the negative of the gain (e.g., if gain is −2 dB, the noise figure is 2 dB).

The mixer and local oscillation (LO) translates the RF frequency (F_{RF}) to the IF frequency (F_{IF}) by multiplying the RF frequency by the local oscillator frequency (F_{LO}). The output of the mixer is the F_{IF}, where:

$$F_{IF} = F_{RF} + F_{LO} \qquad \text{(sum frequency)} \qquad (9.10)$$

and

$$F_{IF} = \left| F_{RF} - F_{LO} \right| \qquad \text{(difference frequency)} \qquad (9.11)$$

The sum frequency ($F_{IF} = F_{RF} + F_{LO}$) shown in (9.10) is eliminated by the low-pass filter Device 5. The mixer loss in this example is assumed to be 8 dB and the noise figure is approximately the negative of the device gain, also 8 dB. The output third-order intercept point of the mixer is assumed in this example to be +10 dBm. It is related to the output power and power capability of the diodes inside the mixer and the local oscillator power.

Table 9.2 calculates the cumulative gain, noise figure, intercept point, and signal level for the system in Figure 9.4. Columns 1 through 5 are the device number, description, gain (dB), noise figure (dB), and output third-order intercept point (dBm), respectively. The accumulated characteristics from the input to the output

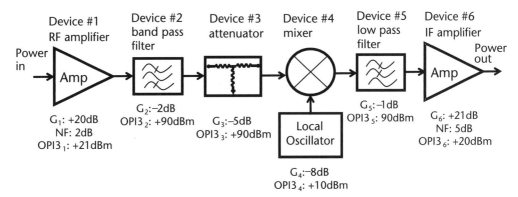

Figure 9.4   Example of a typical receiver, RF input to IF output.

Table 9.2 Cumulative Intercept Point and Noise Figure of Cascaded Devices

		3	4	5	6	7	8	9	10
			Device				*System*		
1 *Device* *Number*	*2* *Description*	*Gain* *(dB)*	*NF* *(dB)*	*OIP3* *(dBm)*	*Gain* *(dB)*	*NF* *(dB)*	*OIP3* *(dBm)*	*IIP3* *(dBm)*	*Each* *Carrier* *(dBm)*
						0.00			−30.0
1	RF amplifier	20	2	21	20.0	2.00	21.00	1.00	−10.0
2	Bandpass filter	−2	2	90	18.0	2.02	19.00	1.00	−12.0
3	Attenuator	−5	5	90	13.0	2.11	14.00	1.00	−17.0
4	Mixer	−8	8	10	5.0	2.77	4.54	−0.46	−25.0
5	Lowpass filter	−1	1	90	4.0	2.95	3.54	−0.46	−26.0
6	IF amplifier	21	5	20	25.0	4.52	18.69	−6.31	−5.0

of the device in each row (1 through 6) are shown in columns 6 through 10. The characteristics calculated are accumulated: gain (column 6), noise figure (column 7), output intermodulation intercept point (column 8), input intermodulation intercept point (column 9), and carrier level (column 10). The total cumulative characteristics of the six devices in series are shown in row 6.

Figure 9.5 is a graph of noise figure and third-order intermodulation intercept point at the output of the referenced numbered device both reflected to the input of the chain. NF and IIP3 are plotted because each respective line clearly shows the system component that provides the most significant degradation to system performance. For the system in Figure 9.5, the graph clearly shows that the output amplifier (Device 6) has the most negative effect on noise figure and intermodulation degradation. Replacing Device 6 with an amplifier that has a lower noise figure and a higher output third-order intercept point will improve the dynamic range of the system. Note that increasing the gain of Device 6 will not affect the noise figure and degrade the third-order intermodulation performance.

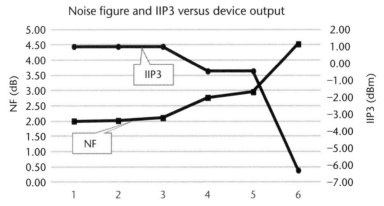

Figure 9.5 Graph of intermodulation intercept point and noise figure at the output of each device in the chain, reflected to the input.

9.6   Out-of-Band Noise Power in a Transmitter: Example

The out-of-band noise problem typically occurs in communication systems and satellite communication Earth station transmitters where the signals transmitted to and received by the satellite are at many different frequencies from many different locations. The Earth station transmission site typically uses a block upconverter and power amplifier configuration to that shown in Figure 9.6. The term block upconverter refers to the fact that the IF section of the converter is the same bandwidth as the RF section of the converter and the entire IF bandwidth is converted to a known and properly allocated satellite RF spectrum.

A signal modulator feeds the IF section of the block converter at a frequency such that the resultant RF signal transmitted to the satellite is at a location and bandwidth allocated to the satellite user. This system is very efficient and cost-effective because the signal transmitted to the satellite can be easily modified with respect to its carrier frequency and bandwidth. A negative issue with this configuration is that the noise produced in the upconverter and the power amplifier emits interference over the entire portion of the allocated satellite receiving spectrum, usually significantly beyond the signal bandwidth. Signals from other satellite Earth stations using different portions of the spectrum will encounter this noise and will experience a slight signal to noise degradation. In the extreme, when there are many diverse users of the same satellite transponder this interfering noise could build up to a level that noticeably degrades the signals to noise of these other usually small bandwidth and lower level carriers. This issue is mitigated by limiting the total noise power spectral density emitted from any single transmitter.

The emitted noise level is shown in Table 9.3, which calculates the cascaded increase in noise figure, noise output power, and total gain. Noise figure is normally reflected to the input of a chain of devices where the input noise (N_{in_n}) reflected back from device n is:

$$N_{in_n} = 10 \cdot Log(k \cdot T \cdot B \cdot F_n) + 30 \qquad (9.12)$$

and the output noise is the input noise multiplied by the gain from the system input to the device output.

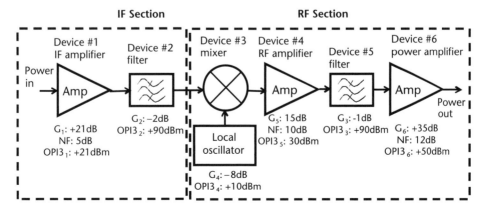

Figure 9.6   Example of a block upconverter and output power amplifier.

Table 9.3 Chain Analysis of a Block Upconverter Feeding a Power Amplifier (The Output Noise Power Is Normalized to a 1-MHz Bandwidth)

		3	4	5	6	7	8	9	10	11
1		Device			System					Output
Device Number	2 Description	Gain (dB)	NF (dB)	OIP3 (dBm)	Gain (dB)	NF (dB)	OIP3 (dBm)	IIP3 (dBm)	Carrier (dBm)	Noise (dBm/MHz)
						0.00			−25.0	−113.86
1	RF amplifier	21	5	21	21.0	5.00	21.00	0.00	−4.0	−87.86
2	Lowpass filter	−2	2	90	19.0	5.01	19.00	0.00	−6.0	−89.85
3	Mixer	−8	8	10	11.0	5.10	7.46	−3.54	−14.0	−97.76
4	RF amplifier	15	10	30	26.0	5.96	21.76	−4.24	1.0	−81.89
5	Bandpass filter	−1	1	90	25.0	5.97	20.76	−4.24	0.0	−82.89
6	Power amplifier	35	12	50	60.0	6.02	48.98	−11.02	35.0	−47.84

$$N_{out_n} = 10 \cdot \text{Log}\left(k \cdot T \cdot B \cdot F_n \cdot G_n\right) + 30 \qquad (9.13)$$

where N_{out_n} Is the noise power at the output of the n device (dBm), k is Boltzmann's constant ($k = 1.38064852 \cdot 10^{-23}$ J/K), B is the noise bandwidth (Hz), F_n is the noise factor from the input through device n, G_n is the gain from the input through device n, and the factor +30 is added to display the result in the more common dBm instead of dBW.

The chain analysis using the input intercept point effectively normalizes the intermodulation degradation and gives an indication of which devices in the chain have the most negative effect on the system linearity. A graph of NF and IIP3 as a function device output from Figure 9.6 is shown in Figure 9.7.

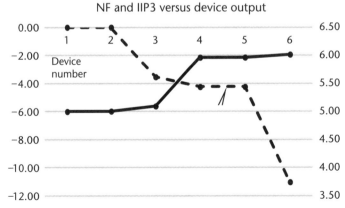

Figure 9.7 NF and IIP3 as the signal progresses to the respective output of each device in Figure 9.7.

It is important to view both factors because a decrease in noise figure can also have a negative effect on the intermodulation distortion. Tables such as these and the graphs derived from these tables allow the design engineer to vary the individual component characteristics and visually see the effects to meet both of the sometimes-conflicting requirements.

9.7   Carrier Triple Beats Interference

When three equal level carriers (F_1, F_2, and F_3) are present in a system, the third-order modulation interference in dBc is 6 dB higher than two carriers' third-order intermodulation interference because the interference from the carriers beating together (carrier triple beats) does not require a second harmonic of any carrier. Figure 9.8 compares the frequency and signal level of two-tone third-order intermodulation interference and carrier triple beat interference for equally spaced carriers.

The third-order intermodulation interference frequencies of two carriers at frequencies, F_1 and F_2, are: $2 \cdot F_1 - F_2$ and $2 \cdot F_2 - F_1$. The third-order intermodulation interference frequencies for three carriers F_1, F_2, and F_3 are: $F_1 + F_2 - F_3$, $F_2 + F_3 - F_1$, and $F_1 + F_3 - F_2$.

If the three carriers in the carrier triple beat example are equally spaced, the third-order intermodulation interference in case 3 ($F_1 + F_3 - F_2$) is at the same frequency as carrier F_2 [8].

9.8   Multiple Carrier (N > 3) Interference

The carrier triple beat discussion can be extended to a greater number of carriers (N) [9]. Assume that the carriers in the band are equal amplitude and equally spaced numbered 1 to N in order of their frequency position in the band. N = 1 is assumed to be the lowest frequency carrier. The worst-case interference occurs near

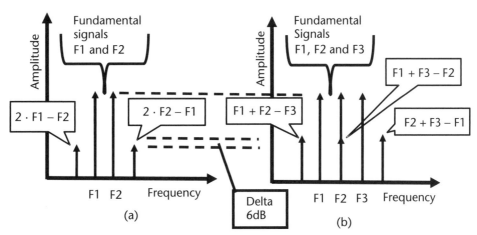

Figure 9.8   (a) Two-carrier third-order intermodulation interference (a) and (b) carrier triple beat of three equally spaced carrier interferences.

the closest carrier to $N/2$ near the center of the band. The maximum number of carrier beats ($Beats_{MAX}$) in a single frequency slot in the center of the band would be:

$$Beats_{MAX} = \frac{3 \cdot N^2}{8} \tag{9.14}$$

Since all the carriers are considered noncoherent, the respective interference level (IM) in dBc would be:

$$IM(dBc) = -2 \cdot \left[OIP3(dBm) \right] - C(dBm)] + 6 \ dB + 10 \cdot Log\left(\frac{3 \cdot N^2}{8} \right) \tag{9.15}$$

where N is the number of carriers for $N > 3$, OIP3 is the output third-order intermodulation intercept point in dBm, and C is the power level of each carrier in dBm.

References

[1]   Browne, J., "Understanding Dynamic Range," *Microwaves and RF*, February 2011.

[2]   Skolnik, M., *Radar Handbook*, New York: McGraw-Hill, 2008.

[3]   Long, S., "Performance Limitations of Amplifiers, 1. Distortion in Nonlinear Systems," UCSB, 2010.

[4]   Watson, R., "Receiver Dynamic Range: Part 1," WJ Communications, Inc., Tech Note.

[5]   Loy, M., "Understanding and Enhancing Sensitivity in Receivers for Wireless Applications," Texas Instruments, Technical Brief SWRA030, May 1999.

[6]   Watson, R., "Receiver Dynamic Range: Part 2," WJ Communications, Inc., Tech Note.

[7]   Karki, J., "Calculating Noise Figure and Third-Order Intercept in ADCs," Texas Instruments Inc., *Analog Applications Journal*, 2003.

[8]   Hausman, H. "Topics in Communication System Design: Carrier Triple Beats," *Microwave Journal*, 2002.

[9]   Hausman, H., "Estimate Multiple Carrier Interference," *Microwaves & RF*, May 2004.

Part II
Designing the Power Amplifier

Defining the Output Power Requirements in a Communication Link and Other Wireless Systems

10.1 Introduction: Power Amplifier Requirements

The mission of the power amplifier is to deliver the signal to the receiver at a high enough level to ensure signal recovery and not interfere with other signals. Specific requirements depend on the application, but generalized common goals for communication systems and radar systems are sufficient output power, an acceptable amount of distortion (nonlinearity), and maximum practical efficiency.

A multitude of specific requirements flow down from these general requirements depending on the complexity of the information to be recovered and proximity of the receiver with respect to the transmitter and the medium between the two respective systems.

The power amplifier is one of the most critical and costly items in a communications or radar system. Undersizing the available power reduces the signal availability and oversizing the available power adds recurring and nonrecurring costs to design and system operation.

10.2 Power Amplifier Requirements in a Wireless Communications Link

Figure 10.1 shows a typical wireless communications link. Baseband information to be transmitted is fed into an IF carrier modulator [1]. The IF signal is upconverted to the required RF and amplified (power amplifier) to the required power for successful signal recovery after the signal travels through the transmit antenna, the path between the transmitter and receiver, the receiver antenna, the low noise amplifier, the downconverter, and the demodulator.

Linearity requirements are important throughout the chain, but typically in a well-designed system most critical in the transmitter power amplifier. It is understood that the required output from the power amplifier is at a linearity level such that all the distortion requirements, typically intermodulation distortion, harmonics, and spectral regrowth, are also met.

In a properly designed communication link, the C/N is primarily determined by the power amplifier and transmit antenna, the path loss, and the receive antenna and the low noise amplifier [2] as shown in Figure 10.2.

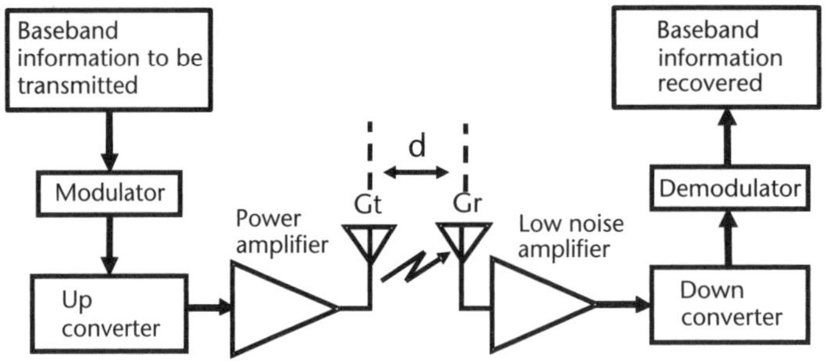

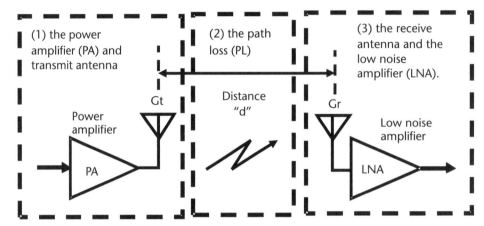

Figure 10.1   Typical block diagram of a wireless communication system.

Figure 10.2   The key components that determine the C/N are the power amplifier (PA) and transmit antenna, the path loss (PL), and the receive antenna and the low noise amplifier (LNA) designated by the respective dashed lines.

Since one of the primary design parameters is the C/N [3], synthesizing a communication link starts with the receive antenna and the low noise amplifier and works backward through the path loss and the power amplifier and transmit antenna. After the transmit antenna characteristics and associated losses are determined, the parameters of the power amplifier can be defined.

10.3   Design a Receiver Input to Meet a Required Minimum C/N in a Communications System

In any wireless communication system, the objective is to recovery the transmitted signal at an acceptable level above the noise (i.e., a minimum C/N). Figure 10.3 is a diagram of a typical wireless receiving system front end. The diagram and example in Table 10.1 describe the C/N to the output of the low noise amplifier. Although the C/N degrades through each component in the receiver chain, a properly designed system will have a minimal degradation after the input low noise amplifier.

The input signal with some added atmospheric noise [usually expressed in degrees kelvin (T_A)] is gathered in an antenna reflector (dish), which focuses the signal to an

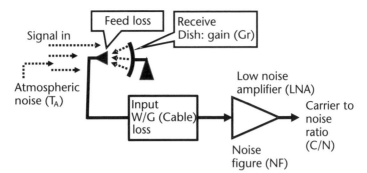

Figure 10.3 Receiving antenna showing where typical gains and losses occur before being amplified in the input low noise amplifier.

antenna feed. The antenna feed sends the signal to the low noise amplifier. When analyzing the receiver input, it is important to separate the noise sources and the signal losses although both add to the effective noise figure of the receiver because only the actual signal loss adds to the system gain [3]. Table 10.1 translates the C/N requirements (row 14) to the actual signal level required to the input of the receive antenna (row 22) given the atmospheric parameters and other characteristics listed in rows 1, 2, 5, 8, 9, 14, 16, and 20 [4]. The other rows are calculated values that are a function of the input data.

Example: Determining Received Input Signal Power from C/N

See Table 10.1. The input data is as follows. Row 1 is the ambient temperature use to calculate the thermal noise; 298K is approximately room temperature. Rows 2 and 3 are the signal frequency and the respective calculated signal wavelength in free space. These parameters are used to calculate antenna gain, but conventionally antenna gain is given with respect to an isotropic radiator and the frequency component of the calculation is handled in the path loss calculation. Row 5 is the signal bandwidth is the minimum bandwidth required to pass the modulated signal with an acceptable amount of distortion. The noise bandwidth given in row 5 is typically 10% to 50% larger than the signal bandwidth. The noise bandwidth is an ideal filter (perfectly square filter characteristic) that contains the same amount of noise power as the real filter containing the same noise spectral density but integrated over the entire spectrum. Row 8 is the atmospheric noise in degrees kelvin. Atmospheric noise includes precipitation noise, solar noise, and galactic noise. This noise parameter is a function of the signal frequency and antenna elevation angle. Row 9 is the noise figure of the input low noise amplifier in degrees kelvin. Row 14 is the required C/N in decibels. Row 16 is waveguide or cable loss between the antenna and the low noise amplifier (dB). Row 20 is antenna gain in dBi. dBi is the gain with respect to that of an isotropic radiator.

For the calculations in Table 10.1, row 6 calculates the thermal noise $(k \cdot T \cdot B)$, where k is Boltzmann's constant, T is the ambient temperature in degrees kelvin (row 1), and B is the noise bandwidth (row 5). Row 10 is the noise figure of the low noise amplifier calculated from the noise temperature in row 9. Row 11 is the system

Table 10.1   Chain Analysis of a Wireless Receiving System

1	Temperature	298	Degrees kelvin
2	Signal frequency	14	GHz
3	Lambda	0.021413747	Meters
4	Thermal noise		
5	Noise bandwidth	30	Megahertz
6	Thermal noise (kTB)	−99.09	dBm
7	Added noise		
8	Antenna noise (T_A)	11	Degrees kelvin
9	Low noise amplifier noise temperature (T_{LNA})	50	Degrees kelvin
10	Low noise amplifier noise figure	0.69	Decibels
11	System noise figure	0.83	Decibels
12	Total system noise	−98.26	dBm
13	Receiver input analysis		
14	C/N	13.56	dB
15	Signal after antenna gain	−84.70	dBm
16	Waveguide loss	0.2	dB
17	Waveguide loss (T_{Loss})	13.667	Degrees kelvin
18	Signal at antenna input		
19	Antenna output signal	−84.50	dBm
20	Antenna gain (G_r)	42.23	dBi
21	G/T (for reference only)	23.50	dB/K
22	Antenna input signal (Sr)	−126.73	dBm

noise figure calculated from the sum of the low noise amplifier noise temperature in row 9 and the antenna noise temperature row 8. Row 12 is the total equivalent system input noise in dBm, $10 \cdot \text{Log}(k \cdot T \cdot B) + NF$ where NF is the system noise figure in row 12. Row 15 is the signal at the low noise amplifier input equal to the sum of the total system noise in dBm (row 12) plus the C/N in decibels (row 14). Row 17 is the loss in row 16 converted to noise temperature in degrees kelvin. This is for reference only since the loss in row 16 is used for these calculations. Row 19 is the antenna output signal in dBm equal to the sum of row 15 plus row 16. Row 21 is G/T, the antenna gain divided by the total noise temperature. In decibels $G/T(\text{dB/K}) = G_r(\text{dBi}) - 10 \cdot \text{Log}(T_A + T_{LNA} + T_{LOSS})$. G/T is a reference parameter used to compare antennas and not used in these calculations. Row 22 is the signal entering the antenna in dBm, the value in row 19 minus the value in row 20.

The results indicate that meeting the carrier-to-noise requirement of 13.56 dB requires a signal into the receive antenna of −126.73 dBm. That signal level is achieved by transmitting a signal high enough to overcome the path loss such that:

Received signal level (−126.73 dBm) = Transmitted signal − Path Loss   (10.1)

10.4 Path Loss Calculation

The path loss between the output of the transmitter antenna and the input of the receiver antenna is a function of the distance between antennas (d) and the frequency of transmission (see Figure 10.4).

A variation of the Friis transmission equation for path loss (PL) [5] is

$$PL = 20 \cdot \log\left(\frac{\lambda}{4 \cdot \pi \cdot d}\right) \tag{10.2}$$

where PL is the path loss excluding the transmitter and receiver antenna gains (dB), λ is in the wavelength of the transmitted signal, and d is the distance between the transmitter and receiver antennas.

Example: Determining Path Loss

See Table 10.2. The input data is as follows. Rows 1 and 2 are the signal frequency and the respective calculated signal wavelength in free space. Row 3 is the distance between the transmitting and receiving antennas. Row 6 is an estimate of atmospheric loss. The transmitted power is required to be sufficient to overcome this loss but typically is backed off (lowered) when lower loss atmospheric conditions are observed.

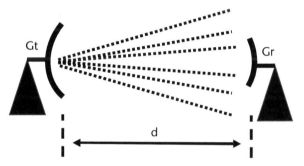

Figure 10.4 Transmit and receive antennas separated by distance d. G_t and G_r are the respective antenna gains.

Table 10.2 Path Loss Analysis

	Path Loss		
1	Frequency	14	GHz
2	Lambda	0.021413747	Meters
3	Distance	35,888.37	km
4	Distance	22300	Miles
5	Path loss, clear sky	206.47	dB
6	Atmospheric effects	3	dB
7	Path loss	209.47	dB

In the calculations in Table 10.2, row 4 is conversion from kilometers to miles for reference only, row 5 is the calculated path loss under clear sky conditions, and row 7 is the calculated path loss under adverse weather conditions.

Notes that on the path loss calculation, for path loss, the transmitted signal is not actually lost. The calculation is based on the signal dispersed in a lossless medium over distance d. Atmospheric conditions such as rain and sleet can increase the loss by increasing the signal dispersing, adding signal attenuation, and causing polarization distortions. The transmit antenna gain (G_t) represents a focusing of the transmitted signal. The receiver antenna gain (G_r) reflects how much of the dispersed signal is collected. The units of length for λ and d in (10.2) must be the same: $F = c/\lambda$ in free space in which F is the signal frequency (Hz) and c is the speed of light (in the same units as the wavelength, λ). Wavelength is not part of the path loss calculation but is typically added because the receiving antenna gain is given with respect to a normalized one meter spherical reflector without considering the frequency relationship.

10.5 Transmitted Power: Equivalent Isotropic Radiated Power

An isotropic antenna radiates power from all directions in three dimensions equally from a point source (see Figure 10.5). Equvalent isotropic radiated power (EIRP) [6] is the amount of power that a theoretical isotropic antenna would emit to produce the same power observed in the direction of a solid angle (Ω) produced by an actual antenna.

10.5.1 Antenna Gain

A typical antenna has a beam primarily radiating in a single direction defined by a solid angle (see Figure 10.6). The actual solid angle (Ω) representing the antenna

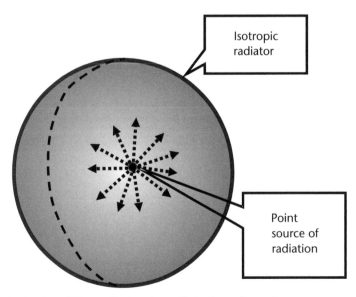

Figure 10.5 Isotropic radiator emits signal equally in three dimensions.

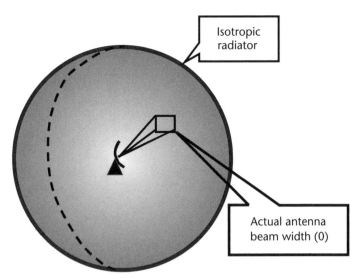

Figure 10.6 Actual antenna beam is a solid angle segment of the surface of an isotropic radiator (i.e., a segment of the surface of a sphere).

beamwidth is given in units of steradians (symbol of a steradian is "sr"). An isotropic antenna theoretically radiates from a point source. The radiation pattern is the surface of a sphere, $4 \cdot \pi \cdot R^2$, where R is the distance from the point source to the surface. The number steradians on the surface of a sphere is $4 \cdot \pi$. The solid angle beamwidth (Ω) projected from the antenna is (Ω) steradians. The antenna gain (G_t) is the number of steradians on a sphere divided by the solid angle of the antenna beamwidth (Ω), also in steradians [7].

Antenna gain (G_t) is

$$\text{Antenna gain } \left(G_t\right) \text{ is } \quad G_t = \frac{4 \cdot \pi}{\Omega} \tag{10.3}$$

The antenna gain (G_t) is usually in decibels (dBi), where dBi is the gain in dB [G_t(dB)] with respect to an isotropic radiator [8].

$$G_t(\text{dB}) = 10 \cdot \text{Log}\left(G_t\right) \tag{10.4}$$

Note that on antenna gain the assumption that the antenna pattern is radiating from a point source to the surface of a sphere with radius d, the distance from the point source to the target is called the far-field approximation. This approximation is based on the assumption that the distance from the antenna d is much larger than the diameter of the antenna.

10.5.2   EIRP

EIRP is the power that an isotropic radiator emits, producing the same power as a directional antenna with a solid angle (Ω) emits in the same direction. EIRP is the antenna gain (G_t) times the actual power from the power amplifier (P_{out}) decreased

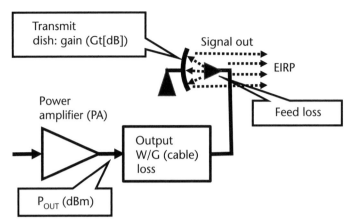

Figure 10.7   Power amplifier feeding a transmit antenna.

by the loss of power from the interface between the power amplifier and the antenna (L_t). EIRP is usually given in decibels with respect to 1 mW (dBm) or 1W (dBW) where the units are the same as the units of output power from the power amplifier.

Figure 10.7 shows an output power amplifier [P_{out}(dBm)] feeding the transmit antenna dish [G_t (dBi)] through a waveguide and antenna feed [L_t(dB)]. The EIRP is calculated in (10.5).

$$EIRP(\text{dBm}) = P_{out}(\text{dBm}) - L_t(\text{dB}) + G_t(\text{dBi}) \qquad (10.5)$$

where EIRP(dBm) is the EIRP in dBm, P_{out}(dBm) is the output of the power amplifier in dBm, L_t(dB) is the loss between the power amplifier and the antenna in decibels, and G_t(dBi) is the transmit antenna gain in dBi.

10.5.3   Example: Determining Power Amplifier Requirements [P_{out}(dBm)]

Assume that the path loss [PL(dB)] between the transmitter and receiver is 209.47 dB and the signal at the receive antenna input signal [Sr(dBm)] must be −126.73 dBm to satisfy the system C/N. The required EIRP is:

$$EIRP(\text{dBm}) = Sr(\text{dBm}) + PL(\text{dB}) = -126.73 \text{ dBm} + 209.47 \text{ dB} = 82.74 \text{ dBm}$$

$$(10.6)$$

The required EIRP(dBm) is inserted into Table 10.3 (row 1) along with the feed loss (F_{LOSS}) in row 2, the transmitter cable loss or waveguide loss (WG_{Loss}) in row 3 and the antenna gain [G_t(dBi)] in row 4. The antenna gain in dBi denotes that the gain is with respect to an isotropic radiator which is consistent with the system EIRP, which is also related to a theoretical isotropic radiator.

From Table 10.3 the P_{out}(dBm) is calculated:

$$P_{out}(\text{dBm}) = EIRP(\text{dBm}) + \left(F_{LOSS}\right) + \left(WG_{LOSS}\right) - G_t(\text{dBi}) = 37.0426 \text{ dBm} \quad (10.7)$$

Table 10.3 Finding Power Amplifier Requirements from EIRP
Information

	EIRP Analysis		
1	System EIRP(dBm)	82.74	dBm
2	Feed loss (F_{LOSS})	0.4	dB
3	Cable or waveguide loss (WG_{Loss})	0.6	dB
4	Antenna gain [G_t(dBi)]	46.7	dBi
5	HPA output [P_{out}(dBm)]	37.0426	dBm
6	HPA output [P_{out}(watts)]	5.0613	Watts

The loss factor L_t(dB) from (10.4) equals the cable or waveguide loss (WG_{Loss}) plus the feed loss (F_{LOSS}).

$$L_t(\text{dB}) = WG_{\text{LOSS}} + F_{\text{LOSS}} \tag{10.8}$$

The output power of the high-power amplifier in watts [P_{out}(watts)] is:

$$P_{\text{out}}(\text{watts}) = \frac{10^{\left(\frac{P_{\text{out}}(\text{dBm})}{10}\right)}}{1{,}000} = 5.0613 \text{ watts} \tag{10.9}$$

10.6 G/T Receiver Comparison Indicator

G/T is a common term that compares the relative sensitivity of receivers. It combines the receiver antenna gain, the loss between the antenna reflector, and the low noise input amplifier and the input amplifier noise figure [9].

In linear terms, G/T is the gain of the receiver antenna divided by a system temperature (T_{Rec}) in degrees kelvin, where the system temperature is:

$$T_{\text{Rec}} = T_A + T_{\text{Loss}} + T_{\text{LNA}} \tag{10.10}$$

where T_{Rec} is the total receiver noise temperature (K → degrees kelvin), T_A is the atmospheric noise (K), T_A is a function of frequency, atmospheric conditions, and antenna elevation, T_{Loss} is the loss between the antenna and the low noise amplifier (K), and T_{LNA} is the input amplifier (low noise amplifier) noise temperature (K).

G/T in decibels [G/T(dB)] is:

$$\frac{G}{T}(\text{dB}) = 10 \cdot \text{Log}\left(\frac{G}{T}\right) = Gr(\text{dBi}) - 10 \cdot \text{Log}(T_{\text{Rec}}) \tag{10.11}$$

where G_r(dBi) is the gain of the receiver antenna in decibels, dBi is the antenna gain with respect to an isotropic radiator, T_{Rec} is the total receiver noise temperature

(K → degrees kelvin), and G/T is a relative comparison between antenna-receiver combinations and not usually directly used in the analysis of communication links.

10.7   Power Amplifiers for Radar Systems

The analysis determining the required output power of the transmitter in a radar system [10] is like the analysis of the communications system with a few notable exceptions.

1. Instead of using C/N and ultimately BER as the criteria for an acceptable signal, radar systems usually use the S/N and the minimum detectable signal as a criterion. S/N is a postdetection ratio, whereas C/N is predetection ratio. In basic radar systems that just detect signal presence, the noise bandwidth of a post detection radar signal is a calculation combining the IF bandwidth and postdetection video bandwidth.
2. The path loss in a radar system is obviously different than the path loss in a communication system (see Figure 10.8). The path loss in a typical radar is the signal loss to the target plus the signal loss back to the radar. Using the radar range equation where R is the distance from the radar to the target, the path loss is proportional to $1/R^4$ to account for the two way travel of the radar signal. In addition to the path loss, the radar signal is reflected off a target that has an effective cross-sectional area (σ). The loss due to the target reflection is added to the path loss.

10.7.1   Radar Equation

The radar equation relates the power received (P_r) into the input low noise amplifier (signal in) to the power transmitted (P_T) [11]; see (10.12).

$$P_r = \frac{P_t \cdot G^2 \cdot \lambda^2 \cdot \sigma \cdot L}{(4 \cdot \pi)^3 \cdot R^4}$$

(10.12)

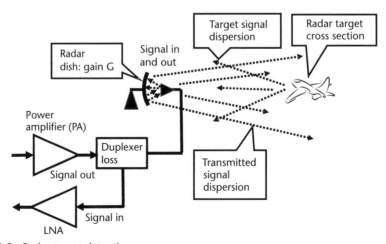

Figure 10.8   Radar target detection.

Separating factors in the radar equation gives a little insight in the represented functions.

$$P_r = P_t \cdot G^2 \cdot L \cdot \left(\frac{\lambda}{4 \cdot \pi \cdot R^2} \right)^2 \cdot \frac{\sigma}{4 \cdot \pi} \qquad (10.13)$$

where P_t is the transmitted power in watts or milliwatts, P_r is the received power into the low noise amplifier in the same units as transmitted power, and G is the gain of the antenna. The gain of the receive antenna is the same as the gain of the transmitter antenna, hence, the term G^2, L is the loss from the power amplifier to the antenna and antenna to the low noise amplifier, including the two-way feed loss, λ is the wavelength of the signal in meters, R is the range from the radar antenna to the target in meters, and σ is the effective radar cross section of the target with respect to a 1-m sphere.

The radar cross section of the target is $\dfrac{\sigma}{4 \cdot \pi \cdot (1 \text{ meter})^2}$. The path loss $(\lambda/(4 \cdot \pi \cdot R^2))$ is squared to account for the signal traveling from the radar to the target (at a range R) and returning to the radar.

10.7.2 Radar Power Amplifier Linearity

Power amplifiers for many radar systems do not require the same linearity criteria as communications systems and therefore can be designed to be highly efficient in terms of the power consumed versus the power delivered.

References

[1]   Raab, F., et al., "Power Amplifiers and Transmitters for RF and Microwave," *IEEE Transactions on Microwave Theory and Techniques*, Vol. 50, No. 3, March 2002.

[2]   Huang, "Microwave Active Circuit Design," Lecture slides, Claremont Graduate University, Claremont, CA.

[3]   Clegg, A., "RF Dialects: Understanding Each Other's Technical Lingo," U.S. National Science Foundation, Third IUCAF Summer School on Spectrum Management for Radio Astronomy, Tokyo, Japan, June 1, 2010.

[4]   Hum, S., "Noise in Radio Systems," Technical note, University of Toronto, ECE422: Radio and Microwave Wireless Systems.

[5]   Torlak, M., "Path Loss," Lecture slides, University of Texas at Dallas.

[6]   Tsai, M., "Link Budget," Lecture slides, Nanyang Technological University, Singapore, 2011.

[7]   Orfanidis, S. J., "Transmitting and Receiving Antennas," from *Electromagnetic Waves and Antennas*, Rutgers University, New Jersey, 2002.

[8]   Cuypers, G., "Noise in Satellite Links," *Belgian Microwave Roundtable*, 2001.

[9]   Nupponen, M., "Satellite Communications," Lecture slides, Post-Graduate Course in Radio Communication, Communications Laboratory, 2006.

[10] Aichele, D., "Next-Generation, GaN-Based Power Amplifiers for Radar Applications," *Microwave Product Digest*, January 2009.

[11] Skolnik, M., *Radar Handbook*, 3rd ed., New York: McGraw-Hill, 2008.

CHAPTER 11

Parallel Amplifier Topology Enhancing SSPA Performance

11.1 Introduction

SSPAs are usually required to produce more power than a single amplification device could deliver. Overcoming this limitation requires amplifier devices to be placed in a parallel configuration [1], as shown in Figure 11.1.

Ideally, the power dividers and power combiners are lossless and the output power capability of the individual power amplifier modules increases by the number of stages in parallel. Understanding the real-world limitations that degrade the ideal output power improvement is necessary to optimize the performance of the output parallel amplifier module.

11.2 Performance of Parallel Amplifiers Using Near-Ideal Perfectly Matched Components

Figure 11.2 shows an output power module with four amplifiers in parallel. The input signal is divided into four outputs each feeding a single amplifier. The outputs of the four amplifiers in parallel are recombined to a single output. Under ideal

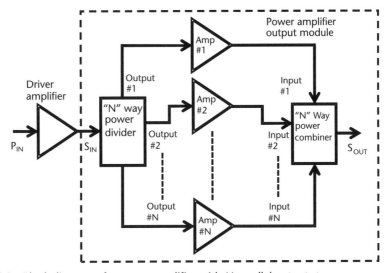

Figure 11.1 Block diagram of a power amplifier with *N* parallel output stages.

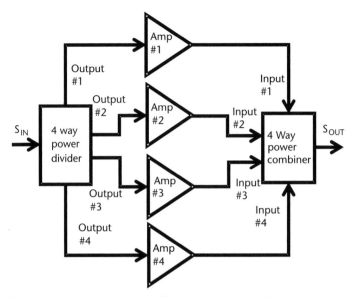

Figure 11.2   Block diagram of a power amplifier output module with 4 parallel output stages.

conditions, the output power capability of this circuit is four times the output power capability of each individual amplifier module.

An ideal parallel amplifier circuit assumes that:

1. All the power dividers and combiners are lossless.
2. The outputs of the power divider are identical in amplitude and phase.
3. The amplifier transfer characteristics (gains and phase) are identical.
4. The power combiner transfer characteristic from each input to the combined output is identical.
5. All inputs and outputs are perfectly matched to a characteristic impedance (Z_0).

Each parallel amplifier in Figure 11.2 has the following characteristics [2]:

- Gain: G(dB).
- Noise figure: NF(dB).
- Output third-order intercept point: OIP3(dBm).

11.2.1   Gain Calculation for an Ideal Parallel Amplifier Module

The input signal in dBm $[S_{IN}(dBm)]$ is divided into four paths such that at each output of the power divider the signal level is one-fourth of the input signal power $[S_{IN}(dBm) - 6$ dB]. The amplifiers have the same gain such each of the amplifier output signal levels are $S_{IN}(dBm) - 6$ dB $+ G$(dB), where G(dB) is the gain of each amplifier in decibels [1]. The four signals from the four parallel amplifiers are coherently summed in the output power combiner. Assuming no loss in the power combiner, the output signal is:

$$S_{\mathrm{OUT}}(\mathrm{dBm}) = S_{\mathrm{IN}}(\mathrm{dBm}) - 6\ \mathrm{dB} + G(\mathrm{dB}) + 6\ \mathrm{dB} = S_{\mathrm{IN}}(\mathrm{dBm}) + G(\mathrm{dB}) \qquad (11.1)$$

From (11.1), it is noted that the gain of a single amplifier stage is identical to the gain of N ($N = 4$ in this example) parallel stages under the assumed ideal conditions.

11.2.2   Noise Figure Calculation for an Ideal Parallel Amplifier Combiner Module

Noise figure is the degradation in signal to noise ratio. Noise can increase when a signal decreases and noise stays constant (e.g., an ideal attenuator), when signal is constant (e.g., noise model at the input of an amplifier), and a combination of both of these (e.g., second-stage additive noise).

In a parallel amplifier configuration determining noise figure is a matter of tracking the signal and noise separately through the topology because the signal is coherent and noise is noncoherent (random). Using parallel amplifier configuration in Figure 11.2 as an example, the analysis steps are:

1. The input signal is attenuated 6 dB through the four-way power divider.
2. The noise figure increases 6 dB because the signal decreases and the thermal noise ($k \cdot T_o \cdot B$) is the same on both sides of the power divider assuming the temperature is constant (standard temperature $T_o = 290\mathrm{K}$).
3. Random noise is added to the signal at each amplifier input ($k \cdot T_o \cdot B \cdot F$), where k is Boltzmann's constant, T_o is temperature in degrees kelvin, B is the system bandwidth in hertz, and F is the noise factor of the amplifier [noise figure (NF) = $10 \cdot \mathrm{Log}(F)$], where noise factor (F) is the noise produced at the amplifier input added to thermal noise and F can represented as an equivalent noise temperature T_e in degrees kelvin such that $F = (1 + T_e/T_o)$; see (11.2).

$$k \cdot T_o \cdot B \cdot F = (k \cdot T_o \cdot B) \cdot \left(1 + \frac{T_e}{T_o}\right) = (k \cdot T_o \cdot B) + (k \cdot T_e \cdot B) \qquad (11.2)$$

where $k \cdot T_e \cdot B$ is the noise added at the input of each parallel amplifier.
4. The signal plus noise increases in level by gain, $G(\mathrm{dB})$ and are added together in the power combiner.
5. At the output of the combiner, the signals are added coherently. Assuming the amplitude and phase through each path is identical, the output signal is increased by 6 dB, recovering the input 6-dB loss.
6. The noise generated in each parallel amplifier is noncoherent. The noise decreases in amplitude by 6 dB with respect to the signal. Signal to noise effectively increases by 6 dB. The increase in signal to noise at the output compensates for the decrease in signal to noise at the input (loss in the input power divider circuit).

The resultant noise increase through the parallel amplifier topology is ($k \cdot T_e \cdot B$), the input noise of a single amplifier, and therefore the resulting noise figure of the parallel circuit is exactly the same as the noise figure of a single parallel amplifier [1].

11.2.3   Available Output Power in an Ideal Parallel Amplifier Module

Available output power is defined in multiple ways depending on the application. Some output power parameters commonly used are: the output level where the gain compresses 1 dB (P1dB), the output level where increasing the input does not increase the output [i.e., saturated power (P_{SAT})], and the output third-order intermodulation intercept point (OIP3), a point where the power output as a function of power input line intersects with the imaginary line depicting the level of the third-order intermodulation interference (see Figure 11.3).

The 1-dB compression point (P1dB) is loosely related to the OIP3, usually 6 dB to 10 dB below OIP3. The power saturation level (P_{SAT}) is typically 1 dB or 2 dB above P1dB. OIP3 is best defined over a wide dynamic range and therefore most often used in calculating the accumulative effects of a systems nonlinearity. P1dB and P_{SAT} are approximated from OIP3 when a direct measurement is not practical.

11.2.3.1   Calculating the OIP3 of a Parallel Amplifier Module

The module in Figure 11.2 has four identical amplifiers in parallel each with an identical output third-order intermodulation intercept point ($OIP3_A$). A signal through the input power divider produces no intermodulation interference products because it is assume to be a linear passive device. Since the parallel amplifiers are identical, each amplifier produces identical third-order intermodulation interference products I(dBc) with respect to the carrier level [C(dBm)]; see Figure 11.4.

The level of the third-order intermodulation interference [I(dBc)] is related to the output third-order intermodulation intercept point [$OIP3_A$(dBm)] and the carrier level [C(dBm)] as shown in (11.3) [1].

$$\left[I(\text{dBc}) \right] = -2 \cdot \left[OIP3_A(\text{dBm}) - C(\text{dBm}) \right] \tag{11.3}$$

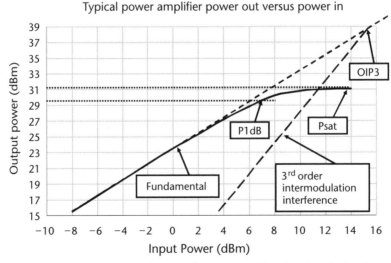

Figure 11.3 Transfer characteristics of a typical power amplifier showing the fundamental response, the 1-dB compression point (P1dB), the saturation level (P_{sat}), the OIP3, and the third intermodulation interference.

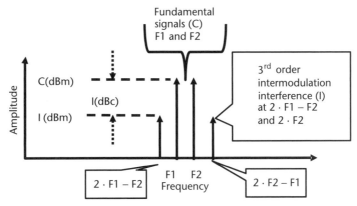

Figure 11.4 Third-order intermodulation products at the output of each parallel amplifier.

Solving for the output third-order intermodulation intercept point,

$$\text{OIP3}_A(\text{dBm}) = C(\text{dBm}) - \frac{1}{2} \cdot \left[I(\text{dBc})\right] \tag{11.4}$$

When the amplifier outputs are combined in the assumed linear power combiner, the carriers [$C(\text{dBm})$] and the interference signals [$I(\text{dBc})$] add coherently. The carrier signal at the output of the power combiner is $C(\text{dBm}) + 6$ dB and the interference levels [$I(\text{dBm})$] are also 6 dB higher, but the relative interference level with respect to the carrier is unchanged [i.e., $I(\text{dBc})$ before the power combiner is the same as $I(\text{dBc})$ after the power combiner].

Solving for the third-order intercept point at the output of the four-way combiner [$\text{OIP3}_B(\text{dBm})$] and noting that the carrier level and the third-order intermodulation level with respect to the carrier are both increased 6 dB.

$$\text{OIP3}_B(\text{dBm}) = \left\{C(\text{dBm}) - \frac{1}{2} \cdot \left[I(\text{dBc})\right]\right\} + 6\,\text{dB} = \text{OIP3}_A(\text{dBm}) + 6\,\text{dB} \tag{11.5}$$

The third-order output intercept point at the of the four-way combiner [$\text{OIP3}_B(\text{dBm})$] is 6 dB or four times higher than the third-order intermodulation intercept point at the output of the individual amplifiers [$\text{OIP3}_A(\text{dBm})$].

Ideally, the output power capability of the four amplifiers in parallel is four times the output power capability of a single amplifier. In general, N identical amplifiers in parallel increase the power of a single amplifier by a factor N or $10 \cdot \text{Log}(N)$ in decibels.

11.2.4   Summary of Parallel Amplifier Combining Using Ideal Devices

The analysis so far was produced using ideal components. The assumptions of ideal characteristics are:

1. Ideal power dividers are lossless (i.e., the insertion loss above the power splitting loss is zero); each output or transfer function is perfectly matched

in gain (or loss) and phase; and input and output impedances are perfectly matched with respect to the characteristic impedance (Z_0).

2. Ideal power combiners are lossless (i.e., the signals add ideally with no additional insertion loss); each input to output transfer function is identical in gain (or loss) and phase; and input and output impedances are perfectly matched with respect to the characteristic impedance (Z_0).

3. The assumption of the amplifier characteristic is that amplifier gains, noise figures, and third-order output intermodulation intercept points are identical and input and output impedances are perfectly matched with respect to the characteristic impedance (Z_0).

Based on these assumptions, the characteristics of the parallel combined amplifier circuits are:

1. The gain of the combined circuit [G_B(dB)] is the equal to the gain of the individual amplifier [G_A(dB)].
2. The noise figure of the combined circuit [NF_B(dB)] is the equal to the noise figure of the individual amplifier [NF_A(dB)].
3. The third-order intermodulation intercept point of the combined circuit [$OIP3_B$(dBm)] is N times the third-order intermodulation intercept point of the individual amplifier [$OIP3_A$(dBm)], where N is the number of parallel amplifiers.

$$G_B(\text{dB}) = G_A(\text{dB}) \tag{11.6}$$

$$NF_B(\text{dB}) = NF_A(\text{dB}) \tag{11.7}$$

$$OIP3_B(\text{dBm}) = OIP3_A(\text{dBm}) + 10 \cdot \log(N) \tag{11.8}$$

11.3   VSWR Mismatch Loss

When the transmission line impedance (Z_0) is not equal to the load impedance (Z_L), there is a mismatch and a portion of the incident signal (V_i) is reflected from the load (V_r) [3]; see Figure 11.5.

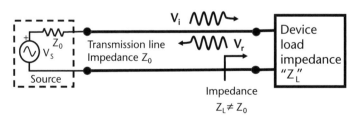

Figure 11.5   A portion of the incident signal (V_i) is reflected from load when the load impedance (Z_L) is not equal to the source impedance (Z_0).

V_i and V_r are voltage waves with their respective normalized power (P_i and P_r):

$$P_i = |V_i|^2 \text{ and } P_r = |V_r|^2 \text{ (the resistance } R = 1\Omega) \qquad (11.9)$$

The total power entering the load is:

$$P_t = P_i - P_r \text{ where } P_t = \text{the signal going into the load} \qquad (11.10)$$

The power loss or mismatch loss in decibels [P_{Loss}(dB)] is:

$$P_{\text{Loss}}(\text{dB}) = 10 \cdot \text{Log}\left(\frac{P_i}{P_t}\right) = 10 \cdot \text{Log}\left(\frac{P_i}{P_i - P_r}\right) \qquad (11.11)$$

An example of power loss due to reflected signals is $P_i = 10\text{W}$, $P_r = 1\text{W}$, $P_t = 9\text{W}$, and P_{Loss}(dB) = 0.457575 dB (note that gain through the interface is −0.457575 dB). Note that on mismatch loss, when a device manufacturer specifies loss (for power dividers and combiners) or gain (for amplifiers), the device is driven and terminated in a near-ideal load (Z_0) and the mismatch loss from or to an ideal load or termination is probably included in the device loss and therefore should not be included in the interface analysis.

11.3.1   Reflection Coefficient

The ratio of the reflected voltage wave to the incident voltage wave is the reflection coefficient (ρ).

$$\rho = \frac{V_r}{V_i} \text{ where } \rho \text{ is a vector quantity} \qquad (11.12)$$

In terms of normalized power:

$$|\rho|^2 = \frac{|V_r|^2}{|V_i|^2} = \frac{P_r}{P_i} \qquad (11.13)$$

From (11.10), the transmitted power (P_t) is $P_t = P_i - P_r$.
Substituting and solving for transmitted power in terms of reflection coefficient (ρ),

$$P_t = P_i \cdot \left(1 - |\rho|^2\right) \qquad (11.14)$$

The mismatch loss in decibels [P_{Loss}(dB)] can be expressed in term of the magnitude of the reflection coefficient ($|\rho|$):

$$P_{\text{Loss}}(\text{dB}) = 10 \cdot \log\left(\frac{P_i}{P_t}\right) = -10 \cdot \log\left(1 - |\rho|^2\right) \tag{11.15}$$

VSWR can be expressed in terms of the reflection coefficient ($|\rho|$):

$$VSWR = \frac{1 + |\rho|}{1 - |\rho|} \tag{11.16}$$

Solving for reflection coefficient in terms of VSWR:

$$|\rho| = \frac{VSWR - 1}{VSWR + 1} \tag{11.17}$$

The mismatch loss in terms of VSWR is found by substituting (11.17) into (11.16).

$$P_{\text{Loss}}(\text{dB}) = 10 \cdot \log\left[1 - \left(\frac{VSWR - 1}{VSWR + 1}\right)^2\right] \tag{11.18}$$

Table 11.1 lists the magnitude of the reflection coefficient, the return loss in decibels and the expected mismatch loss in decibels in terms of VSWR. The listed VSWR is part of a ratio with respect to unity [4].

Mismatch loss occurs at every device interface. The losses primarily affect overall gain distributed along the chain. Losses at the input of a device degrade

Table 11.1 The Magnitude of the Reflection Coefficient ($|\rho|$), the Return Loss in Decibels, and the Mismatch Loss in Decibels as a Function of VSWR

VSWR:1	Reflection Coefficient	Return Loss (dB)	Mismatch Loss (dB)
1.1	0.048	26.444	0.010
1.2	0.091	20.828	0.036
1.3	0.130	17.692	0.075
1.4	0.167	15.563	0.122
1.5	0.200	13.979	0.177
1.6	0.231	12.736	0.238
1.8	0.286	10.881	0.370
2	0.333	9.542	0.512
2.2	0.375	8.519	0.658
2.5	0.429	7.360	0.881
3	0.500	6.021	1.249
3.5	0.556	5.105	1.603
4	0.600	4.437	1.938
4.5	0.636	3.926	2.255

the device noise figure and losses at the output degrade the device power-handling capability.

11.3.2   Effective VSWR at the Interface of Two Nonideal Devices

When the source VSWR ($VSWR_S$) and the load VSWR ($VSWR_L$) (see Figure 11.6) are not ideal, the effective VSWR can degrade significantly or improve assuming there is a minimal transmission line (with ideal characteristic impedance, Z_0) distance (d) between the two devices ($d << \lambda$, where λ is the wavelength of the signal).

The effective VSWR between the two devices can be anywhere between the maximum VSWR ($VSWR_{\mathrm{MAX}}$) and the minimum VSWR ($VSWR_{\mathrm{MIN}}$), where:

$$VSWR_{\mathrm{MAX}} = VSWR_S \cdot VSWR_L \tag{11.19}$$

$$VSWR_{\mathrm{MIN}} = \frac{VSWR_S}{VSWR_L} \text{ for } VSWR_S > VSWR_L \tag{11.20}$$

or

$$VSWR_{\mathrm{MIN}} = \frac{VSWR_L}{VSWR_S} \text{ for } VSWR_L > VSWR_S \tag{11.21}$$

In an example, $VSWR_S = 2.0{:}1$, $VSWR_L = 3.0{:}1$, $VSWR_{\mathrm{MAX}} = 3.0 \cdot 2.0 = 6.0$, and $VSWR_{\mathrm{MIN}} = 3.0/2.0 = 1.5$.

The signal loss due to a poor VSWR between the source and load get considerably worse or can improve. When the source and load reflection coefficients are reactive (a function of frequency), the loss will probably change with frequency and can cause a considerable amount of gain ripple at that interface. The design engineer should attempt to keep the VSWR of the source and load of each device as low as possible, but when this cannot be accomplished a probability should be assigned to the best-case and worse-case scenarios.

11.3.3   VSWR as a Function of Source and Load Impedance

VSWR is relate to the magnitude of the reflection coefficient ($|\rho|$), as shown in (11.16), and is usually specified as a typical or worst-case ratio. The relationship

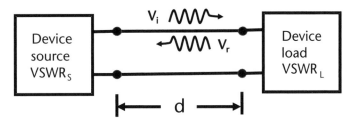

Figure 11.6   The source and load devices each have a nonideal VSWR.

between reflection coefficient (ρ), the source impedance (Z_0) and the load impedance (Z_L) [5] is:

$$\rho = \frac{Z_L - Z_0}{Z_L + Z_0} = \frac{\left(\dfrac{Z_L}{Z_0}\right) - 1}{\left(\dfrac{Z_L}{Z_0}\right) + 1} \tag{11.22}$$

If the real parts of the source impedance (Z_0) and the load impedance (Z_L) are much larger than the respective imaginary parts, the ratio Z_L/Z_0 is approximately real and the reflection coefficient (ρ) is real ($|\rho| \approx \rho$). Under these conditions,

$$|\rho| = \left| \frac{\left(\dfrac{Z_L}{Z_0}\right) - 1}{\left(\dfrac{Z_L}{Z_0}\right) + 1} \right| \quad \text{for } \frac{Z_L}{Z_0} \text{ approximately real} \tag{11.23}$$

Observing (11.17), relating $|\rho|$ and VSWR, the ratio of load impedance (Z_L) to characteristic impedance (Z_0) is approximately equal to the VSWR.

Under the conditions that the ratio Z_L/Z_0 is approximately real,

$$VSWR \approx \frac{Z_L}{Z_0} \text{ if } Z_L > Z_0 \text{ when } Z_L \text{ and } Z_0 \text{ are approximately real} \tag{11.24}$$

or

$$VSWR \approx \frac{Z_0}{Z_L} \text{ if } Z_0 > Z_L \text{ when } Z_L \text{ and } Z_0 \text{ are approximately real} \tag{11.25}$$

The approximations in (11.24) and (11.25) are adequate for many applications.

11.4 Nonideal Power Divider and Power Combiner Losses Effecting Gain, Noise Figure, and Intercept Point

Nonideal components have losses and mismatch impedances, which degrade the performance of the parallel amplifier topology. It is important to identify, quantify, and minimize these effects to optimize the performance. Figure 11.7 shows a two-amplifier parallel architecture. The losses and mismatch effects calculated can be extrapolated to N devices in parallel.

11.4.1 Power Divider Losses

The losses [LPD(dB)] through the power divider (PD1) shown in Figure 11.7 are above the theoretical power divider loss (3 dB),

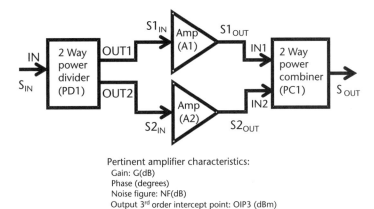

Figure 11.7   Combining two amplifiers in parallel.

11.4.1.1   Power Divider Losses Affecting Noise Figure

All passive losses including the power divider losses before the amplifier input add directly to the noise figure. As an example, if the power divider loss equal on both outputs is 0.4 dB [LPD(dB) = 0.4 dB], the total noise figure (NF_{TOT}) will degrade 0.4 dB. If all other aspects of the circuit in Figure 11.7 are ideal, the total noise figure (NF_{TOT}) of the parallel amplifier circuit is:

$$NF_{TOT} = NF + LPD \text{ (dB)} \tag{11.26}$$

where NF_{TOT} is the total noise figure of the parallel amplifier circuit, NF is the noise figure of the individual amplifier (A1 and A2), and LPD(dB) is the power divider loss equal on both outputs (OUT1 and OUT2).

11.4.1.2   Power Divider Losses Affecting Signal Level

If the power divider loss is 0.4 dB [LPD(dB)] and the loss is equal on both outputs, the signal level into the amplifiers will degrade 0.4 dB plus the theoretical 3-dB divider loss:

$$S1_{IN} = S2_{IN} = S_{IN} - 3 \text{ dB} - LPD \text{ (dB)} \tag{11.27}$$

where $S1_{IN}$ and $S2_{IN}$ are the signals going into amplifiers 1 and 2, respectively. S_{IN} is the input signal and the −3-dB factor is due to the splitting of the input signal into two equal outputs. LPD(dB) is the power divider loss assumed to be the same on both outputs (OUT1 and OUT2).

11.4.2   Power Combiner Losses

The losses [LPC(dB)] through the power combiner (PC1) shown in Figure 11.7 are the difference between the combined amplifier output signals ($S1_{OUT}$ plus $S2_{OUT}$) assuming that the signals ($S1_{OUT}$ and $S2_{OUT}$) are equal in amplitude and phase.

11.4.2.1 Power Combiner Losses Affecting Signal Level

If the power combiner loss is 0.5 dB $[LPC(dB)]$, the output signal (S_{OUT}) is:

$$S_{OUT} = S1_{OUT} + S2_{OUT} - LPC \text{ (dB)} = S1_{OUT} + S2_{OUT} - 0.5 \text{ dB} \quad (11.28)$$

11.4.2.2 Power Combiner Losses Affecting OIP3

The OIP3 is degraded by any losses after the parallel amplifiers. The degradation is decibel for decibel (i.e., 1-dB loss after the parallel amplifier degrades the OIP3 by 1 dB). As an example, if the power combiner loss is 0.5 dB $[LPC(dB)]$, the combined third-order intermodulation intercept point at the output of the combiner $(OIP3_{TOT})$ will be two times the individual amplifier third-order intermodulation intercept point minus the combiner loss $[LPC(dB)]$:

$$OIP3_{TOT} = 2 \cdot OIP3 - LPC \text{ (dB)} = 2 \cdot OIP3 - 0.5 \text{ dB} \quad (11.29)$$

where $OIP3_{TOT}$ is the total output third-order intermodulation intercept point of the parallel amplifier circuit after the output combiner, OIP3 is the output third-order intermodulation intercept point of each parallel amplifier, and $LPC(dB)$ is the power combiner loss equal on both inputs (IN1 and IN2).

11.4.3 Power Divider and Combiner Loss Effects on Gain

Power divider losses $[LPD(dB)]$ and power combiner losses $[LPC(dB)]$ subtract from the circuit gain. If all other aspects of the circuit in Figure 11.7 are ideal, the total gain $[G_{TOT}(dB)]$ of the parallel amplifier circuit is:

$$G_{TOT} \text{ (dB)} = G \text{ (dB)} - LPD \text{ (dB)} - LPC \text{ (dB)} \quad (11.30)$$

where G_{TOT} is the total gain of the parallel amplifier circuit in decibels, $G(dB)$ is the gain of the individual amplifier (A1 and A2) in decibels, $LPD(dB)$ is the power divider loss equal on both outputs (OUT1 and OUT2), and $LPC(dB)$ is the power combiner loss equal with respect to both inputs (IN1 and IN2).

11.5 Losses Due to Amplitude and Phase-Matching Errors in Parallel Channels

The signals out of the parallel amplifiers in Figure 11.7 $(S1_{OUT}$ and $S2_{OUT})$ are added in a power combiner (PC1). Although commonly called power combining, the function (PC1) is actual combining the voltage wave forms, which are vector quantities. The output is the vector sum of signals $S1_{OUT}$ and $S2_{OUT}$. The maximum output level (two times the normalized power level of each input signal) is attained when the signal vectors are in phase and equal in amplitude at the amplifier's maximum level [6].

Practically, the output power will be degraded by amplitude and phase mismatch errors between the signal paths. As the number of parallel signal paths to be combined increases, the difficulty of maintaining minimally acceptable phase and amplitude matching to minimize the summing losses grows substantially. This section discusses the combining of two devices, but the analysis can easily be extrapolated to higher-order combining systems.

The graphs presented relate phase and amplitude signal-tracking errors to combining losses. The plotted curves are a visual aid to help determine the component requirements and trade-offs necessary to meet the applicable amplifier system specification.

The phase and amplitude matching errors tend to be statistical when considering such variables as frequency and temperature so the calculation of combined available power and efficiency is usually related to a Gaussian probability function.

11.5.1   Vector Summing in the Power Combiner

The signals ($V1$ and $V2$) in Figure 11.8 are voltage vectors of signals $S1_{OUT}$ and $S2_{OUT}$. $V1$ and $V2$ have a phase angle θ the angle between the vectors representing the difference in phase between the two parallel signal paths.

Figure 11.9(a) shows two signals ($V1$ and $V2$) going into an ideal summing vector network ($PC1_{IDEAL}$). Vectors $V1$ and $V2$ are the voltage vectors containing

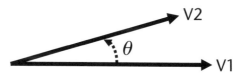

Figure 11.8   Vectors **V1** and **V2** represent the voltage vectors of signals $S1_{OUT}$ and $S2_{OUT}$.

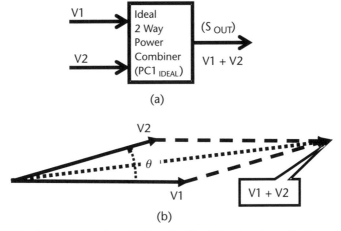

Figure 11.9   (a) Ideal power combiner ($PC1_{IDEAL}$) with all the signal amplitude and phase characteristics acquired through the power divider, amplifiers, and power combiner incorporated into vectors **V1** and **V2**. (b) A diagram of the projection of vector **V2** (the smaller vector) with respect to vector **V1**.

amplitude and phase tracking characteristics coming out of power divider PD1 going through amplifiers A1 and A2 and coming out of the power combiner (PC1) [7].

Figure 11.9(b) shows the two signals **V1** and **V2** at different amplitudes and a differential phase angle θ. The largest signal is usually controlled to limit the third-order and higher-order intermodulation interference; therefore, the desired output power level from the combiner circuit is two times the highest signal level, which in Figure 11.9 is represented as vector **V1**. Due to the amplitude and phase anomalies in the parallel device components, the sum of vectors **V1** and **V2** will be less than the desired two times **V1** (2·**V1**). The difference from the ideal outcome to the actual outcome is a function of the relative amplitude between vectors **V1** and **V2** ($|V2|/|V1|$) and the phase angle between the two vectors (θ).

$$|V1 + V2| = \sqrt{\left[|V1| + |V2| \cdot \cos(\theta)\right]^2 + \left[|V2| \cdot \sin(\theta)\right]^2} \qquad (11.31)$$

11.5.2   Normalized Signal Power at the Output of the Power Combiner

The normalized power of the signal at the output of the power combiner is S_{OUT} is:

$$S_{OUT} = \left(|V1 + V2|\right)^2 = \left[|V1| + |V2| \cdot \cos(\theta)\right]^2 + \left[|V2| \cdot \sin(\theta)\right]^2 \qquad (11.32)$$

Expanding (11.32),

$$S_{OUT} = |V1|^2 + 2 \cdot |V1| \cdot |V2| \cdot \cos(\theta) + |V2|^2 \cdot \left[\cos(\theta)\right]^2 + |V2|^2 \cdot \left[\sin(\theta)\right]^2 \qquad (11.33)$$

Equation (11.33) simplifies to:

$$S_{OUT} = |V1|^2 + 2 \cdot |V1| \cdot |V2| \cdot \cos(\theta) + |V2|^2 \qquad (11.34)$$

The maximum output power is S_{OUTmax}:

$$S_{OUTmax} = \left(2 \cdot |V1|\right)^2 = 4 \cdot |V1|^2 \qquad (11.35)$$

The output power degradation (P_{LOSS}) is:

$$P_{LOSS} = \frac{S_{OUT}}{S_{OUTmax}} = \frac{|V1|^2 + 2 \cdot |V1| \cdot |V2| \cdot \cos(\theta) + |V2|^2}{4 \cdot |V1|^2} \qquad (11.36)$$

Simplifying the power loss equation:

$$P_{LOSS} = \frac{1 + 2 \cdot \dfrac{|V2|}{|V1|} \cdot \cos(\theta) + \left(\dfrac{|V2|}{|V1|}\right)^2}{4} \qquad (11.37)$$

The power loss in decibels $P_{\text{LOSS}}(\text{dB})$ is:

$$P_{\text{LOSS}}(\text{dB}) = 10 \cdot \log\left[\frac{1 + 2 \cdot \dfrac{|V2|}{|V1|} \cdot \cos(\theta) + \left(\dfrac{|V2|}{|V1|}\right)^2}{4}\right] \tag{11.38}$$

11.5.3 Graphical Analysis of Gain Loss and OIP3 Loss as a Function Parallel Path Mismatch

Figure 11.10 is a plotted solution to (11.38). The graphs show the effective loss to the total system gain and the combined OIP3 as a function of differences in gain and phase matching of the two parallel paths shown in Figure 11.7. The 1-dB compression points (P1dB) and saturated power (P_{sat}) degrade one for one with OIP3.

The mismatch gain and phase differences in the parallel paths occur in the power divider outputs, the parallel amplifiers, and the transfer function from the power combiner inputs to the signal output (S_{OUT}). These gain and phase differences are typically frequency-related and therefore systems with larger percentage bandwidths usually experience greater variations than narrowband systems. Power divider and power combiner isolation between ports is also a contributing factor over the frequency band, but usually to a lesser extent and probably included in the manufacturer's amplitude and phase imbalance parameter.

11.6 Mismatch Loss Uncertainty and Phase Uncertainty as a Function of VSWR

As shown in Figure 11.11, when an incident signal (V_i) encounters a load (Z_L) such that ($Z_L \neq Z_0$), a portion of the incident signal is reflected back to the source (V_{RL}).

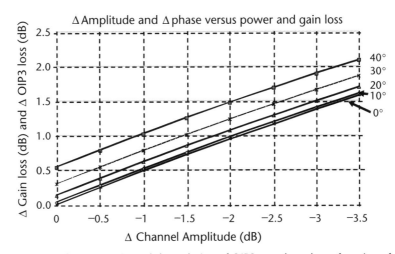

Figure 11.10 Loss of system gain and degradation of OIP3 are plotted as a function of differences in amplitude tracking from 0.0 dB to −3.5 dB and phase tracking (degrees) from 0° to 40° in 10° increments.

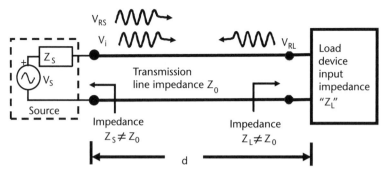

Figure 11.11 The system shows the signal reflections when the source and load impedances are not equal to the characteristic impedance Z_0.

If the source impedance (Z_S) is also not matched to the characteristic impedance ($Z_S \neq Z_0$), the reflected signal (V_{RL}) is rereflected at the source (V_{RS}) (see Figure 11.11).

The rereflected signal (V_{RS}) adds to the incident signal (V_i) resulting in a composite vector ($V_i + V_{RS}$). The resultant vector ($V_i + V_{RS}$) has an uncertainty in phase and amplitude depending on the characteristics of the rereflected signal (V_{RS}) [8] (see Figure 11.12).

The rereflected vector (V_{RS}) creating the uncertainty is a function of the source and load reflection coefficients (ρ_S and ρ_L), which are related to the VSWR at each interface. The relationship between reflection coefficient and VSWR is shown in (11.39).

$$|\rho| = \frac{VSWR - 1}{VSWR + 1} \tag{11.39}$$

The distance that the wave travels [i.e., two times the distance (d) between the source and load] is a factor in the amplitude and phase uncertainty and shows up as an amplitude and phase ripple. The distance between the ripple peaks is a function of the signal wavelength and the distance traveled $2 \cdot d$.

11.6.1 Magnitude Uncertainty Due to Source and Load Mismatch

The magnitude of the rereflected signal ($|V_{RS}|$) is the product of the magnitude source and load reflection coefficients ($|\rho_S| \cdot |\rho_L|$). The loss uncertainty due to

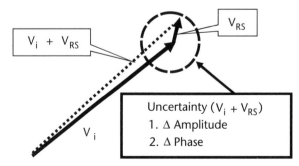

Figure 11.12 The incident vector ($\boldsymbol{V_i}$) sums with the rereflected vector from the source ($\boldsymbol{V_{RS}}$).

the amplitude difference between the incident wave (V_i) and the rereflected wave (V_{RS}) in dB ($LdB+$) and ($LdB-$) is:

$$LdB+ = 20 \cdot \log\left(1 + |\rho_S| \cdot |\rho_L|\right) \tag{11.40}$$

$$LdB- = 20 \cdot \log\left(1 - |\rho_S| \cdot |\rho_L|\right) \tag{11.41}$$

where $LdB+$ is the positive uncertainty in dB and $LdB-$ is the negative uncertainty in decibels. The maximum value of $|LdB+|$ is slightly lower than the maximum value of $|LdB-|$.

11.6.2 Phase Uncertainty Due to Source and Load Mismatch

The phase uncertainty due to source and load mismatch resulting in rereflected vector (V_{RS}) is shown in Figure 11.13. The magnitude of the rereflected wave with respect to the incident wave ($|V_{RS}|/|V_i|$) is equal to the magnitude of the reflection coefficient at the load times the reflection coefficient at the source ($|\rho_L| \cdot |\rho_S|$).

$$\frac{V_{RS}}{V_i} = |\rho_L| \cdot |\rho_S| \tag{11.42}$$

The worst-case phase angle uncertainty (Φ) occurs when the incident vector (V_i) and rereflected vector (V_{RS}) are at right angles as shown in Figure 11.13. The resultant maximum phase angle uncertainty (Φ) [9] is:

$$\Phi = \pm\tan^{-1}\left(|\rho_L| \cdot |\rho_S|\right) \tag{11.43}$$

Using (11.39), the amplitude and phase and uncertainty equations can be written in terms of VSWR.

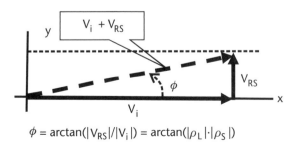

$\phi = \arctan(|V_{RS}|/|V_i|) = \arctan(|\rho_L| \cdot |\rho_S|)$

Figure 11.13 The phase difference between incident vector (V_i) and rereflected vector (V_{RS}).

11.7 Example: Application of Parallel Path Systematic and Random Losses to the Performance of the SSPA Output Stage

Figure 11.14 shows two parallel amplifiers (PATH1 and PATH2) configured to ideally increase their output power capability by two times [10 · Log(2) = 3 dB]. The analysis assumes typical (not necessarily optimum) characteristic of the parallel amplifiers and the power divider/combiner (same device use in a reciprocal configuration). The procedure determines the expected output power degradation from the ideal [10 · Log (2) = 3 dB] when the effects of real (nonideal) device characteristics are considered.

The input source impedance (Z_o) and output load impedance (Z_o) are ideal, but all other interfaces are real with an associated nonideal VSWR. This analysis can be extended to N amplifiers in parallel increasing the available output power of the individual amplifiers N times [10 · Log(N) in decibels].

11.7.1 Characteristics of the Devices Used in the Parallel Amplifier Topology

The device characteristics used in this example are typical (not optimum) of those that might be encountered when designing a parallel amplifier output stage. The specifications represent manufacturing variations, fixed expected losses, and random variations encountered in the manufacturing process. These characteristics usually include variations over the operating frequency range and sometimes over the temperature range. The fixed losses are accumulated arithmetically and the random effects are typically considered a statistically probability.

11.7.1.1 Typical Power Divider/Combiner Specifications

Since most passive couplers are reciprocal devices, the power divider (PD1) and the power combiner (PC1) shown in Figure 11.14 are the same device used in forward and reverse configuration. Table 11.2 lists the specifications relating to this analysis [10].

The specification parameters are:

1. *Insertion loss:* The insertion loss is in decibels and represents the signal loss above the divider theoretical loss (3 dB for a two-way power divider).

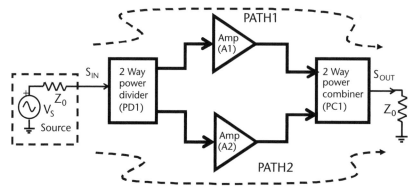

Figure 11.14 Example of two parallel amplifiers and the associated signal paths (PATH1 and PATH2).

Table 11.2   Some Relevant Power Divider and Power Combiner Specifications

Insertion Loss dB (Max.)	VSWR Max.	Amplitude Balance +/– dB (Max.)	Phase Balance +/– Degrees (Max.)
0.45	1.35:1	0.55	6

It is typically given as a maximum although the typical value is usually not more than a few tenths of a decibel below the maximum. The insertion loss is typically the same when the device is configured as a power divider or a power combiner.

2. *VSWR:* The VSWR specification is usually given worst-case and is typically the same at every port. Over the operating frequency, the typical values can vary with the maximum VSWR approaching close to the specified limit. Often, the VSWR is given in terms of the interchangeable parameter, the return loss in decibels.

3. *Amplitude balance and phase balance:* The amplitude and phase balance are the maximum respective difference between the signals at the power divider output ports. As a power combiner, the amplitude and phase balance refers to the maximum difference of each signal at the output if the input signals are identical.

11.7.1.2   Typical Amplifier Specifications

To maximize the required symmetry, amplifiers A1 and A2 are the same device (i.e., the same model number). The amplifiers are typically selected from the same lot or semiconductor wafer to attain the best matching characteristics [11]. Table 11.3 lists the specifications relating to this example.

The following are the specifications listed in Table 11.3:

1. *Gain:* The gain is typically given in decibels and represents a nominal value. For this example, the nominal value will be used for all calculations. In many designs, gain variations over the operating frequency and temperature range can be extensive and critical in the selection of a device.

2. *Noise figure:* Noise figure is usually given as a maximum value, but typical values are usually within a few tenths of a decibel below or at the than the maximum value specified. Noise figure is usually deemed a noncritical parameter in the design of power amplifiers and many times omitted. When the parameter is not given, it is suggested to talk to the manufacturer for a worst-case number because although noise figure is not usually a critical parameter, noise output power (noise figure × gain) is many times specified

Table 11.3   Typical Power Amplifier Specifications

Gain dB (Typical)	Noise Figure dB (Max.)	OIP3 dBm (Typical)	VSWR Input Max.	VSWR Output Max.	Amplitude Balance +/– dB (Typical)	Phase Balance +/– Degrees (Typical)
15	10	50	1.67:1	1.9:1	0.80	10

to prevent adjacent channel noise from accumulating from diverse sources in a common receiver.

3. *OIP3:* The output third-order intercept point for the SSPA is the most critical parameter in a system. Noting the effect of each parameter on the degradation of OIP3 is the primary objective of this example.

4. *Input and output VSWR:* Input and output VSWR are scalar quantities a function of the magnitude of the reflection coefficient. The two aspects of VSWR of concern in this example are: (1) the loss of incident power due to reflections, and (2) the phase and amplitude uncertainty which effects the channel matching. Item 1 is considered a fix loss and usually included as part of the gain specification, while item 2 is considered a random effect.

5. *Amplitude balance and phase balance:* The amplitude and phase balance are the maximum respective difference between the signals traveling through the amplifier. This parameter is usually not specified by the manufacturer but should be measured by the user to insure a controlled design.

11.7.2 Calculating the Fixed Losses Associated with the Parallel Amplifier Topology

The fixed losses identified from the device parameters are: (1) power divider and combiner losses [LPD_{LOSS}(dB) and LPC_{LOSS}(dB)], and (2) losses due to signal reflections (VSWR-related losses at interface paths P1, P2, P3, and P4) shown in Figure 11.15.

11.7.2.1 Power Divider and Combiner Losses

The fixed power divider and combiner losses are LPD_{LOSS}(dB) and LPC_{LOSS}(dB), respectively. Since the power divider and the power combiner are the same device configured in a reciprocal direction the losses are expected to be equal [LPD_{LOSS}(dB) = LPC_{LOSS}(dB)].

11.7.2.2 Summary of Fixed Losses

The other losses indicated in Figure 11.15 are VSWR losses (i.e., loss due to reflected signals). VSWR losses through interface path P3 and P4 may be included in the

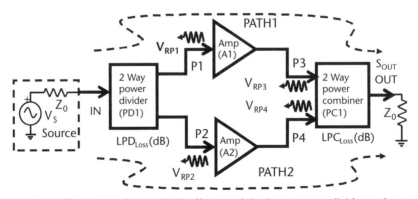

Figure 11.15 The fixed losses due to VSWR effects and the input power divider and output power combiner are shown.

Table 11.4   Fixed Losses for the Two-Amplifier Parallel Topology

Device or Interface	Input or Output	Loss (dB)
Power Divider Loss	Input	0.45
Power Combiner Loss	Output	0.45

power combiner loss [LPC_{LOSS}(dB)] and losses through interface paths P1 and P2 may be included in the amplifier gain. Table 11.4 omits the VSWR losses assuming that they are included in the power combiner loss and the amplifier gain.

11.7.3   Calculating the Uncertainty Loss Associated with the Parallel Amplifier Topology

The uncertainty loss associated with the parallel amplifier topology is caused by phase and amplitude differences at all the interfaces (P1, P2, P3, and P4) and devices in PATH1 and PATH2. These differences are a function of VSWR, frequency, signal level, temperature, DC voltage, and so forth. The effects of these variables are considered random because of the number of variables and the infinite number of increments within each variable.

11.7.3.1   Phase and Amplitude Uncertainty Through the Devices

Table 11.5 lists the amplitude and phase uncertainty for each device in the output chain. The amplitude and phase listed for the amplifiers (A1 and A2) are differential variables and therefore only list one time for both amplifiers.

The power divider and power combiner parameters are given by the device manufacturer. The amplifiers parameters are usually not given by the device manufactures and must be obtained by testing many devices or assumed from prior knowledge of the exact or similar devices.

11.7.3.2   Phase and Amplitude Uncertainty at the Device Interfaces

The device interfaces that effect the parallel channel amplitude and phase uncertainty are labeled P1, P2, P3, and P4 between the power divider (PD1) and the parallel amplifier (A1 and A2) inputs and the parallel amplifier (A1 and A2) outputs and the power combiner (PC1).

The amplitude and phase uncertainty is calculated from the source and load VSWR at each interface (P1 through P4). The VSWR is converted to the reflection

Table 11.5   Typical Amplitude and Phase Uncertainty for Each Device in the Power Amplifier Output Module

Device	Amplitude Uncertainty (dB)	Phase Uncertainty (Degrees)
Power divider (PD1)	0.55	6
Amplifier (A1 and A2)	0.8	10
Power combiner (PC1)	0.55	6

Table 11.6 Amplitude and Phase Uncertainty for Device Interfaces P1, P2, P3, and P4

Path	Load VSWR:1	Load Rho	Source VSWR:1	Source Rho	Mismatch Uncertainty (+dB)	Mismatch Uncertainty (−dB)	Phase Uncertainty (Degrees)	Average Mismatch Uncertainty (+/− dB)
P1	1.671	0.251	1.35	0.149	0.319	−0.331	2.142	0.325
P2	1.671	0.251	1.35	0.149	0.319	−0.331	2.142	0.325
P3	1.35	0.149	1.9	0.310	0.392	−0.411	2.646	0.402
P4	1.35	0.149	1.9	0.310	0.392	−0.411	2.646	0.402

coefficient $|\rho_s|$ and $|\rho_L|$ corresponding to the source and load reflection coefficients, respectively. The amplitude and phase uncertainty is calculated using (11.40) through (11.43). Typically, the amplitude uncertainty is approximated as plus and minus the average of the magnitude of the results of (11.40) and (11.41). Table 11.6 shows the results of the VSWR uncertainty values for device interfaces P1 through P4.

11.7.3.3 Calculating the Standard Deviation of the Phase and Amplitude Differences Between the Two Parallel Channels

Table 11.7 lists the device and interface amplitude and phase uncertainties. The sum of each amplitude and phase parameter (peak-to-peak value) has an unrealistically low probability of occurrence, so it is reasonable to calculate a probability of occurrence using the root sum square (RSS) value that represents one standard deviation (1σ) of a normal probability function.

Equation (11.44) and (11.45) are used to calculate the $RSS_{\Delta\varnothing}$ and $RSS_{\Delta dB}$ (1 · σ) values of phase and amplitude, respectively.

$$RSS_{\Delta\phi} = \sqrt{PD1_{\Delta\phi}^2 + A12_{\Delta\phi}^2 + PC1_{\Delta\phi}^2 + P1_{\Delta\phi}^2 + P2_{\Delta\phi}^2 + P3_{\Delta\phi}^2 + P4_{\Delta\phi}^2} \tag{11.44}$$

where $RSS_{\Delta\varnothing}$ is 13.971°.

Table 11.7 Amplitude and Phase Uncertainty for Devices and Interfaces in the Output Parallel Power Amplifier Module

Device or Path	Amplitude Uncertainty (ΔdB) (+/− dB)	Phase Uncertainty (ΔØ) (+/− Degrees)
Power Divider (PD1)	0.550	6.000
Amplifier (A1 and A2)	0.800	10.000
Power Combiner (PC1)	0.550	6.000
P1	0.325	2.142
P2	0.325	2.142
P3	0.402	2.646
P4	0.402	2.646

$$RSS_{\Delta dB} = \sqrt{PD1^2_{\Delta dB} + A12^2_{\Delta dB} + PC1^2_{\Delta dB} + P1^2_{\Delta dB} + P2^2_{\Delta dB} + P3^2_{\Delta dB} + P4^2_{\Delta dB}} \quad (11.45)$$

where $RSS_{\Delta dB}$ is 1.334 dB.

The calculated RSS value $(1 \cdot \sigma)$ for phase is +/− 13.971° and for amplitude is +/− 1.334 dB. These phase and amplitude results represent one standard deviation and have a 68.2% probability of actual values being within these limits. Two standard deviations $(2 \cdot \sigma)$ +/−2.668 dB for amplitude and +/−27.942° have a 95.4% probability of the occurrence. Many times, in critical applications the $2 \cdot \sigma$ value is used to determine the probable loss (see Table 11.8).

Applying the $1 \cdot \sigma$ and $2 \cdot \sigma$ gain and phase differential to the graph in Figure 11.16 results in a loss probability given in Table 11.9. Uncertainty losses due to phase and amplitude imbalance in the parallel channels effectively occur at the output of the power combiner and directly impact the available output power (OIP3, P1dB, and P_{sat}).

11.7.4   Summary of the Analysis of the Parallel Amplifier Topology

The losses calculated are separated into two categories, losses reflected to the input and losses reflected the output (see Table 11.10).

Table 11.8   Amplitude and Phase Uncertainty for Devices and Interfaces in the Output Parallel Power Amplifier Module with the $1 \cdot \sigma$ and $2 \cdot \sigma$ Parameters Calculated

Device or Path	Amplitude Uncertainty (ΔdB) $(+/- dB)$	Phase Uncertainty $(\Delta \varnothing)$ $(+/- degrees)$
Power divider (PD1)	0.550	6.000
Amplifier (A1 and A2)	0.800	10.000
Power combiner (PC1)	0.550	6.000
P1	0.325	2.142
P2	0.325	2.142
P3	0.402	2.646
P4	0.402	2.646
1σ root sum square (RSS)	1.334	13.971
2σ standard deviations	2.668	27.942

Table 11.9   Summary of the Calculated Amplitude and Phase Mismatch and Its Respective Combining Loss in Decibels for a $1 \cdot \sigma$ and $2 \cdot \sigma$ Probability of Occurrence

Standard Deviation	Δ Amplitude (dB)	Δ Phase (Degrees)	Loss (dB)
1σ (68.2%)	−1.33	13.97	0.706
2σ (95.4%)	−2.67	27.94	1.486

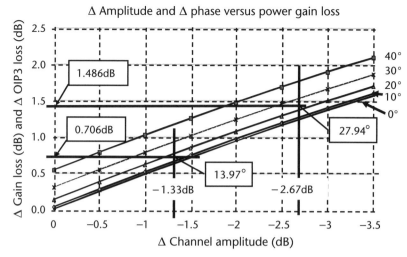

Figure 11.16 Applying the results in Table 11.9 determines the expected (1 · σ and 2 · σ) losses due the amplitude and phase mismatch errors.

Table 11.10 Summary of the Input and Output Losses Expected in the Parallel Power Amplifier Example

Device or Interface	Input Loss (dB)	Output Loss (dB)
Power divider loss	0.45	
Power combiner loss		0.45
1σ phase and amplitude mismatch loss		0.706
Total loss (1σ)	0.45	1.156

The losses reflected to the input (power divider loss) increase the noise figure by the amount of the loss. The other losses in Table 11.10 (combiner loss and phase and amplitude mismatch loss) degrade the output power available by the amount of the output loss. Both input and output losses affect the gain of the parallel amplifiers by the total amount of the loss see (Table 11.11).

The example results, shown in Table 11.11, shows the effects of the assumed losses on the final performance parameters (i.e., noise figure, gain, and output third-order intercept point). This is a tool for the SSPA design engineer to focus on the device parameters that have the most impact on the module performance.

Table 11.11 The Effect of the Input and Output Losses on the Performance of the Parallel Power Amplifier Example

	Loss Origin	Increase	Decrease
Noise figure (dB)	Input	0.45	
Gain (dB)	Input + Output		1.606
OIP3 (dBm)	Output		1.156

References

[1] Hausman, H., "Modeling Parallel Amplifiers," *IEEE International Conference on Microwaves, Communications, Antennas, and Electronic Systems* (COMCAS), November 2015.

[2] Fuentes, C., "Microwave Power Amplifier Fundamentals," Giga-tronics Technical Paper AN-GT101A, October 2008.

[3] Benmoussa, Z., and D. Barrick, "The Effects of VSWR on Transmitted Power," Technical note, April 2006.

[4] Skyworks Solutions, Inc, "VSWR, Return Loss and Transmission Loss vs. Transmitted Power," Technical note, 2003.

[5] Walraven, D., "Understanding SWR by Example," QST, ARRL, Technical note, November 2006.

[6] Hausman, H., "Understanding Mismatch Effects in Power Combining Circuits," *Microwaves & RF*, 2006.

[7] Razavi, B., "Transceiver Architectures," *RF Microelectronics.*

[8] Lymer, T., "Calculating Mismatch Uncertainty," *Microwave Journal*, May 2008.

[9] Ahmad, D., K. Patel, and P. Negi, "Effective Source Mismatch Uncertainty Evaluation Using Resistive Power Splitter Up to 18 GHz," *EDP Sciences*, 2015.

[10] Marki Microwave "Microwave Power Dividers and Couplers Tutorial."

[11] Giga-tronics Incorporated, "GT-1051A Microwave Power Amplifier," Technical Datasheet, 2011.

CHAPTER 12

MMIC Amplifier Modules for Use in Parallel Combining Circuits

12.1 Introduction

The output amplifier of an SSPA chain typically uses MMIC devices configured in a parallel topology; see Figure 12.1.

If the parallel amplifiers are identical and each of the signal paths with respect to its gain and phase are identical, the effective output power of the SSPA is theoretically N times the output power of the individual parallel amplifier modules. Identical amplifiers are theoretically impossible but understanding the parameters that affects the matching of signal paths helps in module selection to approach the goal of maximum effective SSPA output power.

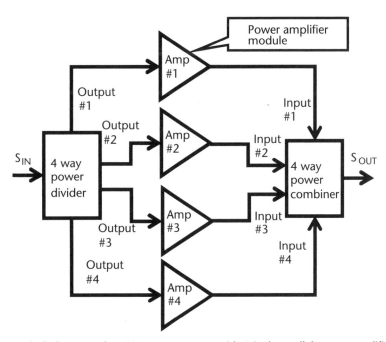

Figure 12.1 Block diagram of an SSPA output stage with 4-inch parallel power amplifier modules.

179

12.2 Criteria for Selecting Parallel Amplifier Modules

Parallel amplifier module selection is a critical function in developing a successful power amplifier. Each parameter can be critical to the design and must be considered independently or dependently in conjunction with other related parameters. The most efficient operation of a parallel amplifier topology occurs when the signals out of all the parallel modules are identical, that is, in phase at the same output level [1].

Key electronic parameters that are typically supplied by the power amplifier module manufacturer are output power, gain, input and output VSWR, frequency flatness, efficiency, and stability [2]. Many times, these parameters are not independent. Knowing the interactive relationship enables the design engineer to optimize the performance in the applicable circuit.

Note that on matching signal level and phase at the output of parallel amplifier modules, the combined output power from parallel amplifier modules degrades when the output signal levels and phase are not identical. The power amplifier transfer parameters are an important factor in meeting identical signal characteristics: signal level and phase matching in the entire parallel signal path must be analyzed and considered, and identical module characteristics by itself are not sufficient to guarantee the desired outcome. Some of the external factors effecting signal level matching are: input signal level and phase, module source and load impedance (e.g., VSWR), and gain and phase matching of the transfer function for each power combiner input to output. Matching of each power amplifier transfer characteristic is not usually provided by the module manufacturer. Some module manufactures will provide this data upon request. Modules with similar transfer characteristics should be selected and used in the same parallel combiner stage.

12.2.1 Output Power

Output power (P_{OUT}), the power delivered to the load is the usually the primary parameter in the selection of a power amplifier module. Output power is usually given in units of watts, dBm, or dBW. When combining parallel amplifier modules, the desired characteristic is equal output power at the maximum available level.

12.2.2 Gain

The minimum gain (G) in decibels is given difference between the output power (P_{OUT}), and the input power (P_{IN}).

$$G(dB) = P_{OUT}(dBm) - P_{IN}(dBm) \tag{12.1}$$

The gain of a power amplifier module should be greater than all signal path losses from the power divider input to the parallel power amplifier module input. This includes but is not limited to the N-way power divider loss [$10 \cdot Log(N)$]. If this condition is not met, the available output power of the amplifier driving the parallel amplifier output stage will be a limiting factor in the overall SSPA output power.

12.2.2.1   Linear SSPA Power Amplifier Module Gain

The power amplifier module gain in linear SSPAs ($Gain_{LINEAR}$) should be much higher than $10 \cdot Log(N)$ such that the driver amplifier does not add a significant amount of third-order intermodulation distortion.

12.2.2.2   Saturated SSPA Power Amplifier Module Gain

Saturated power amplifiers recreate the maximum signal at every saturated stage, so the gain of the saturated parallel amplifier modules ($Gain_{SATURATED}$) should exceed the total losses between the driver stage output and the parallel amplifier module input. Higher gain parallel amplifier modules are desired because they enable the driver to be designed with a lower output power. The gain of an amplifier in saturation is always less than the linear amplifier gain.

$$Gain_{SATURATED} < Gain_{LINEAR} \qquad (12.2)$$

12.2.3   Equal Gain

Equal gain assuming equal input signal levels into each amplifier module in a parallel circuit is important but is not usually a specified parameter.

12.2.3.1   Effect of Unequal Gains on a Saturated SSPA

If the amplifier gain is sufficient to put the output in saturation, the output power should be unaffected, but most likely the phase through the amplifier will vary. Power amplifier transmission phase varies as a function of the depth of saturation (i.e., signal phase is typically a function of the gain compression). Signals with unequal phase will combined with a higher probability of an uncertainty loss at the SSPA output.

12.2.3.2   Effect of Unequal Gains on a Linear SSPA

Equal signal input signal and unequal amplifier gains result in unequal output power. If one output signal is higher than anticipated, the third-order intermodulation interference will increase. If one output signal is lower than anticipated, the total output power will be lower than anticipated.

12.2.4   Input and Output VSWR

Input and output VSWR are a measure of impedance mismatch at the device ports. The effect of mismatch impedances is related to nonoptimum power transfer and an uncertainty loss in the output combiner due to signal amplitude and phase mismatching. High-power amplifier modules with poor VSWRs could present a parallel topology implementation problem.

12.2.5 Frequency Flatness

Frequency flatness or signal variation as a function of frequency usually has two major components, slope and ripple, as shown in Figure 12.2. The frequency flatness is usually statistically different for each device and has the same effect as unequal gains. Minimum slope and ripple is important for an optimum design.

12.2.5.1 Frequency Response Slope (dB/MHz)

Frequency response slope is usually device-related. A decrease in gain as a function of an increase in frequency is usually related to the device parasitic capacitance. The definition of slope can vary, but in Figure 12.2(b) the slope is a line drawn through the frequency response curve such that the area of the ripple above the slope line is equal to the area of ripple below the line. The frequency of concern and areas of the calculation in this example are from F_1 to F_2.

12.2.5.2 Frequency Ripple (dB$_{p-p}$)

Frequency ripple is determined by drawing two straight lines, each parallel to the slope such that the upper straight line touches the gain peaks and the lower straight line touches the gain valley points. The difference between the upper straight line and the lower straight line is the peak-to-peak ripple in decibels (dB$_{p-p}$). Many

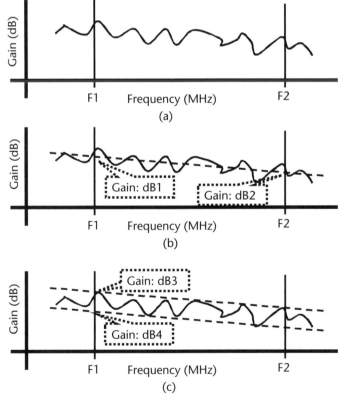

Figure 12.2 (a) An example of a typical SSPA frequency response. (b) The slope of the frequency response (dB/MHz). (c) The frequency response ripple dB$_{p-p}$ (dB peak to peak).

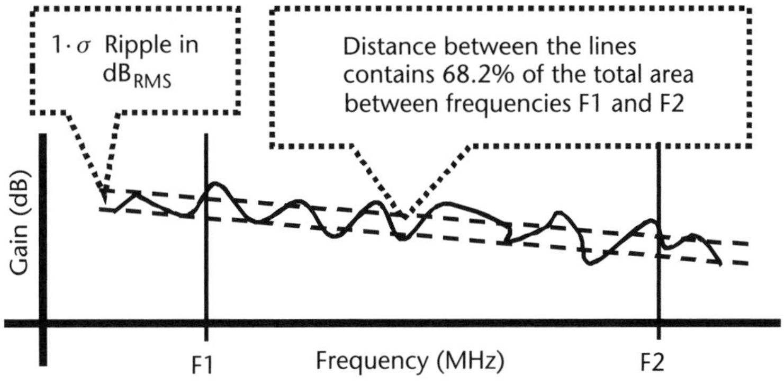

Figure 12.3   Indicates the 1σ ripple probability in dB_{RMS}.

times, the ripple is expressed in statistical terms where the peak to peak ripple is given as 1σ probability. That calculation involves drawing upper and lower lines containing 68.2% of the ripple area above and below the average slope (dB/MHz); see Figure 12.3.

Frequency ripple is often cause by a VSWR mismatch at the device input and output. Improved VSWR helps to reduce the ripple. Since VSWR ripple can be due to multiple reflections, the ripple can be reduced by spacing the individual devices close together (i.e., much less than a wavelength).

12.2.6   Efficiency

Power amplifiers consume a considerable amount of primary power, most of it dissipated in the device [3]. The dissipated power causes a temperature rise that lowers the device reliability or in the extreme causes the device to fail. Thermal design is an important part of designing an SSPA. The other factors associated with high power dissipation are the cost of operating the unit and availability of a high-power DC source. As an example, DC power on a spacecraft is limited and costly. The efficiency of linear power amplifiers and saturated power amplifiers can be very different.

The power-added efficiency (PAE) accounting for the RF input power needed to drive the amplifier is:

$$PAE = \frac{P_{OUT} - P_{IN}}{P_{DC}} \tag{12.3}$$

where P_{OUT} is the output RF power in watts, P_{IN} is the input RF power in watts, and P_{DC} is the DC power in watts.

12.2.6.1   Relative Efficiency of Linear Power Amplifiers

Linear power amplifiers usually operate in a class AB mode [4], accepting a minimal amount of distortion. The higher the amount of acceptable distortion, the more efficient the amplifier. Linear power amplifiers are typically used in communication systems.

12.2.6.2   Relative Efficiency of Constant Wave Saturated Power Amplifiers

Saturated power amplifiers typically operate in a class C mode [i.e., the amplifier is turned on when a portion of a positive or negative signal (not both) is present]. The peak power can be the same as the linear power amplifier, but because the amplifier is not biased on for more than 50% of the time the amplifier operates in a highly efficient mode. Class C amplifier modes are not used in amplitude-modulated communication system and has only limited use in constant wave communication systems due to its inherent intermodulation distortion and associated spectral regrowth. This type of amplifier can be used in some communication systems and radar systems.

12.2.6.3   Relative Efficiency of Pulse-Saturated Power Amplifiers

Pulse-saturated amplifiers operate in an efficiency class C amplifier mode. The signal turning on the amplifier is a pulse modulated RF signal that has a duty factor equal to the pulse width divided by the pulse repetition rate. The efficiency increase over the constant wave saturated amplifier is the inverse of the duty factor.

12.2.7   Amplifier Stability

Instability can occur in any device that has gain and therefore is a critical parameter in the design of SSPAs. An unstable device could generate spurious signal (signals unrelated to the input signal) that stand on their own or modulate signals in the amplifier. Stability is a critical criterion because unintentionally generated signals could severely negatively impact a systems operation.

A commonly used measure of stability is calculated from a device's S-parameters [5, 6].

$$K = \frac{1 - |S_{11}|^2 - |S_{22}|^2 + |\Delta|^2}{2 \cdot |S_{12} \cdot S_{21}|} \tag{12.4}$$

$$\Delta = S_{11} \cdot S_{22} - S_{12} \cdot S_{21} \tag{12.5}$$

When $K > 1$ and $\Delta < 1$, the device is unconditionally stable (i.e., no oscillations are expected under any load conditions).

Note that, with regard to amplifier stability, the K-factor and Δ factor represent a single frequency. To ensure absolute stability, the S-parameters and stability factors must be tested at every frequency up to the frequency where the device gain (G) is less than 1 ($G < 1$). Equations (12.1) and (12.2) are valid over the linear dynamic range of the device. When signals drive the SSPA into a nonlinear mode, the S-parameters could vary significantly. The factors determining stability remain as the K and Δ factors, but because the S-parameters change in the nonlinear mode, a stable device in the linear mode may become unstable when operating in a nonlinear mode.

The basic rules of device stability do not change and therefore good design practices are important to increase the probability of a stable design. Use devices that have a low reverse gain (S_{12}) under all operating conditions. The input source impedance and the output load impedance should be close to the standard transmission line impedance to minimize reflected signals feeding back into the device.

12.3 Interpreting RF Characteristics of MMIC Power Amplifier Devices Using a Typical Example of Manufacturers' Data

MMICs have limited plug-and-play capability depending on the requirements of the application and capabilities of the selected device. Information on the applicability of a device to a specific application can be obtained by investigating the device's published characteristics.

An example of a two-stage GaN 30-W FET amplifier operating in the 9-GHz to 11-GHz region is used to illustrate and interpret typically published RF specifications. The example uses a MMIC designed to operate in a pulsed saturated mode for radar applications as well as a linear CW amplifier for satellite communications applications.

Table 12.1 lists the devices recommended operating conditions. Unless otherwise specified, all parameters are listed under these conditions. Table 12.2 lists typical specifications presented in a data sheet for the MMIC SSPA module used in this example. Since the module operates in the linear mode and saturated mode, examples of specification applicable to both modes are shown.

The following are notes applicable to the specifications listed in Table 12.2.

1. Test conditions for Table 12.2, unless otherwise noted, are recommended operating conditions listed in Table 12.1, device case or heat sink temperature: 25°C, saturated operation with an input pulse modulated signal: pulse width = 10 μS and duty factor = 10%,
2. Row 1 is the frequency limits applicable to all specifications. The MMIC may operate at frequencies beyond these limits with performance not specified.
3. Row 2 is the saturated output power (P_{sat}) specified at a given input power. Note that the small signal gain given in row 4 is approximately 7 dB above the calculated saturated gain indicating the unit is significantly in saturation.
4. Row 3 is the typically expected third-order intermodulation interference level when two closely spaced (in frequency) input carriers are each at an output level of +38 dBm. Note that the signal level of each carrier (+38 dBm) is significantly below the saturation level.

Table 12.1 Typical Recommended Operating Conditions of an MMIC SSPA

Parameter	Value
Drain voltage (VD)	24V
Drain current (IDQ)	1,300 mA
Gate voltage (VG), typical	−1.7V

Table 12.2 Typical RF Specifications for a GaN MMIC Power Amplifier

Electrical Specifications

Item	Parameter	Min.	Typical	Max.	Units
1	Operational frequency range	9		11	GHz
2	Saturated output power (P_{IN} = 25 dBm)				
	9.0 GHz		46		dBm
	11.0 GHz		45.5		dBm
3	Third-order intermodulation level: (P_{OUT}/Tone = 38 dBm) 10.0 GHz		−22		dBc
4	Small signal gain				
	9.0 GHz		29		dB
	11.0 GHz		28.5		dB
5	Small signal input return loss				
	9.0 GHz		15		dB
	11.0 GHz		14		dB
6	Small signal output return loss				
	9.0 GHz		10		dB
	11.0 GHz		14		dB
7	Output power temperature coefficient temperature: 15°C to 75°C, P_{IN} = 25 dBm		0.005		dB/°C
8	Small signal gain temperature coefficient (15°C to 75°C)		0.06		dB/°C

5. Row 4 is the small signal (linear) gain of the device. The output power level is not usually specified but generally assumed to be at least 6 dB below the 1-dB compression point (P1dB) and 8 dB below the saturation level (P_{sat}).

6. Row 5 is the small signal input return loss. When the device is driven at a high level such that the input is in a nonlinear mode, the input return loss could significantly change as a function of input level. Under nonlinear conditions, the return loss should be specified at a particular input level. In this example, row 2 indicates that the output is in saturation with an input level of +25 dBm. The input could also be in a nonlinear mode and the return loss given in row 5 may not be applicable at the level specified in row 2.

7. Row 6 is the small signal output return loss. When the device is in saturation, the output is in a nonlinear mode, which could significantly affect the impedance matching characteristic and thereby alter the output return loss. Output return loss when the device is in saturation is output level-related and therefore typically specified at a referenced output level.

8. Row 7 is the parameter that estimates the variation of output power as a function of temperature when the device is in saturation.

9. Row 8 is the parameter that estimates the variation of linear gain as a function of temperature. In the linear mode at a fixed input power, this parameter directly affects the output power and could affect the level of intermodulation interference.

Note that, on MMIC power amplifier device specifications, all parameters given in Table 12.2 except for row 1, which defines the frequency range, were given as typical values. "Typical" is not a well-defined limit, but unfortunately, it is a common practice. There is an obvious difficulty in designing a system with no parameter limits. This problem is typically mitigated by measuring a sufficient quantity of devices from the same wafer and performing a statistical analysis of the critical parameters. The user can than determine the probability of the usability of the device in an application when all the devices are the measured wafer. When all the devices from the measured wafer are used, the process must be repeated to find another wafer. Some companies take a large amount of data and define the mean as typical and provide an associated standard deviation. These manufacturers should only be selling devices from wafers that match their published statistics.

12.4 Primary DC Voltage and Bias for FET-Based MMIC Power Amplifier Devices

SSPA modules primarily use N-channel depletion-mode FET in a common source configuration. A typical functional circuit is shown in Figure 12.4. The gate to substrate voltage (V_{GS}) induces a field between the gate (G) and substrate (Sub) depleting the number of N-type conductors in the channel between the drain (D) and source (S). Limiting the number of conductors limits the drain current (I_D). Typically, as shown in Figure 12.4, the substrate (Sub) and source (S) are shorted together so the potential between the gate and substrate (V_{GSub}) is identical to the potential between the gate and source (V_{GS}) [i.e., ($V_{GSub} = V_{GS}$)].

Depletion-mode MESFET devices typically have a very low-impedance drain to source when there is no reverse bias gate to substrate or gate to source when the

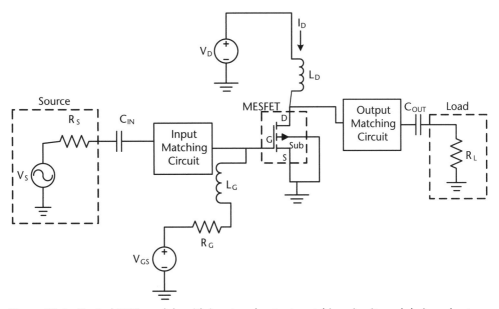

Figure 12.4 Typical SSPA module with input and output matching circuits and drain and gate DC voltages.

source and substrate are shorted together. The drain-to-source impedance when there is no gate-to-substrate voltage is often small enough so that with a fixed normal operating drain voltage (V_D) the current through the drain (I_D) can exceed the maximum allowable current (I_{DMAX}) and power dissipated in the device. Under these conditions, there is a high probability of the device failing.

Note that a depletion-mode device is one where the gate-to-substrate voltage is used to reduce the current within the drain-to-source channel.

12.4.1   RF Signal Path

The signal source (V_S) typically has an internal source impedance (R_S) matched to the characteristic impedance (Z_0). Capacitor (C_{IN}) AC couples the input source voltage to the input matching circuit. Ideally, the MESFET input impedance is reactively matched (lossless impedance transformer) to the characteristic impedance ($Z_0 = R_S$). The matching circuit provides maximum power transfer from the source into the amplifying device (MESFET). The signal at the output of the amplifying device (MESFET) goes through an output matching circuit, an AC coupling capacitor (C_{OUT}) terminating into load resistance R_L. R_L is typically equal to the characteristic impedance (Z_0), maximizing the power transfer from the amplifier device (MESFET) output to the load. The input and output coupling capacitors are usually not part of the matching circuits and designed to appear as RF short-circuits to the signal of interest.

12.4.2   Input DC Bias Circuit

The input DC bias circuit consists of DC voltage source V_{GS}, typically feeding a negative potential through the current-limiting resistor R_G in series with an AC choke, inductor L_G. The gate (G) and substrate (Sub) plates separated by the dielectric material between the drain (D) and source (S) channel behaves essentially as a capacitor. The resistor R_G should be large enough to limit a current surge on turn-on through this capacitor and small enough to have an insignificant voltage drop due to any leakage current through the gate (G). The choke L_G should have a high enough impedance at the operating frequency such that no signal is lost in the input biasing circuit. The self-resonance of the inductor L_G should be greater than the highest operating signal frequency. Sometimes the inductor is in series with a quarter-wave transmission line capacitively shorted at the input to transform the short-circuit to a high impedance at the gate [7].

12.4.3   Output DC Bias Circuit

The output DC bias circuit in Figure 12.4 consists of the voltage source V_D feeding the MESFET drain (D) through an inductor or choke (L_D). The inductor L_D should have a high enough impedance at the operating frequency such that no signal is lost in the output biasing circuit. The self-resonance of the inductor L_D should be greater than the highest operating signal frequency. Sometimes the inductor is in series with a quarter-wave transmission line capacitively shorted at the input to transform the short-circuit to a high impedance at the drain over the operating frequency range [7].

The inductor L_D should have low enough resistance so the resistive loss at the highest operating DC is insignificant. The inductor should also be able to withstand a turn-on current surge with degradation.

12.4.4   Basic Configuration of an *N*-Channel Depletion-Mode MESFET

The basic configuration of an N-channel depletion-mode MESFET is shown in Figure 12.5. The drain and the source are connected through an N-type semiconductor material, which has an excess of negatively charged carriers, effectively creating a low-impedance channel [8]. The gate is separated from the channel by a thin insulator. When a negative potential is established between the gate and substrate (V_{GS} in this configuration), positive charges are attracted into the channel decreasing the number of available electrons for conduction, thereby decreasing the current through the channel.

If there is a sufficient magnitude of the negative gate to source voltage, the channel between the drain and gate could be completely cut off (this is called pinch-off) [9]. An RF signal superimposed on the gate modulates the channel width and changes the channel current. A small change in gate voltage can produce a large change in drain current effectively amplifying the input signal.

12.4.5   Voltage Sequencer

Most MESFET-based SSPA modules operate in the common source configuration (shown in Figure 12.4). The output drain-to-source current (I_D) is controlled by the input voltage (V_{GS}) when the source (S) and substrate (Sub) are shorted. Typically, a negatively biased gate-to-source voltage (V_{GS}) is required to limit the drain-to-source current (I_D). When V_{GS} is zero and the drain voltage (V_D) is at its normal operating level, the drain current is at a maximum and can be high enough to destroy the SSPA module. This situation is avoided by designing a circuit that automatically sequences the turn-on of the gate-to-source voltage (V_{GS}) before the drain voltage (V_D) limiting the available drain current to a safe level [7, 10].

A voltage sequencer is a circuit that controls the switching time of the gate to source voltage (V_{GS}) and drain voltage (V_D) to prevent drain current (I_D) overload [11]. The gate and drain voltages are sequentially switched such that the drain

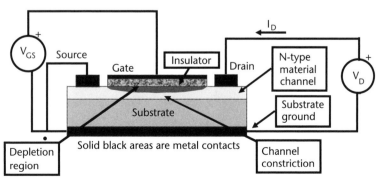

Figure 12.5   Basic configuration of an *N*-channel depletion-mode MESFET.

voltage (V_D) is not turned on until the gate-to-source voltage (V_{GS}) is stable and reverse-biased such that the drain-to-source channel is sufficiently depleted of available carriers to such an extent that the drain current (I_D) and the respective power dissipated in the device is at the recommended operating condition. Figure 12.6 illustrates a typical bias sequence timing diagram.

To turn on a depletion-mode FET-based SSPA module:

1. The V_{GS} voltage is turned on to a negative level sufficient to pinch off the drain-to-source channel.
2. After the gate to source voltage (V_{GS}) is fully charged, the drain voltage (V_D) is turned on.
3. When the drain voltage (V_D) is fully charged, the gate-to-source voltage (V_{GS}) is set to its nominal operating condition.

Note that on the turn-on sequence the gate-to-source voltage (VGS) is initially set to a pinch-off (minimum drain current) position to prevent instabilities and potential lock-up conditions when the charging drain voltage is going through voltages other than its recommended operating value. In pulse applications, if there is a significant amount of drain current when the gate-to-source voltage is in pinch-off, the drain current can be dynamically switched on a pulse-by-pulse basis to save power. Drain voltage switching could cause high transient voltages on the FET (module) output. The maximum voltage rating should be considered when utilizing this technique.

To turn off a depletion-mode FET-based SSPA module:

1. The V_{GS} voltage is set to a negative level sufficient to pinch off the drain-to-source channel.
2. After the gate-to-source voltage (V_{GS}) is fully charged, the drain voltage (V_D) is turned off.
3. After the drain voltage (V_D) is fully discharged, the gate-to-source voltage (V_{GS}) is discharged to 0V.

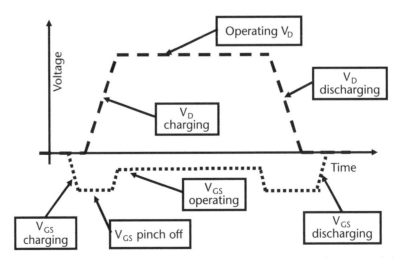

Figure 12.6 Typical FET turn-on and turn-off sequencer timing diagram for gate and drain bias voltages.

Note, on the turn-off sequence, that the gate-to-source voltage (V_{GS}) is set to a pinch-off position (minimum drain current) before the drain voltage is discharged to prevent instabilities during the turn-off process.

12.4.6   Typical Block Diagram of a Depletion-Mode MESFET SSPA Module Bias Voltage Switching Network

Figure 12.7 show a typical block diagram of a depletion mode MESFET SSPA module bias voltage switching network. The input primary positive voltages V_{CC} and primary negative voltage $-V_{EE}$ drive the drain voltage (V_D) switched regulator and the $-V_{GS}$ switch regulator, respectively. The negative voltage ($-V_{EE}$) can be generated from the positive voltage V_{CC} or be an independent input to the network. If the negative voltage ($-V_{EE}$) is an independent input to the network, it must be present before the primary voltage (V_{CC}) is turned on. If the primary voltage (V_{CC}) is sufficiently stable, the primary voltage regulator may just be a voltage-controlled, low-loss switch.

The capacitors (C_D and C_G) at the output of the switched regulators are designed to suppress power-line interference and provide voltage regulator stability as required. These capacitors take a finite time to charge and discharge as shown in the timing diagram in Figure 12.6. It is important that the voltage timing circuit shown in Figure 12.7 [7] include these times in setting the timing sequence indicated in Figure 12.6.

Note on drain voltage switching (Figure 12.7) that the drain voltage switch is only necessary when the FET is driven from a single positive voltage power supply where the negative voltage for gate pinch-off is created from the same DC voltage

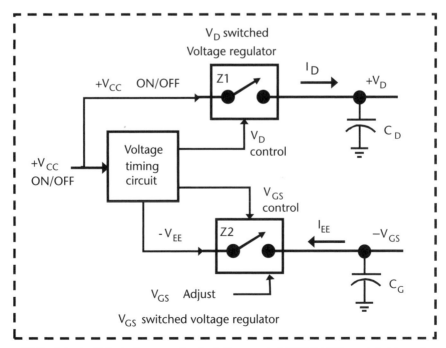

Figure 12.7   A typical block diagram of a depletion-mode MESFET SSPA module bias voltage-switching network.

that supplies the drain power. If the drain and gate voltages are controlled by separate positive and negative DC power supplies, the correct timing sequence can be controlled manually eliminating the need for a drain switch (Z_1). When designing a low duty factor saturated pulse power amplifier and the drain leakage current dissipates a significant amount of power, it might be beneficial to sequence the drain voltage around the RF input pulse. This will lower the total power dissipated.

References

[1]   Hausman, H., "Modeling Parallel Amplifiers," *IEEE International Conference on Microwaves, Communications, Antennas, and Electronic Systems (COMCAS)*, November 2015.

[2]   Huang, C., et al., "Design and Simulation of a Ku-Band MMIC Power Amplifier," Technical paper, School of Aeronautics and Astronautics Zhejiang University Hangzhou, China.

[3]   Frenzel, F., "What's the Difference Between GaAs and GaN RF Power Amplifiers?" *Electronic Design*, October 2012.

[4]   Cripps, S., *RF Power Amplifiers for Wireless Communications*, Norwood, MA: Artech House, 2006.

[5]   Edwards, M., "A Review of Amplifier Stability Analysis Using Modern EDA Tools," AWR Corporation UK (Europe), ARMMS, November 2012.

[6]   Pozar, D., *Microwave Engineering*, 2nd ed., New York: John Wiley & Sons, 1998.

[7]   Ampleon, "Bias Module for 50 V GaN Demonstration Boards," Application Note AN11130, September 2015.

[8]   Colino, S., and R. Beach, "Fundamentals of Gallium Nitride Power Transistors," Efficient Power Conversion Corporation, 2011.

[9]   Pengelly, R., et al., "A Review of GaN on SiC High Electron-Mobility Power Transistors and MMICs," *IEEE Transactions on Microwave Theory and Techniques*, Vol. 60, No. 6, June 2012.

[10]   Kaya, K., "Meeting Biasing Requirements of Externally Biased RF/Microwave Amplifiers with Active Bias Controllers," Analog Devices, Inc., Application Note AN-1363, 2016.

[11]   Nitronex Corporation, "AN-009: Bias Sequencing and Temperature Compensation for GaN HEMTs," Application Note AN-009, October 2008.

CHAPTER 13

Measuring and Matching the Impedance of High-Power MMIC Amplifier Modules

13.1 Introduction

Efficiency, available output power, and phase and amplitude tracking are key elements when designing a parallel amplifier topology SSPA. Optimizing the impedance matching circuit is important to meet these goals because a matched circuit (low VSWR) maximizes power transfer to the load, minimizes reflection losses, and minimizes power loss due to phase and amplitude tracking uncertainties.

13.2 Optimum Impedance Matching

Designing impedance matching circuits for communication amplifiers (i.e., amplifiers that primarily operate in their linear mode) is theoretically easier than designing matching circuits for pulse amplifiers that primarily operate in a saturated mode. The impedance of the amplifiers operating in the linear mode is relatively constant over the linear dynamic range of the amplifier, whereas the impedance characteristics of a device operating in the nonlinear mode change significantly as a function of power.

Many MMIC devices that are designed to operate in the linear mode are internally matched to nominally 50Ω. Impedance matching of these amplifiers attain various degrees of success over the operating frequency range and it is therefore incumbent upon the engineer to decide if the worst-case VSWR and the associated losses meet the system design criteria. Alternate approaches are to search for devices with a better guaranteed impedance match (VSWR), to design an improved impedance matching circuit, or to use passive impedance matching techniques.

Some external passive matching techniques used to improve the VSWR of a power amplifier are adding an isolator to the amplifier input and output or combining two amplifiers in parallel using quadrature (90°) couplers on the input and output one as a power divider and the other as a power combiner.

13.2.1 Using Isolators to Improve Input and Output Amplifier VSWR

Figure 13.1 shows isolators on the input and the output of a power amplifier (MMIC) module. Isolators are three-port directional devices with one port internally

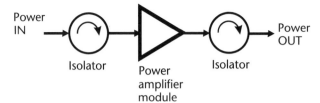

Figure 13.1   Power amplifier module (MMIC) with input and output isolators.

terminated in 50Ω [1]. The device typically has a reverse isolation of 20 dB, a forward loss better than 0.5 dB, and input and output VSWRs of better than 1.25:1. The negative aspects using isolators are the loss (≈0.5 dB), the cost, and the size of the device (lower-frequency units are larger than higher-frequency units).

13.2.2   Using Quadrature Hybrids for Impedance Matching

Quadrature couplers on the input and output of two parallel amplifiers are shown in Figure 13.2. The couplers, shown designed in a branch-line configuration, provide the function of improving the input and output VSWR like that of input and output isolators. The insertion loss is also approximately 0.5 dB, but the added benefit is the combining of two power amplifier modules, theoretically improving the available output power 3 dB [2]. The additional cost is the added power amplifier and added DC to power the parallel amplifier, but this is usually mitigated by the increase in available output power. Many times, the quadrature hybrid can be printed on the substrate adding negligible cost to the unit. The branch-line coupler shown uses a configuration of quarter-wave transmission lines increasing the length of the parallel power amplifier by at least a little over a half-wavelength.

The isolation and return loss performance of the quadrature coupler configuration are usually not as good as the isolator configuration. Typically, the reverse isolation and return loss are greater than 15 dB and are a function of the phase and amplitude matching of the two parallel amplifiers.

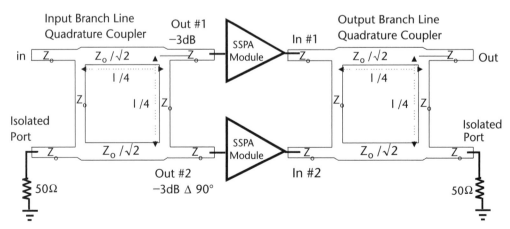

Figure 13.2   Parallel power amplifier modules (MMICs) with input and output quadrature couplers.

13.3   Measuring the Impedance of Linear Power Amplifier Modules

Maximum power transfer occurs when the source impedance and the load impedance are conjugate matches (i.e., the real part of the source is equal to the real part of the load and the imaginary part of the source is the negative of the imaginary part of the load impedance). The external impedance is typically 50Ω, real. The objective is to measure the input and output impedance of the device and then transform that impedance to 50Ω (real with no imaginary part) for maximum power transfer.

Most power amplifiers, to improve their efficiency, lower their cost, and decrease their volume, operate as close as possible to or at their maximum available output power. Linear power amplifiers typically used in communications systems often operate in a class AB mode, trading off an acceptable amount of distortion for added efficiency. Operating in the AB mode, the device S-parameters are affected to a minor extent, related to the amount of signal distortion over the operating dynamic range. Small-signal S-parameter data can usually be used when designing a class AB power amplifier matching network.

Note that with regard to S-parameter measurements of linear power amplifier modules, there are many ways to measure S-parameters. The examples given are functional for instructional purposes and are not necessarily the most efficient means of collecting the data [3].

13.3.1   Measuring S_{11} and S_{21} of a Linear Power Amplifier Module

A functional configuration for measuring S_{11} and S_{21} is [4] shown in Figure 13.3. The test setup is a variation of the typical small signal configuration because the high-power module under test (A5) has a limited amount of gain and the network analyzer usually does not have enough output power to drive the SSPA module to its rated output. A driver stage (A2) and isolator (A3) provide the additional gain necessary and improved source impedance to not have a significant effect on the measurement.

A directional coupler (A4) after isolator (A3) is used to sample and measure the input incident signal (P1) level and the reflected signal (P2) level. The two signals are sent to the network analyzer which performs the S_{11} measurement. The isolator (A3) ensures a matched input to directional coupler (A4), which maximizes the

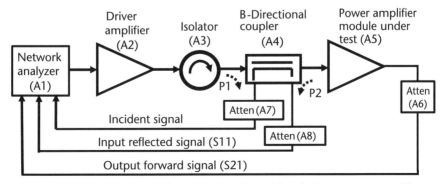

Figure 13.3   A functional test setup for measuring S_{11} and S_{21} of a linear power amplifier.

directivity and separation of the incident and reflected signals. For best results, the forward directivity path (P1) plus attenuator (A7) should be equal to the directivity path (P2) attenuator (A8). Additionally, attenuator (A6) should be equal to the loss through path (P1) plus attenuator (A7), minimizing the error in the S_{21} measurement.

13.3.2　Measuring S_{22} and S_{12} at the Rated Output of a Power Amplifier Module

Measuring the output impedance and the reverse gain of a power amplifier not operating in well-defined linear mode [4] is difficult and more complicated than making the same measurement on a small signal amplifier. The problem is mainly associated with measuring S_{22} and S_{12}, which are a function of the output signal level. A measurement made with the power amplifier emitting a signal at its nominal operating level may burn out the network analyzer input circuit. One method of mitigating this problem is shown in Figure 13.4.

Note that with regard to measuring S_{22} and S_{12} of a power amplifier module, since the power amplifier is typically operating in a nonlinear or quasilinear class AB mode, the output impedance and, by extension, S_{22} and the reverse gain S_{12} will change with output power. This requires the measurement to be made and may only be valid at a specific output power level. In the class AB mode when the signal is minimally distorted, the S-parameter variation from the maximum output power to a lower output powers may be insignificant. This problem with measuring S_{22} has certain similarities to measuring the output return loss of an oscillator.

13.3.2.1　Configuration at the Module (A5) Input When Measuring S_{22} and S_{12} at Rated Output Power

Signal generator (A1) emits a signal at frequency (F_1) in the operating band of the power amplifier under test (A5). The signal is amplified by driver amplifier (A2)

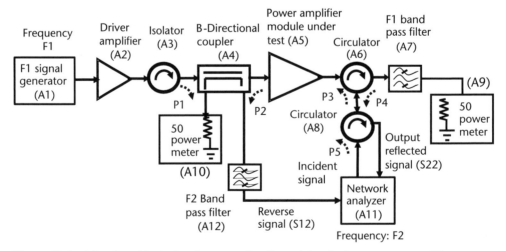

Figure 13.4　A functional test setup for measuring S_{22} and S_{12} of a linear power amplifier.

to a level necessary to produce the rated output power from the device under test (A5). Isolator (A3) improves the driver amplifier (A2) output VSWR. Directional coupler (A4) measures the input power to the device under test (A5) through path (P1) with power meter (A10) and reverse power (S_{12}) through the device under test via path (P2). The bandpass filter (A12) rejects the input signal (F_1) and passes signal (F_2) so the network analyzer (A11) only sees the reverse signal to complete the S_{12} measurement [5].

13.3.2.2   Configuration at the Module (A5) Output When Measuring S_{22} and S_{12} at Rated Output Power

The network analyzer (A11) generates a signal at frequency (F_2) offset from the input signal at frequency (F_1) being emitted from the power amplifier module (A5); see Figure 13.4. The network analyzer signal (F_2) follows path (P5) through the circulator and path (P3) into the output of the power amplifier module (A5) under test. A mismatch impedance with respect to 50Ω at the power amplifier output causes a portion of this signal to be reflected out of the power amplifier adding to the higher-power signal at frequency (F_1) as shown in Figure 13.5.

Both signals (F_1) and (F_2) enter circulator A6 at different ports. Bandpass filter (A7) is centered around frequency (F_1) absorbing the signal and terminating it in power meter (A9). The signal reflected from the output of the power amplifier module (F_2) is in the rejection band of filter (A7) and is reflected from the filter returning to circulator (A6), following path (P4). This signal enters circulator (A8) and is directed back into the network analyzer where its amplitude and phase are compared to the original signal enabling the measurement of S_{22}.

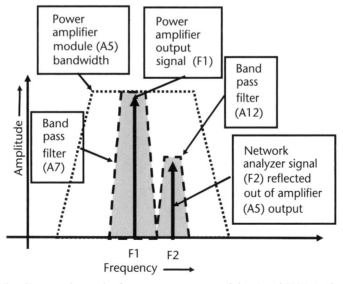

Figure 13.5   The diagram shows the frequency spectrum of the signal (F_1) into the power amplifier (A5), the signal emitted from the network analyzer (F_2), the bandpass filter (A7) response, and the nomial frequency response of the of the power amplifier module under test (A5).

The following are definitions of terms used in this chapter:

- *Circulator:* A circulator is an unterminated isolator, where the signal entering any port travels in the direction of the arrow leaving the device at the next port. If the next port is not terminated in 50Ω, a portion of or the entire signal depending on the reflection coefficient at that port is returned to the circulator and resumes in the direction of the arrow to the next circulator port.
- *Frequencies* (F_1) *and* (F_2): It is important that both frequencies (F_1) and (F_2) are in the passband of the power amplifier module (A5) to get an accurate indication of the S_{22} and S_{12}.
- *Filter (A7):* The bandpass filter (A7) must be capable of absorbing, with minimal reflections, the high-power signal (F_1) emitted from the power amplifier module. If the reflected signal (F_1) from the filter (A7) is at a high level, it can damage the network analyzer or cause the network analyzer to not lock on signal (F_2) making it impossible to make the S_{12} and S_{22} measurement.

13.4 Measuring the Impedance of Large Signal Nonlinear Power Amplifiers Using a Network Analyzer

Power amplifiers for radar and EW systems emitting pulse modulated signals typically operate in a saturated mode, maximizing the output power and efficiency. In these applications, measuring small signal *S*-parameters usually does not accurately portray the power amplifier module in its actual operating mode. There are many techniques and devices used to address this issue. The techniques are described to various levels of detail in application notes from the respective measuring device manufacturers [6].

The impedance of a large signal nonlinear power amplifier can be measured using the technique shown for linear power amplifiers provided that the narrowband filter (A7) in Figure 13.4 can absorb the amplifier output signal such that the reflected signal at frequency (F_1) is below the level that can avoid damage to the network analyzer and low enough that it does not interfere with the network analyzer locking on the signal reflected from the amplifier output (F_2). The bandpass filter (A7) must also have sufficient rejection, that is, reflect all the signals out of the network analyzer (F_2). Any absorption of the analyzer signal (F_2) by the filter (A7) can be misinterpreted as the signal absorption in the output of the amplifier module under test.

Ideally the output impedance of the amplifier is measured at the output frequency (F_1). Using this technique, the output impedance is being measured at frequency (F_2), the signal frequency out of the network analyzer. The error is slight and insignificant if the frequency (F_1) and frequency (F_2) are close together. When the frequencies are too close, the filter (A7) passing signal at frequency (F_1) will have insufficient rejection to network analyzer frequency (F_2), resulting in a measurement error.

The measurement technique shown in Figure 13.4 could have significant error depending on the realizability of the design and manufacture of filter (A7). It is important that the characteristics of the filter as well as the circulators be measure and documented to determine if the theoretical measurement errors are acceptable before configuring and using this test set configuration.

13.5 Load-Pull Impedance Measurements

Another method used to match the input and output impedance of a power amplifier is called load-pull. Load-pull is a technique used to find the optimum impedance seen by a device when the device is operating at its design power level [7]. The technique displays the impedances seen by a device while the device is operating at a specified signal output level. The load-pull device is a microwave tuner that can be used to match a device to 50Ω. The load-pull tuner is calibrated such that each position of its tuning elements is calibrated to reveal its input impedance when terminated in 50Ω. Figure 13.6 is a typical setup with a load-pull tuner on the input and output that matches the devices respective impedances to a 50Ω source and load [8].

Note that load-pull was originally and still being the term used to measure an oscillator's susceptibility to changes in load impedance. The phase and magnitude of an oscillator load is changed and the change in the oscillator frequency is record. The term refers to the pulling of the oscillator frequency as a function of load changes. The variation of the oscillator frequency is caused by the reflected signal adding to the incident oscillator signal resulting in a small change in phase as the vectors add. The phase change is automatically corrected in the oscillator feedback circuit by moving the frequency such that the resonator phase changes, restoring the closed loop phase to 0°, the required phase around the oscillator loop.

13.5.1 Determining the Optimum Input Impedance Using Load-Pull

The directional coupler on the input measures the input power and the return loss from the device under test (DUT). The source tuner varies the input match until the signal into the return loss power meter is a minimum at the expected drive level needed to attain the desired output power. Each position of the load-pull matching device is known as to the impedance at the output with a real 50Ω impedance on the input. When the operator is satisfied that the maximum return loss (minimum

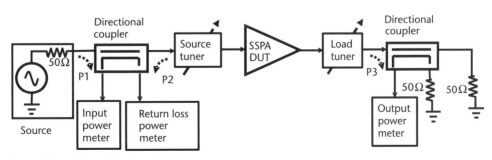

Figure 13.6 Typical load-pull setup.

reflected signal) is attained, the tuner output impedance is read out from a computer that calibrated the load-pull measurement tuner [9].

The readout from the input matching network is not the amplifier input impedance; it is the impedance present to the amplifier input that produces an acceptable input return loss (see Figure 13.7).

13.5.2   Determining the Optimum Output Impedance Using Load-Pull

The output circuit in Figure 13.6 is configured with an output tuner, an output directional coupler, a high-power 50Ω load, and a power meter indirectly measuring the output power through path (P3). Many times, the output power is above the range of a standard power meter, so the measurement is made using a calibrated coupled port out of the directional coupler. The input tuner should be in an optimized position before the output impedance is measured to ensure that the input signal level is known and the input reflected signal losses are minimized. When the input VSWR is high, small changes in the input return loss can translate to significant changes in the signal level going into the amplifier under test.

The load tuner is adjusted until the output power meter displays a maximum power, indicating an optimum output impedance match. It is sometimes necessary to adjust the input power to raise or lower the output power as required. If the input power is adjusted, the load tuner may have to be readjusted. This process may have to be repeated until the desired result is attained [10].

13.5.3   Calibration of the Input and Output Match Networks

The direction of the load-pull measurement impedance is shown in Figure 13.7. The source and load tuners are calibrated such that the setting on these units operating into or from a 50Ω impedance can be converted to a real and imaginary impedance seen by the source or load, respectively.

The diagram shown in Figure 13.7 shows the results of the load-pull measurement, Z_{Source} and Z_{Load}. The resultant impedances are in general complex numbers valid under the specific measurement conditions.

The load-pull measurement conditions are as follows:

1. The matching circuits are terminated in a 50Ω resistive impedance.
2. The measurement is at the singled frequency of the measuring equipment.
3. The measurement is at a single input and output power level.

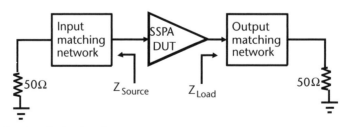

Figure 13.7   Results of a load-pull measurement system.

4. The device is at a specific internal temperature (junction temperature for a bipolar junction transistor (BJT) and a channel temperature for an FET) at the time of the measurement. Ambient temperatures can translate differently to junction or channel temperature depending the thermal conductivity of the device mounting.

Note that load-pull measurement results are the impedance seen by the device and under test, not the impedance of the device. Even when the output power is optimized, the device impedance may or may not be the complex conjugate of the measured values [11]. The output impedance is often, a compromise between maximum output power, best intermodulate distortion (IMD) performance, and best efficiency, into a characteristic 50Ω load.

13.6   Load-Pull Measuring System

The simplified load-pull tuner shown in Figure 13.8 is an adjustable passive network that matches 50Ω to a specific impedance. The basic tuner is an airline conductor between two grounded walls with single or multiple probes whose horizontal position is adjustable along the wavelength of the signal (X direction). The probe also adjusts in the vertical direction (Y direction) changing its depth into the cavity. The horizontal direction primarily adjusts the imaginary part of the transformed impedance and the vertical direction primarily adjusts the real part of the transformed impedance [12]. The Smith chart shown Figure 13.9 is a plot of constant real part lines, constant imaginary part lines, and constant VSWR circles with respect to 50Ω. With 50Ω on one side of the tuner, the impedance seen on the other side of the can be adjusted to different positions on the Smith chart [13].

The tunable passive matching network is calibrated such that, when one port is at 50Ω resistive, the impedance looking into the other port can be reported under actual operating conditions. When the tuner uses a mechanical low-loss device, the reported results are accurate under high-power operations. Accurate high-power results are predicted when minimal heat is generated in the tuner such that changes

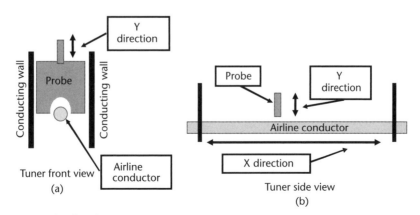

Figure 13.8 A simple microwave tuner.

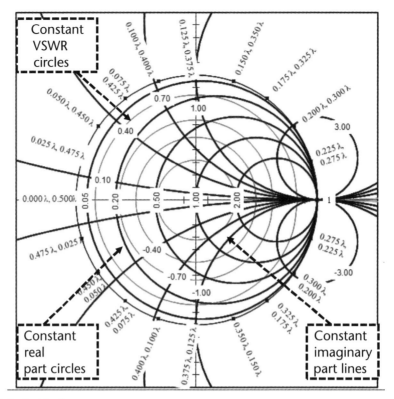

Figure 13.9   Smith chart.

in the mechanical device dimensions are insignificant. Figure 13.10 shows the effect of vertical and horizontal probe positioning on the Smith chart.

Under large signal conditions, the load-pull technique determines the transformed impedance that gives the calibrated result at a specific large signal power output level. The result of this process (achieving the resultant power level measured) could be but is not limited to finding the impedance that is transformed from 50Ω.

The power level measured could represent:

1. A single point of maximum power into a 50Ω load;
2. A contour of points 1 dB, 2 dB, 3 dB, and so forth below the maximum power point;
3. A contour of points representing different efficiency factors (RF output power divided by DC power supplied to the device) for each resulting power level.

Load-pull measurements can characterize impedances seen by a device under different output conditions. This gives the design engineer the ability to select an impedance that weighs the various parameters for an optimized system design.

Note that, with regard to load-pull results, the characteristic input and output impedance of devices operating in the large signal (nonlinear) mode change as a function of the respective input and output power. Many high-power amplifiers have

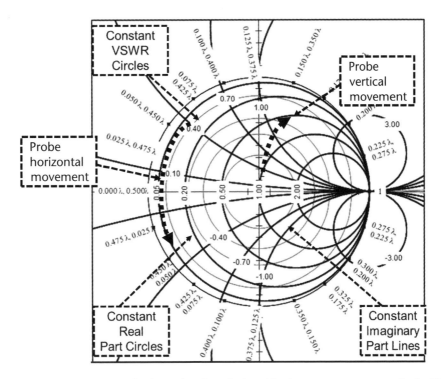

Figure 13.10 Approximate horizontal and vertical positioning movement across the Smith chart.

a very low input and output impedance. These low impedances showing up at the outer edge of the Smith chart could have a large percentage error. This error, when transformed to 50Ω, could result in an unacceptable VSWR. The low-impedance percentage error can be reduced by transforming the impedance with a known network to a value closer to 50Ω. The known transform network has an error that should be considered when evaluating the results.

13.7 SSPA Module Impedances

Solid-state high-power devices inherently have low impedances because the RMS voltage (V_{RMS}) out of the power amplifier module is limited to a large extent by the DC voltage powering the device. As power goes up and the peak-to-peak voltage is limited, current (I_{RMS}) increases and resistance R decreases ($R = V_{RMS}/I_{RMS}$). It is not uncommon for SSPA module impedance to be between 0.5Ω [($|\Gamma|$) $\approx$ 0.99] and 5Ω [($|\Gamma|$) $\approx$ 0.9] depending on the output power of the module.

Power out (P_{out}) of an amplifier is:

$$P_{out} = \frac{\left(V_{RMS}^2\right)}{R} \tag{13.1}$$

Higher DC voltage devices using substrates such as GaN tend to mitigate this issue because the same equivalent power occurs at higher-voltage swings (lower current) resulting in higher impedances. Although mitigated, this issue is not eradicated because GaN has the capability of operating at higher signal power levels than some other substrates such as GaAs. At higher RF powers and higher DC voltages, both available with GaN devices, the problem of low impedance is still an issue.

13.8 Accuracy of SSPA Module Impedance Measurements

The accuracy of impedance measurements decreases significantly when the magnitude of the reflection coefficient ($|\rho|$) is close to unity. This occurs often in solid-state high-power amplifier modules because the input and output impedance (Z_D) tend to be very low compared to the nominal 50Ω characteristic impedance (Z_0).

$$|\rho| = \left| \frac{Z_D - Z_0}{Z_D + Z_0} \right| \tag{13.2}$$

Table 13.1 lists the magnitude of the reflection coefficient ($|\rho|$) and return loss for device impedances (Z_D) from 10Ω to 0.5Ω with respect to the characteristic impedance Z_0.

For reflection coefficients greater than or equal to nine-tenths ($|\rho| \geq 0.9$), the impedance Z_D changes significantly for small changes in return loss.

Example of a Return Loss Measurement Error

If there is a return loss error of 0.7 dB, a device impedance of 2.5Ω could really be 0.5Ω. If a transformer was designed to convert the assumed 2.5Ω to 50Ω (1:20 transformer ratio) to correctly match the device and the return loss measurement was 0.7 dB higher than the true return loss, the device impedance would actually be 0.5Ω. The converted impedance would be 10Ω through the 1:20 transformer instead of 50Ω and the VSWR would be 5:1 instead of the desired 1:1.

Table 13.1 Reflection Coefficient ($|\rho|$) and Return Loss (dB) as a Function of Device Impedance (Z_D)

$Z_0\ \Omega$	$Z_D\ \Omega$	Rho $|\rho|$	Return Loss (dB)	VSWR:1
50	10.00	0.67	3.52	5
50	5.00	0.82	1.74	10
50	2.50	0.90	0.87	20
50	1.25	0.95	0.43	40
50	1.00	0.96	0.35	50
50	0.75	0.97	0.26	66.67
50	0.50	0.98	0.17	100

13.9   Matching the Impedance of Power Amplifiers

Matching impedances of linear and saturated-mode power amplifiers is similar with one important exception; the matching circuit in linear devices is applicable over the linear signal dynamic range, in contrast to devices operating in the nonlinear mode where the matching circuit is applicable only at specified input and output signal levels. The actual dynamic range of a matching circuit for a device operating in the nonlinear mode is determined by extensive testing of the device impedance at various power levels and deciding the accepting level of power matching degradation.

Since input and output impedances are partially reactive, they can vary over the operating frequency range. Matching these impedances over a narrow band is therefore easier than matching over a wide band [14]. Typically, narrow bandwidths are <10% of the center frequency. As the bandwidth increases, the problem of designing an optimum impedance matching becomes harder to resolve.

13.9.1   Impedance Matching Procedure Using RF/Microwave Simulation Program

Matching power amplifier devices is theoretically a straightforward procedure. It starts with measuring the input and output impedance at the operating power and continues with the selection of an impedance matching topology that best fits the measured data.

Typically, the device impedance and the matching network are entered into a computer simulation program to analyze the performance of resulting network [15]. The matching network can be adjusted and reanalyzed until an acceptable circuit is found. More sophisticated computer simulation programs have an optimization routine that iteratively performs this process, sometimes doing thousands or millions of calculations until it converges on a solution and sometimes there is no solution. In the case of no solution, the optimization criteria or the frequency range is usually modified until a compromise solution is found.

Faster convergence always depends on how close the initial matching circuits are the final design. There are many standard matching lumped constant and transmission line circuits available that can be used as a starting point for the design that can be found in technical papers and textbooks focus on that issue. Usually a good place to start is data supplied by the power amplifier module manufacturer. They typically have designed a circuit to test their devices and usually make that circuit available. Many times, the device manufacturer has evaluation printed circuit boards that can be used to test the applicability of the module to an application.

13.9.2   Using a Smith Chart to Select an Optimum Impedance, Power, and Efficiency

Output impedances in nonlinear or saturated amplifiers vary as a function of the respective power level. Many times, it is informative to see how the matching impedance affects the available output power. This information is attained by taking load-pull data at the maximum power level and power levels a few decibels below

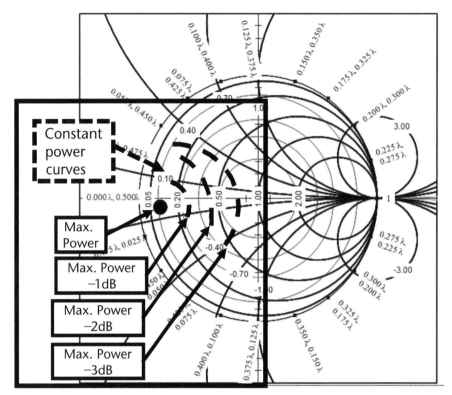

Figure 13.11   Constant power contours plotted on a Smith chart.

the maximum level. This data, when plotted on a Smith chart (see Figure 13.11), gives the design engineer a region of operation at an acceptable output power level.

The contours in Figure 13.11 are plotted in 1-dB increments. Contours that are further apart indicate less output power degradation as a function of impedance variations.

Many times, load-pull data and the respective Smith chart contours are taken and plotted as a function device efficiency. When plotted on the same Smith chart as power contours a decision can be made as the best matching impedance to satisfy output power and efficiency simultaneously (see Figure 13.12).

13.10 An Example of a Computer Simulation of a Matching Network

Output load-pull data can be measured when the power out (P_{out} into load resistor R_L) of the amplifier module is at its maximum power level is shown in Figure 13.13.

In this example, $Z_{Load} = 0.35 + j \cdot 2.025$ at maximum output power (P_{out}) and an operating frequency of 1,525 MHz. The designed frequency band is 50 MHz around the center frequency (3.3% bandwidth), considered narrowband. The measured load impedance is not expected to change significantly over the 1,500-MHz to 1,550-MHz bandwidth and for this example the load impedance is considered constant.

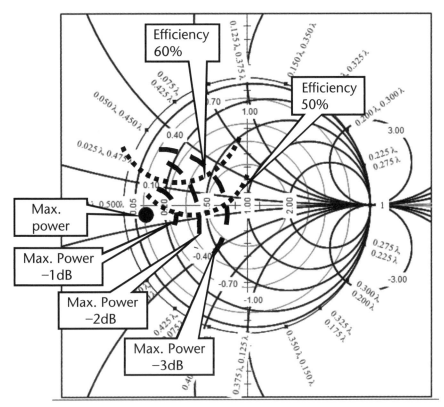

Figure 13.12   Maximum power and efficiency contours plotted on a Smith chart.

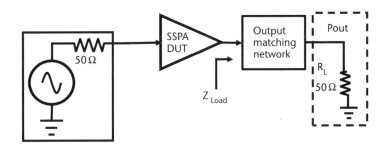

Z_{Load} is the impedance looking into the output circuit

Figure 13.13   The load-pull data indicates the impedance that is seen by the device under test (Z_{Load}).

13.10.1   Computer Simulation Setup

The load pull test at 1,525 MHz determined that the maximum power into the load $(R_L = 50\Omega)$ occurred when transforming the 50Ω load impedance (R_L) to Z_L (see Figure 13.14) where:

$$Z_L = 0.35 + j \cdot 2.025 \ \Omega \tag{13.3}$$

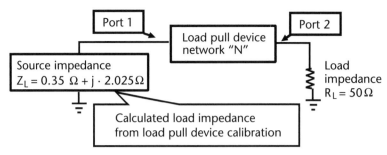

Figure 13.14   Load-pull test setup.

Note that Z_L is the device output impedance (i.e., the Thevenin equivalent impedance). It does not refer to the source of an FET. The objective of the simulation program is to determine the network (N) and measure the expected effects over the frequency range 1,500 MHz to 1,550 MHz.

13.10.2   Computer-Generated Simulation of Network *N*

Figure 13.15 is the simulation of matching network *N* incorporating microstrip transmission lines, TL1 through TL8. The microstrip transmission lines are a series of copper traces on substrate (Subst) MSub1 with each segment (TL1 through TL8) having a width *W* and a length *L*, respectively, given in mils (1 mil = 0.001 inch). The substrate MSub1 is described in Table 13.2. The units in Table 13.2 are: mil = one-thousandth of an inch (0.001 inch), S/m = Siemens per meter, and μm = micrometers.

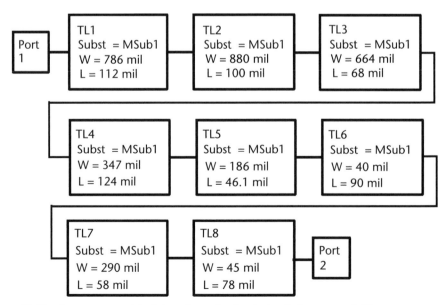

Figure 13.15   Computer model of the microstrip circuit simulation network *N*.

Table 13.2 Characteristics of the Microstrip Substrate MSub1 Used to Model Network N

Parameter	Value	Description
H	25 mil	Height of the substrate
ε_r	10.2	Relative dielectric constant
μ_r	1	Relative permeability constant
Conductivity	5.96×10^7 S/m	Conductivity of copper
Hu	1,000 mil	Free space above the substrate
T	0.6889 mil	Thickness of the copper trace
Roughness	0.4 μm	Roughness of the copper trace

The physical layout of network N is depicted in Figure 13.16. The sequential stepping down from wide trace widths to narrow trace widths raises the relatively low output impedance of the SSPA to the 50Ω load impedance for optimal power transfer [16].

13.10.3 Results of the Computer Simulation for Network N

The results of the computer simulation of the impedance matching network N converting the impedance $Z_L = 0.35 + j \cdot 2.025\Omega$ seen by the SSPA to the load impedance $R_L = 50\Omega$ in terms of S-parameters S_{21}, S_{11}, and S_{22} are shown in Figures 13.17 to 13.19.

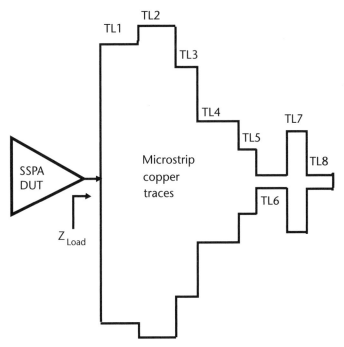

Figure 13.16 Actual SSPA microstrip matching circuit (network N) analyzed by the computer simulation program.

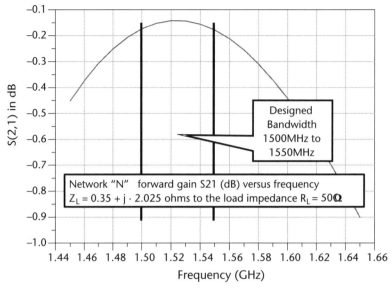

Figure 13.17   *S-parameter S(2,1) for network N is plotted from 1.44 GHz to 1.66 GHz.*

Note that with regard to the resulting *S*-parameter data, since the network is passive bidirectional, S_{21} is expected to equal S_{12} and is therefore not plotted. The graphical plot shows the response from 1,440 MHz to 1,660 MHz. The band of interest is from 1,500 MHz to 1,550 MHz, a 50-MHz bandwidth denoted by two vertical straight lines on the graphs.

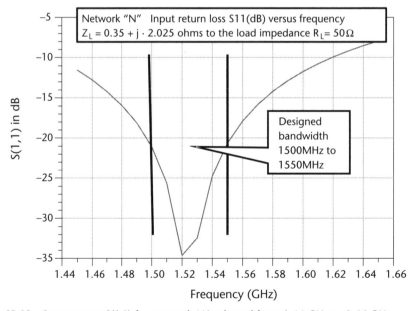

Figure 13.18   *S-parameter S(1,1) for network N is plotted from 1.44 GHz to 1.66 GHz.*

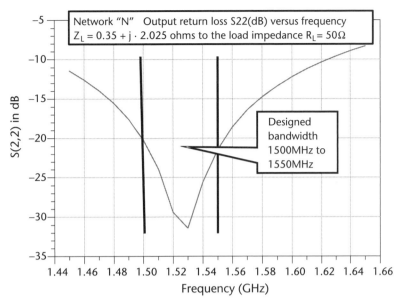

Figure 13.19 S-parameter S(2,2) for network N is plotted from 1.44 GHz to 1.66 GHz.

13.10.3.1 Forward Transfer Power S_{21}

The forward transfer power of network N is shown in Figure 13.17. The simulated insertion loss over the designed 50-MHz bandwidth (1,550 MHz–1,500 MHz) is approximately 0.14 dB.

13.10.3.2 Input Return Loss S_{11} (Figure 13.18)

The input return loss in dB of network N is shown in Figure 13.18. The simulated return loss over the designed 50-MHz bandwidth (1,550 MHz–1,500 MHz) is greater than 20 dB.

13.10.3.3 Output Return Loss S_{22} (Figure 13.19)

The output return loss in decibel power of network N is shown in Figure 13.19. The simulated return loss over the designed 50-MHz bandwidth (1,550 MHz–1,500 MHz) is greater than 20 dB.

13.11 Summary

Measuring and matching the impedance of an SSPA module are critical to the design of a power amplifier. When used in parallel, the available output power and gain of each device should be ideally identical with deviations degrading the system performance. The level of acceptable deviations correlates with the level of acceptable performance from optimum.

The manufacturer's power amplifier module performance, promised or assumed, should be verified on an individual basis and a manufacturing lot basis. The units should be binned so that a similar device can be installed in the same lot of parallel amplifiers. Product consistency is a key factor in producing the product in a predictable time and budget.

References

[1]   MECA Electronics, Inc., "Isolator & Circulator Basics," *Microwave Product Digest*, June 2010.

[2]   Corsini, J., J. Malaver, and S. Lushllari, "90 Degree Hybrid Coupler," Worcester Polytechnic Institute, April 2013.

[3]   Wood, A., and B. Davidson, "RF Power Device Impedances: Practical Considerations," Freescale Semiconductor Application Note AN1526, December 1991.

[4]   Hirato, T., "Basics of RF Amplifier Test with the Vector Network Analyzer (VNA)," Agilent Technologies, March 2012.

[5]   Keysight Technologies, "Understanding the Fundamental Principles of Vector Network Analysis," Application Note, May 2015.

[6]   Keysight Technologies, "High Power Amplifier Measurements Using Nonlinear Vector Network Analyzer," Application Note.

[7]   Simpson, G., "A Beginner's Guide to All Things Load Pull," *Microwaves & RF*, December, 2014.

[8]   Di Falco, S., "Design Techniques for High-Efficiency Microwave Power Amplifiers," Ph. D thesis, Università degli Studi di Ferrara, 2011.

[9]   Ferrero, A., "Overview of Modern Load-Pull and Other Non-Linear Measurement Systems," *ARFTG Nonlinear Measurements Workshop*, San Diego, CA, November 2001.

[10] Focus Microwaves Inc., "Accuracy and Verification of Load Pull Measurements," Application Note 18, September 1994.

[11] RFMD, "Output Return Loss of High Power Class AB Amplifiers," Application Note AN056, 2002.

[12] Ghione, G., "Nonlinearity Characterization and Modelling," Politecnico di Torino Microwave & RF Electronics Group, June 2004.

[13] Cripps, S., *RF Power Amplifiers for Wireless Communications*, Norwood, MA: Artech House, 2006.

[14] Kim, J., et al., "6–18 GHz Reactive Matched GaN MMIC Power Amplifiers with Distributed L-C Load Matching," *Journal of Electromagnetic Engineering and Science*, Vol. 16, No. 1, January 2016, pp. 44–51.

[15] Boshnakov, I., and J. Divall, "Two RF/Microwave Software Programs Used in Tandem Streamline the Design of Power Amplifiers," Aerial Facilities Ltd., Aerial House, Chesham, Bucks, UK.

[16] Lia, X., et al., "Application of Load-Pull in Power Amplifier Design," *Chemical Engineering Transactions*, Vol. 51, 2016.

CHAPTER 14

Power Dividers and Combiners Used in Parallel Amplifier SSPAs

14.1 Introduction

Parallel power amplifier combining (see Figure 14.1) ideally increases the available output power from a single amplifier by N times, where N is the number of amplifiers in parallel.

The selection of the N-way power dividers and combiners are critical to ensure minimal losses and results in output powers as close as possible to the ideal desired level. Amplitude and phase matching of the signals in each parallel path is important so it is common to use the same device in the forward and reversed direction (i.e., power divider and power combiner, respectively).

In this chapter, all the power dividers are reciprocal and can be used as power combiners so whether the device is called a power divider or power combiner, it is understood that the device can be used for both applications.

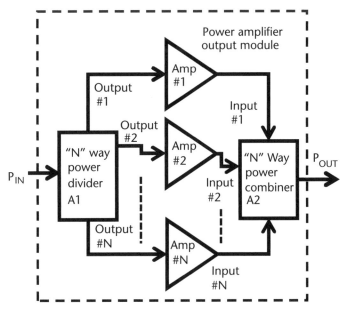

Figure 14.1 Block diagram of the output stage of a power amplifier.

14.2   Factors to Consider in Power Divider Selection

The primary factors to consider in the selection of the power divider are:

1. *Insertion loss:* Insertion loss on the power combiner side is critical; it directly subtracts from the SSPA available power. On the input side (power divider), the loss can be made up with additional gain and is usually not usually considered a significant issue.

2. *Isolation between output ports:* Isolation between ports can affect the signal phase differently at each port. This parameter on the power divider side and the power combiner side are not necessarily the same. Phase differences developed in the signal paths translate into additional losses at the output, negatively impacting the available power from the SSPA.

3. *Size of the power divider:* Power divider topologies have different spatial requirements. Most power divers used various configurations of quarter-wave transmission lines, which limit the spatial efficiency. It is important to note that the power dissipated in an SSPA can be significant and real estate available for heat spreading is not necessary a negative aspect of the design.

4. *Usable bandwidth:* The usable bandwidth is not definitive, it depends on the system requirements and how much power loss is acceptable when all the parallel power amplifier modules are combined. Typically, a 0.5-dB ($\approx 10\%$) loss as a function of frequency is acceptable.

5. *Power-handling capability:* The power-handling capability of microstrip transmission lines is restricted by heating caused by line resistance, dielectric losses, and dielectric breakdown. In addition, the any resistor terminations should be able to withstand the peak and average power that may be reflected from the load [1].

14.3   Two-Way Wilkinson Power Dividers

The Wilkinson power divider shown in Figure 14.2 consists of two quarter-wave transmission lines connected at the input Port 1. Each transmission line, TL_1 and TL_2, has a characteristic impedance of 70.7Ω ($\sqrt{2} \cdot 50\Omega = 70.7\Omega$) feeding an output 50Ω load. Placed between the two outputs is a 100Ω resistor providing a 50Ω termination for in-band even-mode and odd-mode return signals at the frequency where the transmission lines (TL_1 and TL_2) are a quarter-wavelength ($\lambda/4$) [2].

Due to the symmetry of the device, the signal splits evenly and in phase over a large bandwidth. The usable bandwidth of the Wilkinson power divider can be an octave or greater depending on the application. Typically, the bandwidth is limited by the VSWR and isolation between ports [3].

14.3.1   An Example of a Computer Simulation of a Wilkinson Power Divider

A computer simulation of a 17.5-GHz Wilkinson power divider designed on a microstrip substrate, RO4350 [4], was performed to show the theoretical

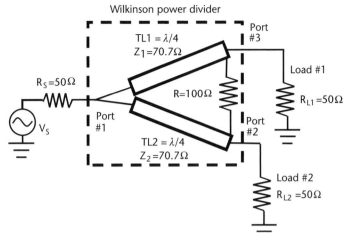

Figure 14.2 Wilkinson power divider.

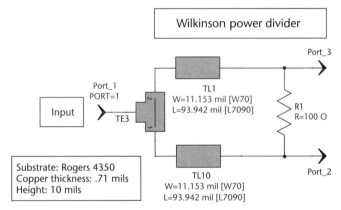

Figure 14.3 Computer simulation of a Wilkinson power divider.

characteristics of the device. Figure 14.3 shows the computer model of the power divider, used to derive the expected performance graphs shown in Figure 14.4.

In this example, the device bandwidth was approximately 11.9 GHz to 22.8 GHz, limited by the input VSWR (S_{11}), which has a return loss greater than 15 dB. The isolation between the output ports 2 and 3 is similar to the results obtained for S_{11} and not shown.

Note that the bandwidth limited by the 15-dB return loss is chosen arbitrarily in this example. Actual return loss specifications are set by the applicable system requirements.

14.3.2   Wilkinson Power Divider: Characteristics Pertinent to Parallel Amplifier SSPA Topologies

The Wilkinson power divider presented in this section is a symmetrical two-way power splitter. The device symmetry with respect to the output ports gives it its wide bandwidth characteristic.

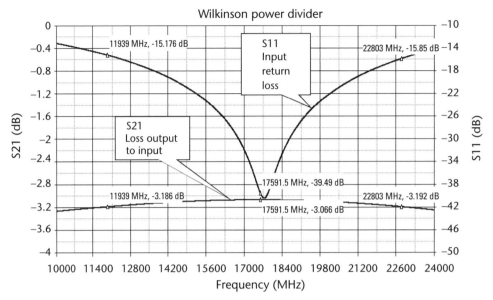

Figure 14.4 Results of the computer simulation of the circuit in Figure 14.3.

The bandwidth limiting factors, the VSWR and the port-to-port isolation, are due to the impedance transformation characteristics of the frequency-dependent quarter-wave transmission lines from the input to the two outputs.

The 100Ω resistor between the power divider outputs under ideal source and load conditions does not see any signal or dissipate any power. The power loss at the center frequency in the simulations is theoretical two-way signal splitting (−3 dB) and the conductor and substrate losses. Losses across the operating band are caused frequency dependence of the conductor (skin effect), the substrate, and the VSWR (see Figure 14.4). The reflected signal at the output is terminated in the source impedance. Additional losses are due to path differences and different load VSWR.

14.4 Quadrature Power Dividers

Quadrature power dividers and couplers are widely used because of their inherent property of directing reflected signals to an isolated load [5]. This effectively makes poorly matched identical parallel devices immune to the consequences of reflected signals [6]. The reflected signal power is directed to an isolated port and dissipated in a 50Ω load placed at that port. The isolated port is not located near the signal ports making it easier to implement a high-power load resistor as necessary.

14.4.1 An Example of a Computer Simulation of a Quadrature Power Divider

A computer simulation of a 16.8-GHz quadrature power divider (see Figure 14.5) was performed to show the theoretical characteristics of the device. Figure 14.6 is the computer model of the quadrature power divider used to derive the expected performance graphs shown in Figure 14.7.

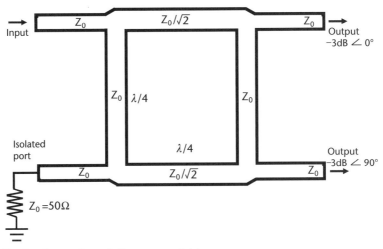

Figure 14.5   Quadrature branch-line power divider.

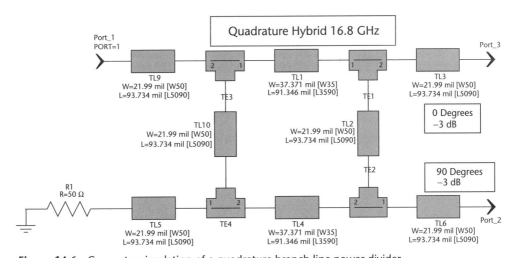

Figure 14.6   Computer simulation of a quadrature branch line power divider.

In this example, the device bandwidth was approximately 15 GHz to 19 GHz, limited by device loss increasing (S_{21} and S_{31}). The 90° phase difference between ports 2 and 3 is maintained throughout the bandwidth and typically holds quadrature over a wider bandwidth than the 0.5-dB output port insertion loss set in this example.

The isolation between the output ports 2 and 3 is similar to the results obtained for the S_{11} and is not shown. Note that on the symmetry of a branch-line quadrature coupler, the branch-line coupler is not a symmetrical design; therefore, the theoretical performances of S_{21} and S_{31} are not the same.

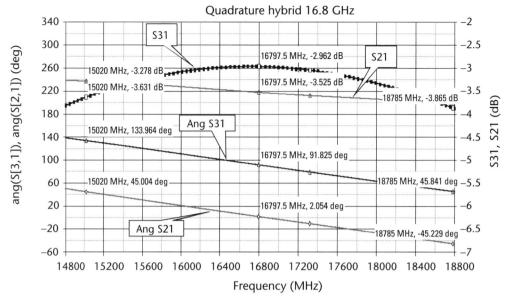

Figure 14.7 Results of the computer simulation of the quadrature branch line power divider circuit in Figure 14.6.

14.4.2 Quadrature Power Divider Characteristics Pertinent to Parallel Amplifier SSPA Topologies

The quadrature power divider presented in this section is a branch-line, two-way power splitter with the phase of the signals at the two outputs 90° out of phase. The bandwidth limiting factors are the VSWR and port-to-port isolation due to the impedance transformation characteristics of the frequency-dependent quarter-wave

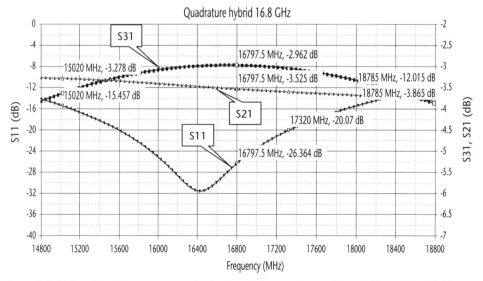

Figure 14.8 Results of the computer simulation of the quadrature branch line power divider circuit in Figure 14.6 showing the input return loss S_{11}.

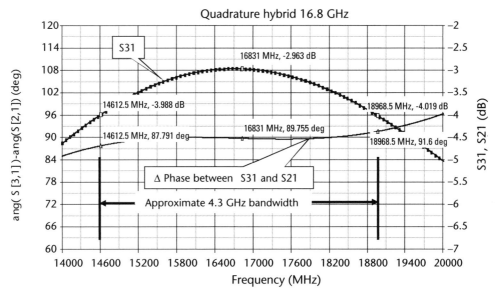

Figure 14.9 Results of the computer simulation of the quadrature branch-line power divider circuit showing the amplitude of the port 3 and the difference in phase of output port 3 with respect to the phase of output port 2.

transmission lines. The 50Ω resistor at the isolated port under ideal source and load conditions does not see any signal or dissipate any power. The 1-dB bandwidth of the branch-line coupler is approximately 4 GHz at a 6.8-GHz center frequency (≈25% bandwidth) (see Figure 14.9). The insertion loss and phase deviation from 90° start to quickly degrade as the bandwidth increases. Alternative, more complex techniques are utilized to design wider bandwidth quadrature couplers.

14.5 Higher Order In-Phase (Wilkinson) Power Dividers

There are many ways to design higher-order ($N > 2$) power dividers, with each technique having different issues and order of difficulty. In general, the higher-order dividers and combiners have practical configuration issues that degrade the performance of that observed with binary (two-way) dividers and combiners.

14.5.1 Three-Way Wilkinson Power Divider

One form of a Wilkinson three-way power divider is shown in Figure 14.10. In this topology, the resistors are connected in a delta configuration (i.e., each resistor is connected to ends of a pair of transmission lines). The difficulty in implementing this three-way Wilkinson power divider is that the resistors (R_1, R_2, and R_3) are not in the same plane. Theoretically, this is not a problem, practically the mechanical configuration of the resistors with respect to the transmission lines are not identical resulting in degraded VSWR and isolation between ports.

An alternate form of resistor placement in the Wilkinson three-way power divider is shown in Figure 14.11.

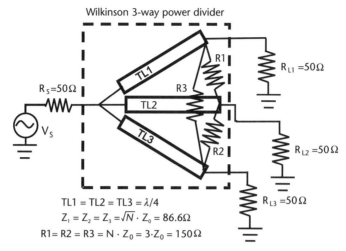

$$TL1 = TL2 = TL3 = \lambda/4$$
$$Z_1 = Z_2 = Z_3 = \sqrt{N} \cdot Z_0 = 86.6\Omega$$
$$R1 = R2 = R3 = N \cdot Z_0 = 3 \cdot Z_0 = 150\Omega$$

Figure 14.10   Three-way Wilkinson power divider with isolation resistors in a delta configuration.

In this configuration, all of the output resistors are tied together in a star configuration [7], each emanating from a single point at the end of the quarter-wave transmission line.

In both these configurations, all the outputs are in phase performing well over a wide bandwidth.

14.5.2   Example of a Simulation of a Three-Way Wilkinson Power Divider

A simulation of a 17-GHz Wilkinson three-way power divider is shown in Figures 14.13 and 14.14. The simulation was performed using the delta resistor configuration (see Figure 14.12). The results theoretically should be the same as the star, but

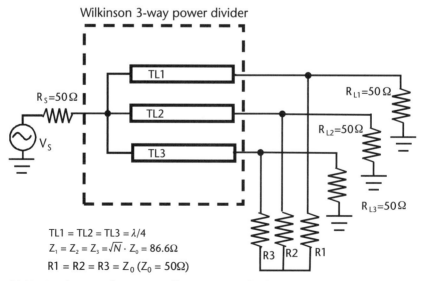

$$TL1 = TL2 = TL3 = \lambda/4$$
$$Z_1 = Z_2 = Z_3 = \sqrt{N} \cdot Z_0 = 86.6\Omega$$
$$R1 = R2 = R3 = Z_0 \ (Z_0 = 50\Omega)$$

Figure 14.11   An alternate three-way Wilkinson power divider configuration with isolation resistors in a star configuration.

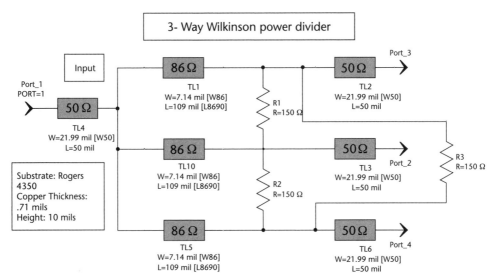

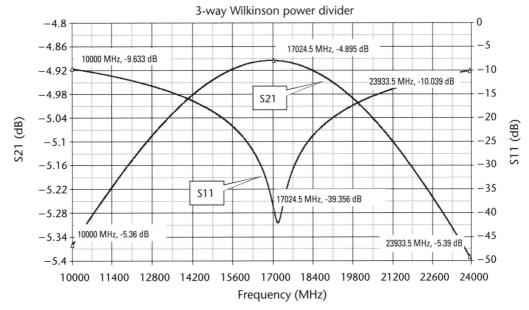

Figure 14.12 Simulation model for a three-way Wilkinson power divider with the isolation resistors in a delta configuration.

practically the performance of the delta configuration is usually superior because there are fewer nonplanar resistors (planar configurations are easier to construct and usually result in fewer parasitic effects).

Figure 14.13 shows the magnitude of the transfer function (S_{21}) and the input return loss (S_{11}). The S_{21} parameter includes the theoretical power divider loss [$10 \cdot \text{Log}(N)$, where $N = 3$ in this example] of 4.77 dB. Losses at the center frequency greater than the 4.77-dB theoretical divider loss are due to the substrate

Figure 14.13 Simulation results of the magnitude of S_{11} and S_{21} for the three-way Wilkinson power divider shown in Figure 14.12.

and conductor loss. Losses away from the center frequency where the transmission lines deviate from a quarter wavelength are due to VSWR mismatch losses in addition to the conductor and substrate losses. The forward transmission losses reach approximately 0.5 dB as the VSWR decreases to a 10-dB return loss. The 0.5-dB bandwidth is approximately 10 GHz and 24 GHz, which is over an octave. Because the device is symmetrical, each of the three outputs are in phase over this entire range.

Figure 14.14 shows the isolation between ports 3 and 2 and ports 4 and 2. The theoretical isolation between the all ports is identical.

Isolation is a function of the 86Ω quarter-wave transmission lines. As the length of the line deviates from a quarter-wave (λ/4 at 17 GHz) the isolation degrades. At the edge of the approximate 0.5-dB bandwidth (10 GHz to 24 GHz), the theoretical isolation degrades to ≈15 dB. This three-way Wilkinson power divider is a symmetrical structure, so all the outputs with respect to the input are identical (i.e., $S_{21} = S_{31} = S_{41}$).

14.5.3 Example of a Three-Way Wilkinson Power Divider in a Planar Configuration

It is often mechanically and electrically advantageous to have all components on the same plane. This is accomplished with the three-way Wilkinson power divider by removing the resistor between output ports 3 and 4 (see Figure 14.15).

The magnitude simulation results S_{21} and S_{11} shown in Figure 14.16 are almost identical to the non-planer simulation results shown in Figure 14.13 because when all output ports are matched into a 50Ω load the isolation resistors R_1 and R_2 do not see any signals and the electrical model looks virtually identical to the nonplanar

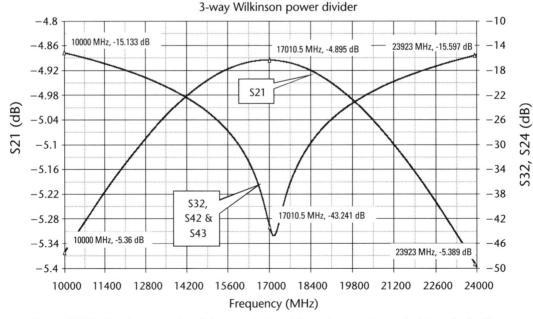

Figure 14.14 Simulation results of the magnitude of S_{21} and output to out isolation S_{21} for the three-way Wilkinson power divider shown in Figure 14.12.

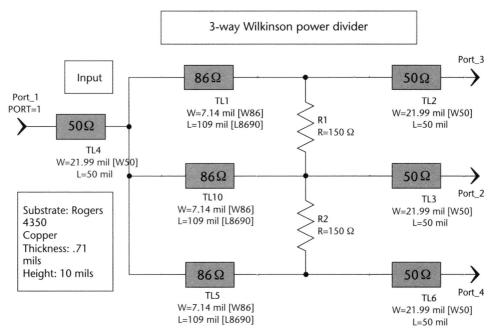

Figure 14.15   Simulation of a three-way Wilkinson power divider with the nonplanar resistor eliminated.

three-way power divider. The 0.5-dB bandwidth of this 17-GHz planar three-way power divider spans the same 10-GHz to 24-GHz frequency range as the nonplanar model. The 0.5-dB bandwidth approximately correlates with the input return loss (S_{11}) reaching 10 dB at the band edge. When all output ports are matched into a 50Ω load, the isolation resistors R_1 and R_2 do not see any signals and the planar

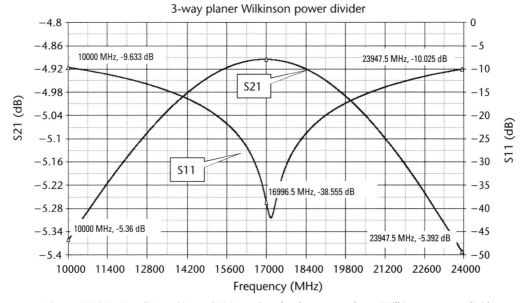

Figure 14.16   Simulation (S_{11} and S_{21}) results of a three-way planar Wilkinson power divider.

three-way power divider appears to be a perfectly symmetrical structure. Under these conditions, the forward signal loss and frequency characteristics of the output ports with respect to the input port are identical (i.e., $S_{21} = S_{31} = S_{41}$).

Figure 14.17 shows the magnitude of the isolation between all the output ports and the magnitude of the transfer S_{21}. The magnitude of the isolation between ports 3 and 2 and ports 4 and 2 of the planar three-way Wilkinson power divider in Figure 14.15 are similar to that simulated for a nonplanar three-way Wilkinson power divider in the delta configuration shown in Figure 14.12. The isolation between ports 3 and 4 where there is no direct isolation resistor connection is degraded significantly in the center of the frequency band. The simulation shows an approximate 11-dB to 12-dB loss across the 10-GHz to 24-GHz band. For most SSPA parallel amplifier topology applications, this level of degraded isolation may be acceptable considering the fact the signal emitted from each output is in phase and return loss from each output is minimal.

14.5.4 An Example of a Three-Way Wilkinson Power Divider with No Isolation Resistors

Figure 14.18 shows a three-way Wilkinson power divider with no isolation resistors between outputs. A quarter-wave transformer was inserted at each output of the power divider (TL_2, TL_3, and TL_6) to match the output VSWR and minimizes the loss at the center frequency. Figures 14.19 is a simulation of the output of the magnitude of S_{21} and the magnitude of the input return loss and S_{11}. The results are similar to that of the three-way Wilkinson dividers with output isolation resistors, with the exception of a small degradation in the operating bandwidth due

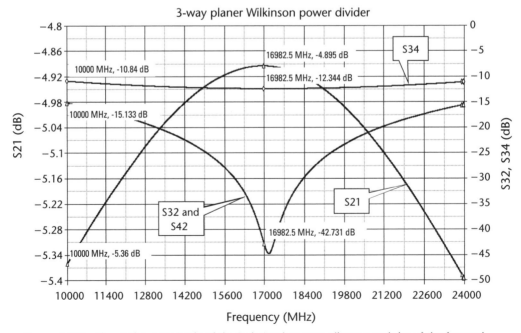

Figure 14.17 Shows the magnitude of the isolation between all ports and the of the forward transfer function S_{21} for the circuit shown in Figure 14.15.

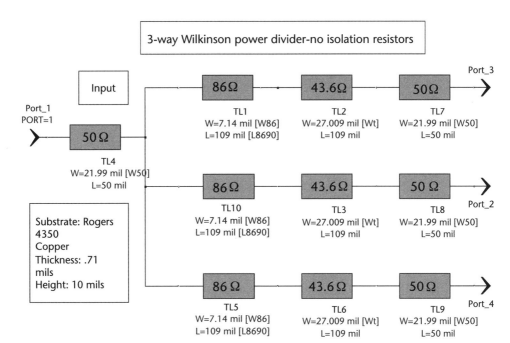

Figure 14.18 A three-way Wilkinson power divider with no isolation resistors.

the addition of the output quarter wave transformers. The 0.5-dB bandwidth is approximately 12.3 GHz to 22.2 GHz, less than an octave.

Figure 14.20 shows the magnitude of the forward loss, S_{21}, S_{31}, and S_{41}, and the isolation between ports, S_{23}, S_{24}, and S_{34}. The simulation has a symmetry between

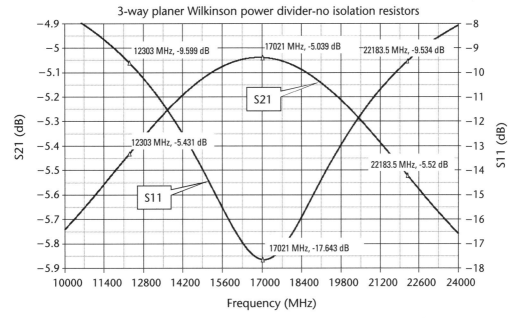

Figure 14.19 S_{21} and S_{11} of a three-way Wilkinson power divider with no isolation resistors.

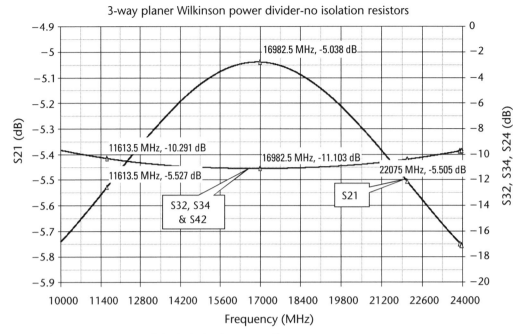

Figure 14.20 S_{21} and the isolation between all output ports.

each output with respect to the input; therefore, the forward transmission loss and the isolation between output ports are identical (i.e., $S_{21} = S_{31} = S_{41}$ and $S_{23} = S_{24} = S_{34}$). Since the output signals are in phase and the parallel amplifiers at the outputs are theoretically identical the minimum 10-dB isolation between channels is typically acceptable.

14.5.5   Higher-Order Wilkinson Power Dividers

The analysis of three-way Wilkinson power dividers and combiners can be extended to N-way power dividers and combiners as shown in Figure 14.21. When $N > 3$, the star configuration is the only practical way to implement the design when maximizing the isolation between ports is a key design parameter.

Each transmission line emanating from the input is designed for an impedance Z_N, where:

$$Z_N = \sqrt{N} \cdot Z_0$$

(14.1)

Each output of the power divider is tied together through an isolation resistor $(R_1, R_2, ..., R_N)$. All the isolation resistors are designed for 50Ω equal to the characteristic impedance at the input and output of the device, $Z_0 = 50\Omega$. The transmission lines are a quarter wavelength long $(\lambda/4)$ where λ is the wavelength at the designed center frequency. The inputs and all the outputs of the device are designed to match into the characteristic impedance $(Z_0 = 50\Omega)$. As the N increases the N-way power divider structure becomes increasingly impractical.

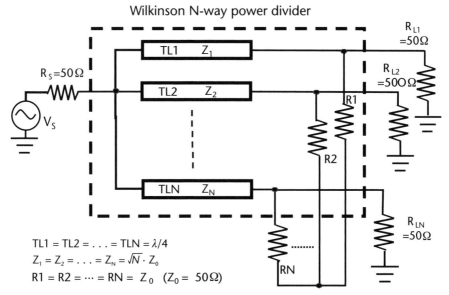

Figure 14.21   *N*-way Wilkinson power divider.

The effective limits on N with respect to the mechanical structure and the microwave design are:

1. *Mechanical structure limits:* Implementing the complex mechanical structure with good microwave design practices becomes more difficult as the number of lines emanating from a single input increases.
2. *Microstrip design limits:* The impedance of each transmission line emanating from the input (Z_N) is $Z_N = \sqrt{N} \cdot Z_0$. As the transmission line impedance gets larger the width of the transmission line gets smaller. Practical microstrip transmission line impedances are generally limited to less than 120Ω [8].

$$N_{\text{MAX}} = \left(\frac{120}{50}\right)^2 = 5.76 \tag{14.2}$$

This effectively limits the practical maximum *N*-way microstrip Wilkinson divider to $N = 5$.

Note that on *N*-way Wilkinson dividers the power dividers are reciprocal and therefore can configured as power dividers or power combiners.

14.6 Higher-Order Radial and Spatial Power Dividers and Combiners

Radial and spatial combining systems for parallel SSPA modules have a single input, a single output, and N-parallel elements. The loss associated with these structures is independent of the number of N-parallel amplifiers (elements). Typically, the combining loss for a Wilkinson power divider is 0.15 dB and the combining loss for a spatial or radial combiner is 0.5 dB. A three-stage binary combining system

(8 parallel amplifiers) loss will be approximately 0.45 dB making the crossover point where radial or spatial combining becomes more efficient is approximately 10 outputs [9].

Note that on combiner system losses, since efficiency depends on the loss after the output, only the loss in the power combiner is considered critical. The power divider loss does not directly affect the available output power.

14.6.1 Radial Power Dividers and Combiners

Radial power dividers are an extension of the N-way Wilkinson power divider with no output isolation resistors [10]. The input port is at the center of a circle and the output ports are radial extensions symmetrically emanating to the circumference (see Figure 14.22).

The radial combiner shown is a reciprocal device and can be used as a power divider or power combiner. These structures provide low loss and excellent phase and amplitudes balance at all outputs. As the number of outputs increases the input impedance decreases. This requires impedance transformers that limit the device bandwidth. The isolation is less than that of a two-way power divider but usually sufficient to be used in parallel amplifier topologies desired for a high-powered SSPA.

Radial combiners have a constant loss independent of the number of parallel elements assuming the mechanical configuration does not increase the path loss in the combiner network.

In an example of the mechanical issue with radial combiners, each element in a 10-element combining system is spaced of 36° apart, and each element in a 20-element combining system is spaced 18° apart. At some point the number of elements

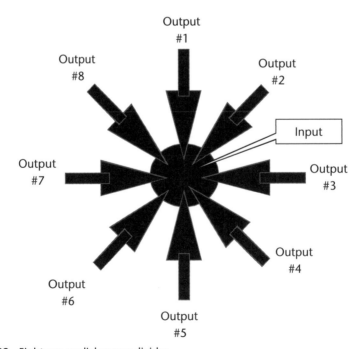

Figure 14.22 Eight-way radial power divider.

N will increase beyond the limit where the outputs are far enough apart to prevent interaction and occurrence of isolation issues. Separating the outputs requires a larger radius leading to a larger mechanical footprint and more microstrip transmission line length resulting in higher combiner losses.

14.6.2 Spatial Power Dividers and Couplers

As previously stated, for a small number of parallel amplifier elements binary combining is most efficient. Radial combining becomes more efficient when the number N exceeds eight elements. Radial combining, although the loss is independent of the number parallel amplifiers, is limited by the mechanical structural requirements. Overcoming the structural requirements requires additional transmission-line length and associated increases in the device loss.

Spatial combining (shown in Figure 14.23) utilizes a transmit antenna at the input, a receive antenna at the input of every parallel amplifier, a transmit antenna at the output of every parallel amplifier, and a receive antenna at the out collecting all the signals from the amplifier outputs.

The advantages of spatial combining are that the loss associated with this technique is independent of the number of parallel amplifiers and, depending on the antenna structure, the usable bandwidth could be very wide.

The disadvantages of spatial combining are a complicated mechanical assembly, issues with heat sinking the amplifier containing in the parallel structure since the parallel amplifiers are contained in an isolated waveguide assembly, and spatial combiners, which can have poorer combining efficiency due to diffraction loss and variations in amplitude and phase through the spatial structure [11].

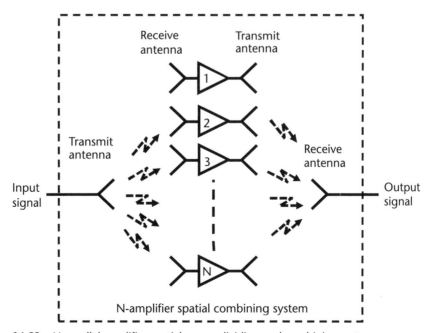

Figure 14.23 *N-parallel amplifier spatial power dividing and combining system.*

14.6.2.1 Tile Spatial Arrays

In a tile array, all the elements are in a single plane [12]. Each tile consists of a receive antenna, a power amplifier, and an output amplifier. The receive and transmit antennas are typically cross-polarized to minimize interference. The array is typically designed to be placed in closed waveguide, where the diffraction losses and focusing errors are minimized, because the energy is confined by the metallic waveguide. Figure 14.24 shows a planar N by N array.

In this configuration, the individual amplifiers and antennas are very small so the module can be monolithically fabricated. Some issues with this design are the nonuniform field in the waveguide. The coupling of the edge elements is not as efficient as the elements in the center of the array. If the array is very large and gain of each element is limited, the array driver may need to have a significant amount of output power. A second smaller array may be necessary to drive the output array. A large array may not fit in a standard waveguide, forcing the waveguide walls to be enlarged possibly introducing harmonic mode problems. Although in a large array each of the amplifier elements are operating at a relatively lower power, the accumulated heat dissipation can be a significant issue. A stable thermal design requires a low-loss heat transfer path from the array to the walls of the waveguide to dissipate the thermal energy while not interfering with the radiated signals.

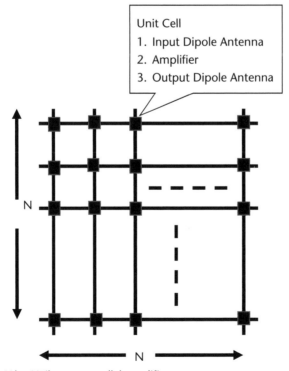

Figure 14.24 *N* by *N* tile array parallel amplifiers.

14.6.2.2 Tray Spatial Arrays

Tray spatial arrays typically use MMIC amplifier elements mounted on a printed circuit board with the input and output antennas either etched into the board or coaxially connected to the amplifier module. The printed circuit boards collect, amplify, and transmit amplified signals to an output collector antenna.

The individual parallel amplifier modules give the design engineer more flexibility than that available with the tile spatial configuration. The power amplifier modules configured in the longitudinal direction can grow only limited by the available system space. With the additional available length, the parallel amplifier module can be designed with multistage high gain and higher-power MMICs. Higher gain power amplifier modules have the advantage of requiring less output power from the stage driving the output parallel amplifier assembly. Each amplifier module can also be mounted on a metal carrier, providing a path for conducting the heat away from the amplifier module. The power amplifier modules are typically constructed on a printed circuit board which can be configured in a rectangular tray or aligned in a circular topology perpendicular to the face of the array to fit in a circular waveguide structure (see Figure 14.25).

The challenge of using a rectangular waveguide combiner is the nonuniform illumination from the source or nonuniform radiation into the load. This problem is sometimes resolved by configuring the N amplifier and antenna trays equally spaced radially in a cylindrical resonant cavity. The symmetry of the configuration minimizes the phase and amplitude differences from each of the parallel amplifier modules [13] and maximizes the probability of attaining optimum output power. The maximum number of radially configured trays is limited by the mechanical configuration of the trays and the size of the resonant cavity.

14.7 Higher-Order Corporate Power Dividing and Combining Techniques

Corporate power dividing is a topology that uses a cascade levels of devices to achieve the required multiple outputs [14]. Figure 14.26 shows three levels of two-way power dividers to split the input signal into eight outputs. The individual power dividers could have greater than two outputs with degraded performance.

Binary combining, a special case of corporate combining, is a multilevel topology using a basic two-way power divider to split power to 2^N sources, where N is the number of levels necessary to achieve 2^N outputs. Figure 14.26 is a three-level

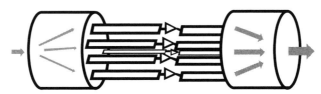

Figure 14.25 Cylindrical tray array of parallel amplifiers.

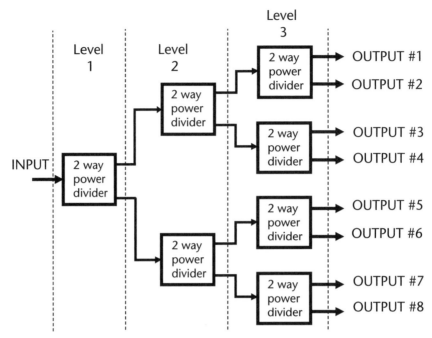

Figure 14.26   Three-level ($N = 3$) binary power divider.

($N = 3$) binary power divider with eight theoretically identical outputs. The configuration is reciprocal and can be reversed to be configured as an eight-way combining network.

Table 14.1 lists a typical arrangement of corporate dividers for two to eight parallel amplifier configurations using two-way or three-way power dividers. In the table power dividers using primary numbers above 3 use a combination of two-way and three-way power dividers attaining one more than the required number of outputs. The unused output is terminated resulting in a loss through the assembly greater than the ideal $10 \cdot \text{Log}(N)$ loss in decibels.

Table 14.1 configures the three-way power dividers at the input and the binary (two-way) power dividers towards the output. Quadrature couplers (binary devices)

Table 14.1   A Typical Corporate Power Divider Configuration for Up to Eight Outputs

	Number of Outputs			*Total*			
Dividing Ratio N	*Level 1*	*Level 2*	*Level 3*	*Outputs Not Used*	*Actual Ideal Loss (dB)*	*Theoretical Loss (dB)*	*Delta Loss (dB)*
2	2			0	3.01	3.01	0.00
3	3			0	4.77	4.77	0.00
4	2	2		0	6.02	6.02	0.00
5	3	2		1	7.78	6.99	0.79
6	3	2		0	7.78	7.78	0.00
7	2	2	2	1	9.03	8.45	0.58
8	2	2	2	0	9.03	9.03	0.00

can optionally be used at the output to provide a matched source for the parallel power amplifiers. The mirror image of the power dividers shown in Table 14.1 are power combiners. Using quadrature power combiners at the input present a matched load to the output of the individual parallel power amplifiers.

Up to three levels of binary power dividers (2, 4, or 8) provide the most efficient matched design. The loss (termination of unused outputs) associated with nonbinary dividers numbers could be very high and is usually not recommended.

14.8 Summary of Power Divider and Power Combiner techniques

The number of parallel amplifiers N designed into the system structure is a function of the system output power requirements and the maximum power available of each individual amplifier. Binary power dividers are typically used for N less than or equal to eight parallel amplifiers. Wilkinson two-way power dividers are used for wide bandwidth. Quadrature two-way power dividers are used to improve the interface match (lower VWSR). Wilkinson power dividers can be used for $N > 2$, but the losses and interface issues increase. Radial power dividers are typically used for $N > 8$ and typically less than $N = 20$. Spatial tray power dividers are used in approximately the same range as radial power dividers (i.e., $N \geq 8$ and $N \leq 20$). The loss is typically less than radial dividers but the construction is more complicated. Tile spatial dividers can be used for combining power amplifiers when $N \gg 20$. The individual tile power amplifiers are lower in power and the design cost can be very high. Corporate (multilevel) power dividers can be constructed on a printed circuit board, making the design very cost-effective. This construction is usually limited to eight or fewer parallel power amplifier modules.

Note that, on power dividers and power combiners, all the power dividers mention in this chapter are reciprocal and therefore can be used as power combiners with the same performance level described herein. Phase array antennas are a form of spatial combiners. The signal from each element is focused on a small section of a spherical solid surface designated by the spherical solid angle Ω. Power from multiple sources is added in the solid angle (Ω). The units of the solid angle are steradians. Note that a steradian is $(180/\pi)^2$ square degrees or about 3,282.8 square degrees [15].

References

[1]   Browne, J., "Choosing to Combine or Divide Power," *Microwaves & RF*, July 2004.

[2]   Pozar, D., *Microwave Engineering*, New York: Wiley, 2012.

[3]   Maloratsky, L., "The Basics of Print Reciprocal Dividers/Combiners," *Microwave Journal*, September 2000

[4]   Rogers Corporation, "RO4000® Series High Frequency Circuit Materials."

[5]   M/A-COM Division of AMP Incorported, "RF Directional Couplers and 3dB Hybrids Overview," Application Note M560.

[6]   Raab, F., et al., "RF and Microwave Power Amplifier and Transmitter Technologies—Part 3," *High Frequency Electronics*, September 2003.

[7]   Nassiri, A., "Power Dividers and Couplers," Massachusetts Institute of Technology, RF Cavities and Components for Accelerators, USPAS, 2010.

[8]   Maloratsky, L., "Reviewing the Basics of Microstrip Lines," *Microwaves & RF*, March 2000.

[9]   York, R., "Some Considerations for Optimal Efficiency and Low Noise in Large Power Combiners," *IEEE Transactions on Microwave Theory and Techniques*.

[10] Ghanadi, M., "A New Compact Broadband Radial Power Combiner," Doctorate Thesis, Elektrotechnik und Informatik der Technischen Universität Berlin, 2012.

[11] Bryerton, E., M. Weiss, and Z. Popovic, "Efficiency of Chip-Level Versus External Power Combining," *IEEE Transactions on Microwave Theory and Techniques*, August 1999

[12] Tsai. F., and M. Bialkowski, "Spatial Power Combiner Using a Planar Active Transmit Array of Stacked Patches," *Journal of Telecommunications and Information Technology*, February 2005.

[13] Bakhsh, J., and F. Dana, "A Wideband Twenty-Element Microwave Spatial Power," Technical paper, Sharif University of Technology, Transactions D: Computer Science & Engineering and Electrical Engineering, October 2013.

[14] Ozbey, B., "Design of an S-Band Power Combiner System with Two Parallel Power Amplifiers and Phase Shifters," Masters Thesis, Graduate School of Engineering and Science of Bilkent University, August 2011.

[15] Pérez, E., "What Are Radians and Steradians," http://4DLab.info.

Power Amplifier Chain Analysis

15.1 Introduction

SSPA systems consist of many devices connected in tandem (typically called a chain) that amplify the input signal to the required output power level. The devices in the chain have various functional applications consistent with the system requirements [1]. The main amplifier sections are typically the preamplifier, the driver amplifier, and the high-power amplifier. Other functions are inserted into the chain as required. Typical ancillary functions in the SSPA chain might be power monitors, return loss monitor points, filters, modulators, attenuators, and isolators. Figure 15.1 shows an SSPA chain of devices amplifying a 1-mW (0-dBm) signal to 40W (+46 dBm).

Each of the functions play a significant role in the system design, but the output parallel amplifier section is typically most critical function delivering the required output power. The preamplifier is usually a high gain stage increasing the input signal to a level sufficient to power the driver amplifier without adding a significant amount of signal distortion. The driver amplifier supplies the signal to the high-power amplifier enabling it to ultimately reach its rated output level.

A chain analysis typically identifies the gain, noise figure, and output power of each device and the accumulative effects of those devices through the SSPA system.

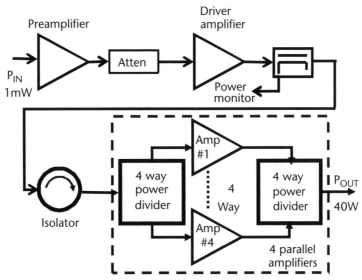

Figure 15.1 Block diagram of a power amplifier showing some typical functions applicable to the system requirements.

The analysis of an SSPA operating in the linear mode (satellite communications, terrestrial communications, and so forth) is a little different than that of an SSPA operating in a saturated mode (radar, EW, and so forth). A linear SSPA requires the analyzing the accumulative effects of every device in the chain while SSPAs operating in a saturated mode function efficiently when each device in the chain is saturated. The analysis of a saturated SSPA can be separated into pieces from one saturated section to the next saturated section typically without considering the accumulative effects.

15.2 Example: Chain Analysis of a Linear Power Amplifier System

A chain analysis is critical in the design and evaluation any microwave system. The primary system parameters calculated for linear power amplifiers are the gain, noise figure, and third-order intermodulation intercept point [2]. These parameters are noted separately for each device in the chain. The cumulative effects of each of the individual components are added a single component at a time so the performance degradation of that component can be seen and adjusted as required to better optimize the system performance.

Additional parameters such as carrier-to-noise ratio, noise output power, and third-order intermodulation interference can be added to the basic chain analysis to give a more in-depth understanding of the system performance.

Parameters used to define SSPAs such as 1-dB compression point (P1dB typically stated in dBm) or saturated power (P_{sat} typically stated in dBm) can be approximated from the output third-order intermodulation intercept point (OIP3, typically stated in dBm). Typical approximations are:

$$\text{P1dB} \approx \text{OIP3} - 6 \text{ to } 10 \text{ dB} \tag{15.1}$$

$$\text{P}_{sat} \approx \text{P1dB} + 1 \text{ dB to } 2 \text{ dB} \tag{15.2}$$

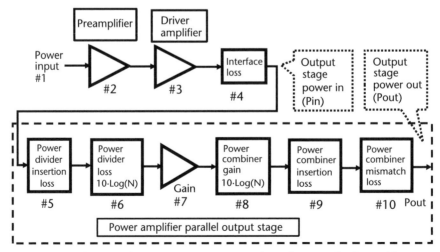

Figure 15.2 Block diagram of an SSPA system separated into functional gains and losses each designated by an item number (1 to 10).

15.2.1   Chain Analysis Spreadsheet for Linear Power Amplifiers

Figure 15.2 is block diagram of the output stage of an SSPA fed from a preamplifier and a driver amplifier stage. The output parallel amplifier module is modeled with respect to its gain and loss functions relating to the real (nonideal) devices used in building the power amplifier system and the parasitic effects resulting from the interfaces and the parallel channel signal phase and amplitude mismatch losses.

The cumulative output power, gain, and noise figure, the chain analysis, can be analyzed using a multitude of software programs. In this example, a simple spreadsheet program (see Table 15.1) gives insight into the effects of each block in Figure 15.2 on the overall system performance [2].

The individual device and parasitic parameters (columns 1 through 5) referenced to Figure 15.2 are as follows:

- *Column 1:* The item number.
- *Column 2:* Item description.
- *Column 3:* Item gain in decibels (loss is a negative gain).
- *Column 4:* Item noise figure (NF) in decibels. The noise figure of the ideal 8-way divider loss (row 6) is 9 dB, but this quantity is recovered in the ideal 8-way combiner gain (row 8), so both are listed as 0 dB for convenience.
- *Column 5:* OIP3 in dBm. Passive devices are assumed to be linear and are listed for calculation purposes as they have a very high OIP3 of 99 dBm.

The calculated accumulated effects (columns 6 through 11) are as follows:

- *Column 6:* Cumulative gain in decibels at the output of the item in the respective row.

Table 15.1   Analysis of the Devices and Parasitic Effects for a Linear SSPA Shown in Figure 15.2

		3	4	5	6	7	8	9	10	11
		Device			System					
1	2	Gain	NF	OIP3	Gain	NF	OIP3	Carrier Power	Carrier Power	Each IM
#	Description	dB	dB	dBm	dB	dB	dBm	dBm	Watts	dBc
1	Input power					0.00		−6.45		
2	Preamplifier	17	10	30	17.0	10.0	30.0	10.6	0.0	−38.9
3	Driver amplifier	12	10	45	29.0	10.1	40.2	22.6	0.2	−35.4
4	Interface loss	−0.3	0.3	99	28.7	10.1	39.9	22.3	0.2	−35.4
5	8-way divider loss	−0.9	0.9	99	27.8	10.1	39.0	21.4	0.1	−35.4
6	Divider loss	−9	0	99	18.8	10.1	30.0	12.4	0.0	−35.4
7	Parallel amplifier	24	10	50	42.8	10.1	48.6	36.4	4.3	−24.4
8	8-way gain	9	0	99	51.8	10.1	57.6	45.4	34.3	−24.4
9	8-way loss	−0.9	0.9	99	50.9	10.1	56.7	44.5	27.9	−24.4
10	Mismatch loss	−0.8	0.8	99	50.1	10.1	55.8	43.7	23.17	−24.4

- *Column 7:* Cumulative noise figure in decibels [3] at the output of the item in the respective row, reflected to the input of the chain of devices.
- *Column 8:* Cumulative OIP3 [4] in dBm at the output of the item in the respective row.
- *Column 9:* Carrier power in dBm at the output of the item in the respective row assuming the input power is the power level in dBm listed in row 1, column 9.
- *Column 10:* Carrier power in watts at the output of the item in the respective row assuming that the input power to item 1 is the power level in dBm listed in row 1, column 9.
- *Column 11:* Level of each intermodulation interference signal with respect to a single carrier (dBc) when two closely spaced carriers each at a level given in column 9 are present in the SSPA bandwidth.

15.2.2 Interrupting the Information in the Chain Analysis Spreadsheet in Table 15.1 Relating to a Linear Power SSPA

Table 15.1 is an example of a spreadsheet describing the chain of devices making up an SSPA used in a communications system (i.e., a linear mode power amplifier [5]). The chain of devices and associated parasitic effects starts from the input of a preamplifier and ends at the output of a parallel amplifier section (eight parallel power amplifier modules in this example).

The output power amplifier section shown in Figure 15.2 (inside the dashed line) is the most critical item in the SSPA system design. It is desired ideally that the OIP3 of the individual parallel power amplifier modules (row 7, column 5, written as R7, C5) multiplied by the number of amplifier modules ($N = 8$ converted to decibels) is equal to the final OIP3 of the amplifier system (R10, C8) in dBm:

$$\mathrm{OIP3(R10,C8)} = 10 \cdot \mathrm{Log}(N) + \mathrm{OIP3(R7,C5)} \quad \text{(ideal goal)} \quad (15.3)$$

In a linear power amplifier, the final OIP3 determines the maximum output power that attains an intermodulation interference level shown in column 11 with respect to a single carrier (dBc). As stated in previous chapters, the third-order intermodulation interference relates to other important factors in communication systems such as spectral regrowth when a single carrier is present and noise power ratio when many carriers are present.

15.2.3 Output Power Degradation in a Linear Power Amplifier Chain

The ideal output power, in the example shown in Table 15.1, related to the third-order intermodulation intercept point is the value of a single parallel output amplifier +50 dBm (row 7, column 5) + 9 dB (row 8, column 3) when the outputs of the eight parallel amplifiers are combined (+59 dBm). Equation (15.4) is used to calculate the third-order intermodulation interference level [IM(dBc)].

$$IM(\mathrm{dBc}) = -2 \cdot [\mathrm{OIP3(dBm)} - C(\mathrm{dBm})] \quad (15.4)$$

and solving for carrier level [C(dBm)]:

$$C(\text{dBm}) = \text{OIP3}(\text{dBm}) + \frac{1}{2} \cdot IM(\text{dBc}) \qquad (15.5)$$

Ideally, the maximum output power of the carrier with an intermodulation interference of −24.4-dBc maximum is:

$$59 \text{ dBm} + \frac{1}{2} \cdot (-24.4 \text{ dBc}) = +46.8 \text{ dBm } (47.8\text{W}) \qquad (15.6)$$

Equation (15.6) assumes that there are no losses after the parallel power amplifier module (row 7) and third-order intermodulation interference (column 11) produced before the parallel power amplifier module (row 7) is insignificant to the results. Under these conditions, the actual output power of +43.7 dBm (row 10, column 9) is degraded by 3.1 dB. This calculation assumes that −24.4 dBc is the maximum acceptable intermodulation interference level.

15.2.4   The Cumulative Effect of Intermodulation Distortion Through the SSPA Chain of Devices on the Available Output Power

Linear SSPAs must achieve the required output power with a specified minimum amount signal distortion. The primary measure of distortion in a linear amplifier is the third-order intermodulation interference, IM(dBc), produced by two equal amplitude carriers [6]:

$$IM(\text{dBc}) = -2 \cdot \left[\text{OIP3}(\text{dBm}) - S_{\text{OUT}}(\text{dBm}) \right] \qquad (15.7)$$

where IM(dBc) is the intermodulation interference in decibels with respect to a single carrier, OIP3(dBm) is the output third-order intermodulation intercept point in dBm, and S_{OUT}(dBm) is the level of a single carrier in dBm.

The third-order intermodulation interference is produced by every component in the chain with the active components contributing most of the distortion. Typically, the output amplifier in a parallel amplifier topology (the parallel power amplifier module, Table 15.1, row 7) adds the most distortion and the amplifier driving the parallel chain (row 3) contributes the second highest amount of distortion. Figure 15.3 shows the third-order intermodulation distortion produced by an output amplifier stage and a driver stage. Δ dB is the level difference between the intermodulation interference produced by the output stage and the intermodulation interference produced by the driver stage. Intermodulation interference produced at each stage in the chain adds coherently.

Table 15.2 shows the total added intermodulation distortion (added dB) as a function Δ dB (labeled "Intermods"). Δ dB is the level of intermodulation that would come out of the final amplifier (row 7) if there was no intermodulation distortion entering the amplifier minus the intermodulation distortion entering the final amplifier (row 6, column 11).

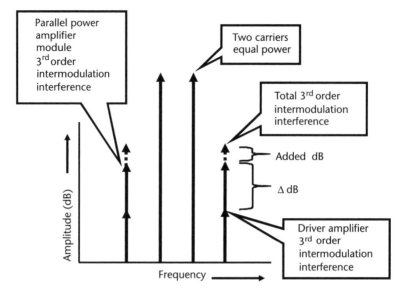

Figure 15.3 The effects of third-order intermodulation interference produced by two amplifier stages.

Example of Driver and Preamplifier Intermodulation Distortion Adding to the Distortion Produced in the Final Amplifier

In Table 15.1, the final parallel amplifier in the chain has an OIP3 of +50 dBm (row 7, column 5). The expected signal level out of each amplifier is +36.4 dBm (row 7, column 9). The intermodulation distortion (IM in dBc) produced in this amplifier is calculated using (15.7). Substituting the example values the resultant IM(dBc) is:

$$IM(\text{dBc}) = -2 \cdot [50 \text{ dBm} - 36.4 \text{ dBm}] = -27.2 \text{ dBc} \qquad (15.8)$$

Table 15.2 The Additional Signal Added dB to the Higher of Two Coherent Signals Added Together (The Separation in Amplitude of the Two Signals Is Δ dB)

Intermods Δ dB	Added dB	Intermods Δ dB	Added dB	Intermods Δ dB	Added dB
0	6.02	12	1.95	24	0.53
1	5.53	13	1.75	25	0.48
2	5.08	14	1.58	26	0.42
3	4.65	15	1.42	27	0.38
4	4.25	16	1.28	28	0.34
5	3.88	17	1.15	29	0.30
6	3.53	18	1.03	30	0.27
7	3.21	19	0.92	31	0.24
8	2.91	20	0.83	32	0.22
9	2.64	21	0.74	33	0.19
10	2.39	22	0.66	34	0.17
11	2.16	23	0.59	35	0.15

The intermodulation distortion entering the parallel amplifiers (row 7) is given in row 6, column 11 (–35.4 dBc).

The Δ dB in Table 15.3 is:

$$\Delta \text{ dB} = (-27.2 \text{ dBc}) - (-35.4 \text{ dBc}) = 8.2 \text{ dB} \tag{15.9}$$

The added intermodulation distortion from the preamplifier and driver stages in rows 2 and 3 from Table 15.1, respectively, is 2.85 dB. From (15.8), it can be seen that the degradation of the effective available output power is half the increase in intermodulation distortion or 1.43 dB. It is obvious that the lower the intermodulation interference produced from the driver stage, the smaller the degradation in the output signal distortion.

15.3 The Effect of *N* Parallel Stages on the Intermodulation Distortion in a Linear SSPA

A simplified amplifier chain indicating the relationship between the power output of the driver (PD_{OUT}), the parallel amplifier module power out (SM_{OUT}), and the SSPA power out (P_{OUT}) is shown in Figure 15.4.

Table 15.3 The Effective Power Degradation of Adding Intermodulation Distortion 8.2 dB (Δ dB) Below the Distortion Created in the Output Stage of an SSPA

Intermods		Power Output Degradation
Δ dB	Added dB	dB
8.2	2.85	1.43

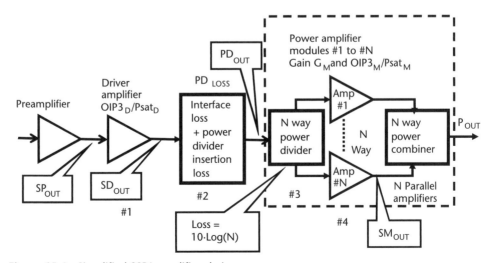

Figure 15.4 Simplified SSPA amplifier chain.

The number of parallel amplifier stages in the output module is related to the capability of the driver stage and, to a lesser extent, the preamplifier, assuming that the preamplifier has no significant intermodulation distortion.

Determining an acceptable value for the number of parallel stages (N) is limited by:

1. Analyzing the driver amplifier output signal with respect to the driver amplifier output signal capability (SD_{OUT} and OIP3$_D$).
2. The interface loss between the driver amplifier and the loss in the power divider (PD_{LOSS}).
3. The theoretical loss in the divide by N circuit [$10 \cdot \text{Log}(N)$].
4. The gain of the power amplifier module (G_M).
5. The signal output of the parallel power amplifier module (SM_{OUT}) and the respective output power capability (OIP3$_M$).

As N increases, the respective divider loss (PD_{LOSS}) plus the $10 \cdot \text{Log}(N)$ increases. Maintaining the same intermodulation distortion level requires that the available output power of the driver stage proportionally increases.

15.3.1 Example: Output Parallel Power Amplifier Third-Order Intermodulation Interference in a Linear SSPA with N Parallel Amplifiers

SSPAs designed for linear operation primarily focus on the third-order intermodulation distortion, column 11 in Table 15.1. The intermodulation distortion in column 11 is 24.4 dBc (dBc is with respect to the carrier level). After the parallel amplifier module (row 7), the virtual gain of the combiner and the actual combiner losses and the signal amplitude and phase mismatch losses are passive and do not affect the intermodulation interference. These losses drop the output power of the SSPA dB for dB without increasing the relative intermodulation distortion. This impact is critical to the design because the carrier power level is degraded [5]. Table 15.4 shows the results when the number of parallel amplifiers is doubled to increase the output power of the SSPA by 3 dB from +43.7 dBm to +46.7 dBm.

15.3.1.1 Analysis of the Effect of Increasing the Number of Parallel Amplifiers

The theoretical gain of the SSPA remains the same. The input signal was increased to attain an output level +46.7 dBm, a 3-dB improvement over the topology with eight parallel amplifiers. Assume that the power divider loss, the power combiner loss and the mismatch loss have not changed. In a real system, these losses will probably all increase. The intermodulation interference increased by 2.2 dB, effectively decreasing the available output power by 1.1 dB. The effective output power therefore increased by 1.9 dB maximum instead of the theoretical increase of 3 dB.

15.3.1.2 Increasing the Output Power of the Drive Stage

The driver amplifier has an OIP3 of +45 dBm. Assuming that the parallel power amplifier has an OIP3 of +50 dBm because that is the highest power available, the

Table 15.4 Modified Table 15.1: The Number of Parallel Power Amplifier Modules Increased from $N = 8$ to $N = 16$

		3	4	5	6	7	8	9		10	11
		Device			System						
1	2	Gain	NF	OIP3	Gain	NF	OIP3	Carrier		Carrier	Each IM
#	Description	dB	dB	dBm	dB	dB	dBm	dBm		Watts	dBc
1	Input power					0.00		−3.40			
2	Preamplifier	17	10	30	17.0	10.0	30.0	13.6		0.0	−32.8
3	Driver amplifier	12	10	45	29.0	10.1	40.2	25.6		0.4	−29.3
4	Interface loss	−0.3	0.3	99	28.7	10.1	39.9	25.3		0.3	−29.3
5	16-way loss	−0.9	0.9	99	27.8	10.1	39.0	24.4		0.3	−29.3
6	Divider loss	−12	0	99	15.8	10.1	27.0	12.4		0.0	−29.3
7	Parallel amplifier	24	10	50	39.8	10.2	47.5	36.4		4.4	−22.2
8	16-way gain	12	0	99	51.8	10.2	59.5	48.4		69.2	−22.2
9	16-way loss	−0.9	0.9	99	50.9	10.2	58.6	47.5		56.2	−22.2
10	Mismatch loss	−0.8	0.8	99	50.1	10.18	57.78	46.7		46.77	−22.2

driver amplifier can be increased up to a maximum of +50 dBm for the OIP3. The results are shown in Table 15.5.

In the analysis of increasing the available output power of the driver stage, the intermodulation interference increased by 1.3 dB from the original interference

Table 15.5 The Number of Parallel Power Amplifier Modules Increased from $N = 8$ to $N = 16$ and the OIP3 of the Driver Amplifier Increased to +50 dBm

		3	4	5	6	7	8	9		10	11
		Device			System						Each
1	2	Gain	NF	OIP3	Gain	NF	OIP3	Carrier		Carrier	IM
#	Description	dB	dB	dBm	dB	dB	dBm	dBm		Watts	dBc
1	Input power					0.00		−3.40			
2	Preamplifier	17	10	30	17.0	10.0	30.0	13.6		0.0	−32.8
3	Driver amplifier	12	10	50	29.0	10.1	41.4	25.6		0.4	−31.5
4	Interface loss	−0.3	0.3	99	28.7	10.1	41.1	25.3		0.3	−31.5
5	12-way loss	−0.9	0.9	99	27.8	10.1	40.2	24.4		0.3	−31.5
6	Divider loss	−12	0	99	15.8	10.1	28.2	12.4		0.0	−31.5
7	Parallel amplifier	24	10	50	39.8	10.2	47.9	36.4		4.4	−23.1
8	12-way gain	12	0	99	51.8	10.2	59.9	48.4		69.2	−23.1
9	12-way loss	−0.9	0.9	99	50.9	10.2	59.0	47.5		56.2	−23.1
10	Mismatch loss	−0.8	0.8	99	50.1	10.18	58.24	46.7		46.77	−23.1

level of −24.4 dBc. This effectively decreased the available output power from the expected +46.7 dBm was 0.65 dB. The effective output power therefore increased by 2.35 dB, 0.65 dB short of the theoretical increase of 3 dB.

On the input side of the parallel power amplifiers (item 4 of Figure 15.4) increasing the number of parallel stages (N) requires an increase in the driver stage (item 1) to maintain the required intermodulation interference level and converge on the desired increase in output power. The losses in the power dividers, the power combiners, and the mismatch losses are expected to increase to a lesser extent, also degrading the output power.

15.3.2 Other Parameters Affecting the Output Third-Order Intermodulation Interference and Available Output Power

The relationship between the signal out of the driver stage (item 1) and the output parallel amplifier stage (item 4) in Figure 15.4 is:

$$SM_{OUT} = SD_{OUT} - PD_{LOSS} - 10 \cdot Log(N) + G_M \qquad (15.10)$$

where SD_{OUT} is the output power of the driver stage in dBm (assuming that there is minimal intermodulation distortion from the preamplifier stages), SM_{OUT} is the output power from the parallel amplifier modules in dBm, G_M is the gain of the parallel amplifier modules in dB, PD_{LOSS} is the power divider loss above the theoretical $10 \cdot Log(N)$, and N is the number of parallel power amplifier modules.

From (15.10), if the gain of the parallel amplifier module equals the loss in the power divider the equation can be rewritten as:

$$PD_{LOSS} + 10 \cdot Log(N) = G_M \rightarrow SM_{OUT} = SD_{OUT} \qquad (15.11)$$

If the output third-order intercept point of the driver amplifier ($OIP3_{Driver}$) is equal to the output third-order intercept point of the parallel power amplifier module ($OIP3_{PPA}$), the third-order intermodulation inference produced in amplifier under the conditions stated in (15.11) will be equal (Δ dB = 0) and the total intermodulation interference will rise by $\approx$ 6 dB.

$$OIP3_{Driver} = OIP3_{PPA} \quad \text{and} \quad PD_{LOSS} + 10 \cdot Log(N) = G_M \qquad (15.12)$$

The output power capability of the system will be degraded by 3 dB (half the power in watts). The intermodulation interference performance can be improved by raising the parallel power amplifier module gain (G_M). Alternatively, the driver amplifier power can be increased by using a parallel topology in the driver stage, but this usually decreases the efficiency and complexity of the SSPA.

If the gain of the parallel amplifier module (G_M) is increased, the signal level out of the driver amplifier will decrease by the same amount. This will tend to mitigate many of the driver amplifier issues.

15.3.3 A Summary of Some Design Practices Concerning Intermodulation Interference

Assuming that the $OIP3_{Driver} = OIP3_{PPA}$, the gain of the parallel amplifier module (G_M) should be greater than the loss in the power divider such that the added intermodulation interference (Added dB from Table 15.2) calculated from the difference between the driver amplifier intermodulation interference (Δ dB from Table 15.2) is at an acceptable level.

$$G_M = PD_{LOSS} + 10 \cdot \text{Log}(N) + \Delta \text{ dB} \tag{15.13}$$

Many times, the driver amplifier output power capability is less than that of the driver amplifier where the difference between the two modules ($\Delta OIP3$) is:

$$\Delta OIP3 = OIP3_{PPA} - OIP3_{Driver} \tag{15.14}$$

Under these conditions, the gain of the parallel amplifier module (G_M) should be:

$$G_M = PD_{LOSS} + 10 \cdot \text{Log}(N) + \Delta \text{ dB} + \Delta OIP3 \tag{15.15}$$

15.4 Chain Analysis of a Saturated Power Amplifier System

The chain analysis of a saturated power amplifier is similar to that of a linear power amplifier with the exception that intermodulation interference and OIP3 is not the governing criteria. A saturated output level (P_{sat}), the highest or near highest available power from an amplifier, is the desired outcome under all conditions. P_{sat} is not a well-defined level and is typically assumed to occur when the linear gain of the device is compressed by 3 dB, (P3dB) [7].

15.4.1 Example of a Chain Analysis Spreadsheet for Saturated Power Amplifiers

The same block diagram (Figure 15.2) for a linear power amplifier is representative of a typical saturated power amplifier. Functionally, the SSPA has a preamplifier and a driver amplifier that feeds the output power stage. In this example, the output stage is configuration with eight parallel power amplifiers modules.

The primary difference between the linear power amplifier and the saturated power amplifier is that the preamplifier and driver stages, as well as the output stage, perform better when each of the stages is in saturation. Table 15.6 is an example of a power amplifier operating in the saturated mode.

The device and parasitic effects (columns 1 through 7) are as follows:

- *Column 1:* The item number referenced to Figure 15.2.
- *Column 2:* Item description referenced to Figure 15.2.

Table 15.6 Analysis of the Devices and Parasitic Effects for a Saturated SSPA Shown in Figure 15.2

| 1 Device Number | 2 Description | Device | | | | | System | | |
		3 Gain dB	4 OIP3 dBm	5 P_{sat} dBm	6 P_{sat} Watts	7 Actual Gain dB	8 Gain dB	9 Carrier dBm	10 Carrier Watts
1	Input power							0.0	0.00
2	Preamplifier	25	30	22	0.16	22.0	22.0	22.0	0.16
3	Driver Amplifier	20	45	37	5.01	15.0	37.0	37.0	5.01
4	Interface loss	−0.3	99	91		−0.3	36.7	36.7	4.68
5	8-way divider loss	−0.9	99	91		−0.9	35.8	35.8	3.80
6	Divider loss	−9	99	91		−9.0	26.8	26.8	0.48
7	Parallel amplifier	24	50	42	15.85	15.2	42.0	42.0	15.85
8	8-way gain	9	99	91		9.0	51.0	51.0	125.89
9	8-way loss	−0.9	99	91		−0.9	50.1	50.1	102.33
10	Mismatch loss	−0.8	99	91		−0.8	49.3	49.3	85.11

- *Column 3:* Item gain in decibels (loss is a negative gain).
- *Column 4:* OIP3 in dBm. Passive devices are assumed linear and are listed for calculation purposes as have a very high OIP3 of 99 dBm.
- *Column 5:* P_{sat} is the saturation level of the device in dBm. P_{sat} for this example is set at a maximum level of 8 dB below the OIP3. This is an approximation and not necessarily related to a particular device.
- *Column 6:* P_{sat} in watts, which was converted from the value in dBm shown in column 5 in the respective row.
- *Column 7:* The actual gain of the device. If the device is in saturation, the device gain is compressed.

Note that in columns 4 and 5 device manufacturers typically supply maximum power information in terms of OIP3, P1dB, or P_{sat}. Since all of these parameters are not usually supplied, it is common to use approximations to fill in the necessary information. Restated from Section 15.2, common approximations are:

$$P1dB \approx OIP3 - 6 \text{ to } 10 \text{ dB} \tag{15.16}$$

$$P_{sat} \approx P1dB + 1 \text{ dB to } 2 \text{ dB} \tag{15.17}$$

The calculated accumulated effects (columns 8 through 10) are as follows:

- *Column 8:* Cumulative gain in decibels at the output of the item in the respective row. The actual gain in column 7 used to calculate the accumulative gain.

- *Column 9:* Carrier power in dBm at the output of the item in the respective row assuming the input power is the power level in dBm listed in row 1, column 8.
- *Column 10:* Carrier power in watts at the output of the item in the respective row.

Note that in column 9 the maximum carrier in the respective row cannot exceed the saturated power (P_{sat}) of the respective device listed in column 5. If the specified gain applied to the device input power would exceed the saturated power of the device, the saturated power was substituted for the input power times the respective device gain. In saturated pulse-modulated amplifiers, this a desired condition because it provides some operating margin [Gain (dB) is greater than P_{out}(dBm) − P_{in}(dBm)] tending to stabilize the maximum output power.

15.4.2   Interrupting the Information in the Chain Analysis Spreadsheet in Table 15.6 Relating to a Saturated SSPA

Table 15.6 is an example of a spreadsheet describing the chain of devices making up an SSPA used in a saturated mode for applications such as radar, EW, and so forth. These applications typically operate with pulse modulated signals. The critical characteristic of these signals is peak power with little or no consideration to linearity. Under these conditions all the amplifier stages are driven in an efficient class C operating mode, turning on only when a pulse is present.

Figure 15.5 is a pulse modulated RF signal with pulse width (PW) in seconds and pulse repetition frequency (PRF) in hertz. Pulse repetition interval (PRI) in seconds is the inverse of PRF. The percentage of time that the pulse is on is called the duty factor (*DF*).

$$PRI = \frac{1}{PRF} \tag{15.18}$$

$$DF = \frac{PW}{PRI} \tag{15.19}$$

The average power transmitted (P_{AVE}) is the peak power (P_{PEAK}) times the duty factor (*DF*).

$$P_{AVE} = P_{PEAK} \cdot DF \tag{15.20}$$

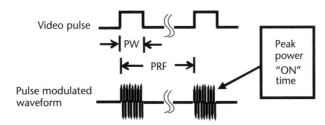

Figure 15.5 Typical pulse modulated waveform.

Since the power amplifier is only on when the signal is present, the average power dissipated in the amplifier is also a fraction of the power dissipated during pulse transmission increasing the efficiency of the SSPA.

In the example of peak and average power calculations, the signal shown in Figure 15.5 has the following characteristics:

- Peak power (P_{PEAK}): 100W;
- Pulse width (PW): 5 microseconds;
- Pulse repetition frequency: 2 kHz;
- Pulse repetition time (PRI) = 1/PRF: 500 microseconds.

$$\text{Duty Factor (DF)} = \frac{PW}{PRI}: \ 0.01 \tag{15.21}$$

Average power (P_{AVE}) equals the peak power (P_{PEAK}) times the duty factor (DF).

$$P_{AVE} = P_{PEAK} \cdot DF = 1\text{W} \tag{15.22}$$

The average power relates to the heat dissipated in the device. The constrictions of heat generate is often a limiting factor in the available output power of the power amplifier.

Unlike transmitting communication signals, pulse modulation focuses on the transmission of the pulse in the time domain and is less concerned about the linearity of the preamplifier and driver stage. The important criteria of the preamplifier stage and the driver stages are to ensure that the final high-power output amplifier is and remains in saturation under all signal and operating environmental conditions. It is also desirable if possible to keep all the stages before and including the output stage in saturation.

The advantages of each stage in saturation are:

1. *Efficiency:* Each stage can be designed to operate more efficiently when it is in a highly nonlinear state.
2. *Signal renormalization:* If each preamplifier and driver stage is in saturation, the levels in each section of the power amplifier will tend to remain constant for a range on input powers, depending on the depth of saturation. Each amplifier stage effectively reconfigures itself to the proper level. This tends to ensure that the next stage is in saturation. This configuration lessens the chance of the output power changing under differing operating conditions.

Note that for average power and peak power dissipation [8, 9] the junction or channel temperature of an MMIC is a critical in the evaluation of a devices reliability. Semiconductor junctions or channels have a much shorter thermal transient time constant than the metal chassis or substrate used to mount the MMIC device. Typical thermal transient time constants of a power amplifier MMIC could be in the range of 1 to 50 microseconds. For long pulses, the device junction or channel temperature will rise during the application of the pulse. Under long pulse applications, the thermal analysis will have to be performed on the peak as well as the average power to determine the reliability of the device.

15.5 Example: Parasitic Issues Associated with the Parallel High-Power Output Amplifiers

A typical power amplifier block diagram is shown in Figure 15.6. The output assembly with eight parallel amplifiers is shown within the dotted line (#5 through #10).

The output power under ideal conditions is eight times the output power of each identical power amplifier modules (#7). The power combiner loss (#9) and the statistical uncertainty loss of phase and amplitude matching of the each of the eight signal paths (#10) have a significant effect on the power output (P_{out}) of the entire SSPA. These losses are shown to be the same value in rows 9 and 10 of the chain analysis, Table 15.1, for the linear power amplifier and in Table 15.6 for the saturated power amplifiers. They represent a one to one degradation (in decibels) to the available output power of the SSPA assembly. Row 9 represents the loss in the output combiner and row 10 represents the statistical uncertainty loss due to the phase and amplitude matching of the parallel signal paths. This loss is reflected to the output of the combining stage.

15.5.1 Output Combiner Loss

The eight-way power divider in this example is made up of three levels of binary dividers (seven power dividers) configured as shown in Figure 15.7. This binary

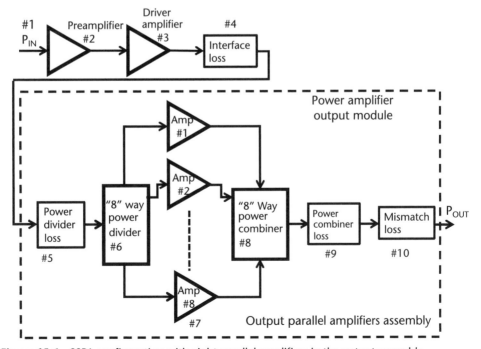

Figure 15.6   SSPA configuration with eight parallel amplifiers in the output assembly.

configuration has some properties different than spatial and radial eight-way power dividers that make it advantageous to use in many applications. The reciprocal of this configuration is an eight-way power combiner, Figure 15.8. Table 15.7 shows a typical specification for a two-way power dividers/combiner. Since there are three layers of binary dividers/combiners in the eight-way configuration, the insertion loss is three times the specified 0.3 dB equal to a total insertion loss of 0.9 dB (row 9 in Tables 15.1 and 15.6).

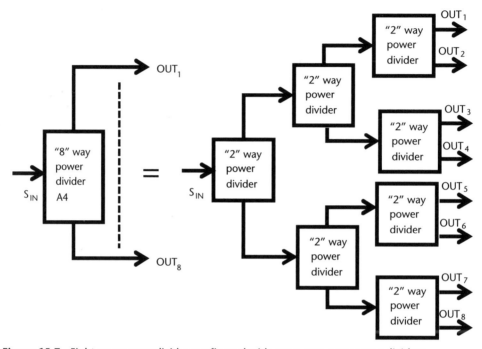

Figure 15.7 Eight-way power divider configured with seven two-way power dividers.

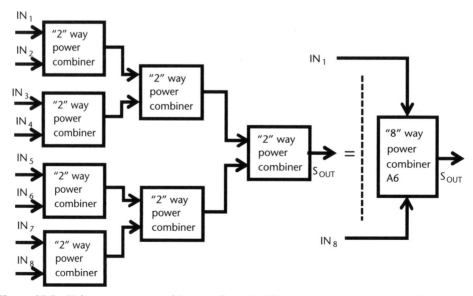

Figure 15.8 Eight-way power combiner configured with seven two-way power combiners.

Table 15.7   Typical Binary Power Divider and Combiner Specifications

Typical Two-Way Divider/Combiner Specifications

Insertion Loss dB (Max.)	VSWR Max.	Amplitude Balance +/– dB (Max.)	Phase Balance +/– Degrees (Max.)
0.3	1.35:1	0.55	6

15.5.2   Phase-Amplitude Parallel Channel Matching Loss

The phase and amplitude channel matching through N parallel path, (eight parallel paths in this example) is critical to minimizing and the loss reflected to the output in row 10 of Tables 15.1 and 15.6 for linear and saturated SSPAs, respectively. The expected phase and amplitude error is obtained from the specifications of the power dividers and combiners (Table 15.7) and the specifications of the amplifiers. An example is given in Table 15.8.

With regard to power amplifier specifications, the specifications listed in Table 15.8 are rarely mentioned by the power amplifier module manufacturer; they usually can be obtained when requested or the user can make the measurements, and when measured data is available, it is generally a good practice to bin the devices in lots of N (eight in this example) to used amplifier modules with the closest amplitude and phase match in the same SSPA.

15.5.3   VSWR Mismatch Loss and Phase and Amplitude Uncertainty

VSWR mismatch losses occur when the source and load are not conjugate matches (real parts are equal and imaginary parts are the negative of each other). These losses are assumed to be included in the divider and combiner loss and amplifier gain. An additional factor added to the amplitude and phase mismatch is the VSWR uncertainty.

The phase and amplitude uncertainty due to VSWR mismatch between devices is caused by the reflected signal from the load going back to the source, reflected at the source and vectorially adding to the original incident signal. If either the source or the load is matched to the characteristic impedance, typically 50Ω, there is no phase or amplitude uncertainty at the respective interface.

VSWR mismatch uncertainty in Figure 15.6 exists between the power divider output (#6) and the amplifier module input (#7), and the amplifier output (#7) and the power combiner input (#8).

Table 15.8   Specification of a Parallel Amplifier Used in This Example

Typical Parallel Power Amplifier Specifications

Linear Gain dB (Typical)	Noise Figure dB (Max.)	OIP3 dBm (Typical)	VSWR Input Max.	VSWR Output Max.	Amplitude Balance +/– dB (Typical)	Phase Balance +/– Degrees (Typical)
24	10	50	1.67:1	1.9:1	0.50	10

Table 15.9 Phase and Amplitude Uncertainty at the Parallel Amplifier Input Interface

Source VSWR :1	Load VSWR :1	Average Mismatch Uncertainty +/– dB	Phase Uncertainty Degrees
1.35	1.67	0.3247733	2.1403474

Table 15.10 Phase and Amplitude Uncertainty at the Parallel Amplifier Output Interface

Source VSWR :1	Load VSWR :1	Average Mismatch Uncertainty +/– dB	Phase Uncertainty Degrees
1.9	1.35	0.3251002	2.1424975

In the example of VSWR amplitude and phase uncertainty, for the example shown in Figure 15.6 and Tables 15.7 and 15.8, the VSWR phase and amplitude uncertainty is shown in Tables 15.9 and 15.10. The combined uncertainty is the statistically the root sum square (RSS) of the input and output uncertainties.

$$\text{Phase Uncertainty} = \sqrt{\left(2.14034^2\right) + \left(2.14249^2\right)} = 3.02843° \tag{15.23}$$

$$\text{Amplitude Uncertainty} = \sqrt{\left(0.3248^2\right) + \left(0.3251^2\right)} = \pm0.4595 \text{ dB} \tag{15.24}$$

Note that on phase and amplitude uncertainty the uncertainty increases as the number of parallel channels (N) increases. The uncertainties due to the VSWR mismatch increases with the number interfaces in the parallel channels. The total uncertainty is found by calculating the RSS of all the interfaces in the parallel amplifier chain and the phase and amplitude tracking in the signal paths. Transmission-line interfaces should be included if the design suggests that these differences are significant enough to affect the result.

15.6 Summary of Losses in an *N* Parallel Amplifier SSPA

Parallel amplifier-combining circuits ideally increase the available output power by a factor $10 \cdot \text{Log}(N)$ in decibels. This factor assumes that the combiner loss is zero and the signals being combined are in phase and equal in amplitude. The combiner loss was discussed in other sections as well as the loss associated with the mismatch of signals being combined. As the number of parallel channel (N) increases, the losses and uncertainties due to phase and amplitude increase. Table 15.11 lists some issues causing degradation in the SSPA system output power related to increasing the number of parallel amplifiers (increasing N) and some possible mitigation techniques.

Table 15.11 Typical List of Reasons That Available Output Is Degraded and Possible Solutions to Mitigate These Issues

Issues	Reasons for Degradation of the Available Power	Corrective Actions to Mitigate the Available Power Degradation
Losses after the parallel power amplifier module	All SSPAs: Power combiner loss	The loss in the output combiner is difficult to mitigate. Careful selection of the microwave substrate and choosing a low-loss combiner topology helps reduce the loss.
Losses after the parallel power amplifier module	All SSPAs: Signal level and phase mismatch loss	The mismatch uncertainty can be mitigated by improving the source and load VSWR at each interface and carefully matching the gain and phase through the parallel paths.
Intermodulation interference created in the preamplifier and/or driver amplifier adding to the intermodulation interference created in the parallel power amplifier module.	Linear SSPAs: This added intermodulation interference forces the output power to be lowered.	Increasing the gain of the output amplifier. Increasing the OIP3 of the preamplifier and/or the driver amplifier may mitigate this issue.

References

[1]   Von Mehlem, U., and R. Wallis, "Solid-State Power Amplifiers for Satellite Radar Altimeters," *Johns Hopkins APL Technical Digest*, Vol. 10, No. 4, 1989.

[2]   Hausman, H., "Modeling Parallel Amplifiers," *Microwaves, Communications, Antennas and Electronic Systems (COMCAS), IEEE International Conference*, November 2015.

[3]   Poshala, P., K. Rushil, and R. Gupta, "Signal Chain Noise Figure Analysis," Texas Instruments Incorporated, Application Report SLAA652, October 2014.

[4]   Karki, J., "Calculating Noise Figure and Third-Order Intercept in ADCs," Texas Instruments Incorporated Data Acquisition, *Analog Applications Journal*, 4Q 2003.

[5]   Hausman, H., "Analyzing MMIC Device Specifications for SSPA Applications-Submitted," *Long Island RF/Microwave Symposium, Trends in Microwaves*, 2017.

[6]   Khalil, M., "Characterizing Intermodulation Distortion of High-Power Devices," *High Frequency Electronics*, July 2007

[7]   Smith, J., and P. Malloy, "Input Power Requirements for AR RF/Microwave Instrumentation's Amplifiers," AR Application Note #45, June 13, 2007.

[8]   Prejs, A., et al., "Thermal Analysis and Its Application to High Power GaN HEMT Amplifiers," *IEEE MTT-S International Microwave Symposium (IMS)*, June 2009.

[9]   Joh, J., et al., "Measurement of Channel Temperature in GaN High-Electron Mobility Transistors," *IEEE Transactions on Electron Devices*, December 2009.

Part III
Designing the Power Amplifier System

CHAPTER 16

RF Signal Monitoring Circuits

16.1 Introduction

SSPAs operate at stress levels above those of most components in a microwave system. Some of the stresses are inherent in the design and some are caused by external events. Remotely monitoring can help to warn of impending situations and aid the alignment of the critical parameters. Some common RF monitorings are RF/microwave signal levels at critical interfaces and reflected signals at critical points, indicating an interface problem.

Signal power monitoring is typically separated into two distinct functions, sampling the signal and measuring (detecting) the signal level. Sampling the signal is an inline function; therefore, care must be taken to avoid SSPA performance degradation. Signal-level detection converts the RF information into baseband data. Detection methods are related to the expected signal modulation technique employed.

16.2 Monitoring RF/Microwave Power Levels at Critical Interfaces

Monitoring the power level at critical points in an SSPA is always a good practice. Figure 16.1 is a typical SSPA showing suggested points of power monitoring.

The lineup of devices in Figure 16.1 is:

1. *Preamplifier:* The preamplifier increases low-level signals to those required by the driver amplifier input.
2. *Driver amplifier:* The driver amplifier is usually a medium-power or high-power amplifier supplying a sufficient signal level to drive the output amplifier modules through the respective *N*-way power divider.
3. *Isolator:* An isolator is often used between the driver amplifier and the high-power amplifier section to ensure that the power divider is being driven by a low VSWR source impedance. Power divider port-to-port isolation and output impedance are a function of the input driving impedance. These parameters affect the parallel signal channel amplitude and phase ripple and uncertainty [1].
4. *Directional coupler sampling forward and reverse power:* Monitoring forward and reverse power at this interface checks that the driver amplifier is delivering the correct power and that the output amplifier module is not reflecting a significant amount of that power. A change in the ratio of

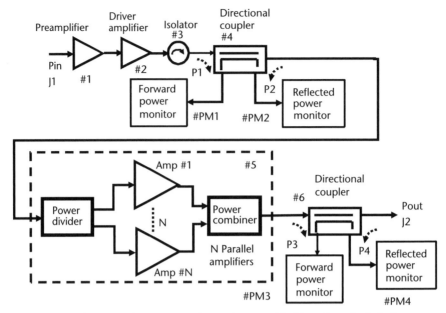

Figure 16.1 Typical SSPA with power monitor points at critical interfaces included.

reverse power to incident power is a possible indication that one or more of the parallel amplifiers is not functioning properly [2, 3].

5. *High-power output amplifier section:* The high-power output amplifier section consisting of *N* parallel amplifiers delivering the required SSPA output signal level.

6. *Directional coupler sampling forward and reverse power:* Monitoring forward and reverse power at this interface between the high-power output amplifier section and SSPA load impedance is critical to determine that the correct power is being delivered and the load impedance is not reflecting a significant amount of that power.

Excessive reflected power can damage the high-power amplifier output module [4]. High-power semiconductor devices exhibit increase stress under severe mismatch conditions. High-impedance mismatches increase the RF voltage and low-impedance mismatches increase the RF current [5].

Many times, reflected signals from the output load are monitored in real time. When a threshold level deemed to be dangerous is encountered, the RF signal is shut down within microseconds of the detection of a mismatch.

With regard to the output isolator, an isolator is desired at the output of the high-power output amplifier section to improve the output matching VSWR but is not used many times because of its loss and sometimes size. The added isolator loss directly (dB for dB) degrades the available output power. The large size is generally due to the microwave resistor on the terminated port that must be capable of absorbing all the output signal if the J2 port is looking into a poor VSWR.

With regard to the configuration shown in Figure 16.1, the array of *N* parallel power amplifier modules (item 5) is the most critical component in the SSPA and therefore is most important to monitor. The configuration shown monitors the input and output forward and reverse power and is an optimum suggestion but not a mandatory requirement.

16.3 RF/Microwave Bidirectional Coupler for Forward and Reverse Power Sampling

There are many configurations of RF power monitoring devices. A commonly used device is a bidirectional coupler as shown in Figure 16.1, with items 4 and 6 placed at the respective input and output of the high-power parallel amplifier output stage.

The major requirements of the device are:

1. Low loss in the signal path. This is especially true for the output bidirectional coupler (item 6) where loss directly effects the SSPA output power.
2. Good directivity such that there is enough separation of the forward and reflected signals at the coupled ports. Poor directivity will result in incorrect power monitoring.
3. Compatibility with the spatial requirements of the system. Directional couplers typically employ as a minimum a quarter-wave transmission line, which adds length to the SSPA.

A bidirectional coupler is schematically shown in Figure 16.2.

The basic bidirectional coupler is made up of a pair of parallel transmission lines (see Figure 16.3), spaced "s" inches apart. The length of the transmission lines is typically a quarter-wave and the width determines the characteristic impedance, typically 50Ω. The spacing "s" between the transmission lines determines the coupling coefficient. The input of the four-port device is Port 1. The main line (low loss) output is Port 2. There are two directional ports (Ports 3 and 4) coupling power from signals going into Ports 1 and 2, respectively. The device is passive, symmetrical, and reciprocal so Ports 1 and 2 and the respective coupled Ports 3 and 4 can be reversed. Since the device is symmetrical in both the horizontal and vertical directions, Port 3 could be the input port with Port 4 as the low-loss output port and Ports 1 and 2 would be the respective coupled ports.

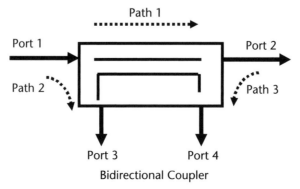

Figure 16.2 Schematic of a bidirectional coupler.

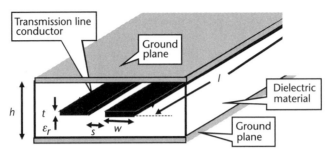

Figure 16.3   The parallel transmission lines of a stripline coupler.

16.3.1   Example of a Typical Bidirectional Coupler Used for an SSPA

The example presented is a 20-dB stripline coupler shown in Figures 16.2 and 16.3.

1. In this example, the main transmission line (the line that passes the in-line signal) is designated from Port 1 to Port 2, where Port 1 is the main signal input and Port 2 is the main signal output.
2. The main line (Port 1 to Port 2) is typically lightly coupled to a parallel transmission line (Port 3 and Port 4) such that the loss through the main line is minimal. Low loss is a primary requirement for the output bidirectional coupler (item 6).
3. Port 1 is coupled to Port 3 with a fixed designed signal loss and ideally completely isolated from Port 4. A 20-dB coupler has an ideal main line loss of 0.04 dB. Table 16.1 lists the ideal main line loss as a function the device coupling coefficient.
4. The device is reciprocal so; Port 2 is coupled to Port 4 with the same fixed design loss as Port 1 to Port 3. Port 2 is ideally completely isolated from Port 3.
5. With this configuration, Port 3 is monitoring the power going into Port 1 (forward signal) and Port 4 is monitoring the power going into Port 2, the signal reflected from the device after the directional coupler.
6. In a nonideal world, the isolation Port 1 to Port 4 and Port 2 to Port 3 is not infinite but must be high enough so the forward and reflected signals are distinguishable and measurable. As an example, if the isolation between Port 1 to Port 4 is 40 dB, a reflected signal 10 dB below the incident signal will look approximately 0.5 dB higher than the actual reflected power.

$$\text{Incident signal Port 1 at Port 4 (P41) is:} \quad P42 = -40 \text{ dBc} \qquad (16.1)$$

$$\text{Reflected signal into Port 2 (PR2) is:} \quad PR2 = -10 \text{ dBc} \qquad (16.2)$$

$$\text{The coupling from Port 2 to Port 4 (P42) is:} \quad P41 = -20 \text{ dBc} \qquad (16.3)$$

The signal at Port 4 (from Port 2) with respect to Port 1 (PR21):

$$PR42 = PR2 + P42 = -30 \text{ dBc} \qquad (16.4)$$

The difference between the desired signal (PR42) and the spurious signal (P41) is 10 dB. If the two signals at the same frequency were in phase, the signal-level monitoring error would be 0.83 dB. Statistically the phase can be random, and the two signals can be modeled as adding as two random signals that cause a statistical error of 0.41 dB (one standard deviation).

7. The high isolation requirement (Port 1 to Port 4 and Port 2 to Port 3) precludes the use of a microstrip design because the signals radiating off the transmission line inherently limit the isolation between parallel paths to unacceptable levels.

To attain the required high isolation between Port 1 to Port 4 and Port 2 to Port 3, bidirectional couplers are usually designed in a stripline configuration. The stripline configuration shown in Figure 16.3 has a ground plane above and below the transmission line, which terminates the fields, thereby preventing unintentional cross-coupling.

16.3.2 Example: A Computer Simulation of a 1-GHz to 2-GHz Bidirectional Coupler

A directional coupler designed to operate in the 1-GHz to 2-GHz frequency range was simulated using the model shown in Figure 16.4. The transmission lines were modeled on a Rogers RO4350 substrate, 0.04 inch high with copper traces 0.00142 inch thick [6]. The port numbers in Figure 16.2 (schematic) and Figure 16.4 (simulation model) are consistent. Figure 16.5 is the simulated response over the 1-GHz to 2-GHz frequency range. The three response curves shown in Figure 16.5 are:

1. The forward path, db{S[2,1]} (db{S[2,1]} = $|S21|$ in dB): Loss ≈ 0.2 dB.
   a. The forward path is the signal-through path (Path #1).
   b. This path should be designed for low loss.

Table 16.1 Ideal Main Line Coupling Loss as a Function of Parallel Line Coupling

Coupling dB	Main Line Loss dB	Coupling dB	Main Line Loss dB
−3	−3.02	−12	−0.28
−4	−2.20	−13	−0.22
−5	−1.65	−14	−0.18
−6	−1.26	−15	−0.14
−7	−0.97	−16	−0.11
−8	−0.75	−17	−0.09
−9	−0.58	−18	−0.07
−10	−0.46	−19	−0.06
−11	−0.36	−20	−0.04

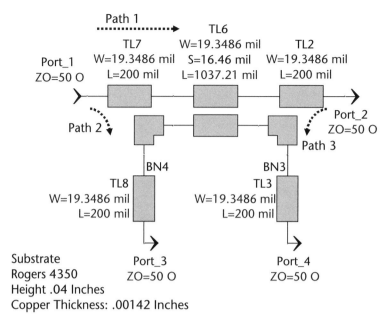

Figure 16.4 Simulation model of a stripline bidirectional coupler.

2. The forward-coupled path, db{S[3,1]} (db{S[3,1]} = |S31| in dB): Loss ≈ 20 dB.
   a. S[3,1] is a sample of the forward (incident) signal, path #2.
   b. The 20-dB loss is a design parameter of the directional coupler.
3. The reverse coupled path db{S[4,1]} (db{S[4,1]} = |S41| in dB): (S41): Loss ≈ 50 dB.
   a. S[4,1] is an isolation path.
   b. It is the amount of incident signal at Port 1 showing up at Port 4
   c. Ideally there should be no forward signal at this port if Port 2 is matched.
   d. Practically, the Loss |S[4,1]| should be much greater than the Loss |S[3,1]|

Since the directional coupler is passive, reciprocal, and symmetrical:

$$S[4,1] = S[2,3] \tag{16.5}$$

where S[2,3] is the reflected signal entering Port 2 and showing up at Port 3. This unwanted signal adds linearly to the desire signal S[3,1], where S[3,1] is a known sample of the forward incident signal. The monitoring point operates correctly when:

$$\left| S[2,3] \right| + \left| S[3,1] \right| = \left| S[4,1] \right| + \left| S[3,1] \right| \approx + \left| S[3,1] \right| \tag{16.6}$$

16.3.3 Summary of the Bidirectional Coupler Function and Example

Due to the symmetry of the device the coupled port S[3,1] (a sample of the forward signal, Path 2) is equal to the coupled port S[4,2] (a sample of the signal reflected by the load) and measured at Port 4. In summary,

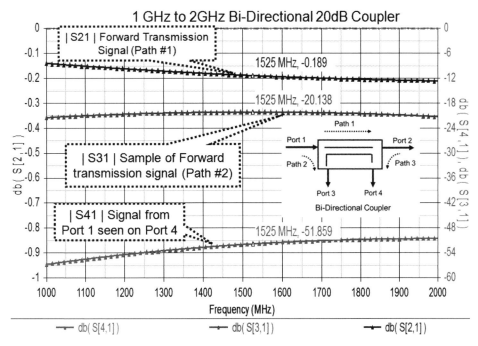

Figure 16.5   Simulation results of a stripline bidirectional coupler.

1. The primary signal travels from Port 1 to Port 2 (Path 1) with minimum loss.
2. Port 3 is a sample of the signal entering Port 1 (Path 2) with a fixed attenuation, in this example designed to −20 dBc, where 0 dBc is the signal level entering Port 1.
3. Port 4 is a sample of the signal entering Port 2 (Path 3) with a fixed attenuation, in this example designed to be −20 dBc, where 0 dBc is the signal level entering Port 2). A signal entering Port 2 is the incident signal entering Port 1 reflected by the load connected to Port 2.
4. The ratio of the signal level at Port 3 with respect to the signal level at Port 4 is the return loss, due to the mismatch of the load connected to Port 2. Return loss can be converted to the VSWR.

16.4   RF/Microwave Signal Power Level Detection

Power (watts) is the transfer of energy per unit time (1W is 1 J/s) [7]. In microwave systems, the pertinent information is the power is transferred into the desired load. The average power dissipated in the load can be computed by measuring the actual heating effect, that is, temperature rise of the load [8] or the potential heating effect, that is, analyzing the voltage waveform seen by the load impedance. Both methods and variations of these techniques have their advantages and limitations depending on the application.

Systems employing pulse microwave signals such as radar or navigation applications require peak power measurements in addition to average power information.

Peak power measurements typically employ indirect power measurements techniques that allow for a faster measurement typically limited to a designated time frame. Signal power detection methods are determined by the characteristic of the carrier modulation and the desired or pertinent signal-level information.

16.4.1 Root Mean Square Power

RMS is a mathematical quantity used to equate the power dissipated by a time-varying current or voltage waveforms to that of a DC or voltage passing through the same resistor or the REAL part of a complex impedance (R_Z).

RMS power (P_{RMS}) does not actually have a direct definition, it is the RMS voltage (V_{RMS}) and RMS current (I_{RMS}) with respect to the resistance (R_Z) that defines the power (P_{RMS}).

$$P_{RMS} = V_{RMS} \cdot I_{RMS} = \frac{\left(V_{RMS}\right)^2}{R_Z} = \left(I_{RMS}\right)^2 \cdot R_Z \tag{16.7}$$

The RMS voltage (V_{RMS}) is:

$$V_{RMS} = \sqrt{\frac{1}{T} \cdot \int_0^T \left[v(t)\right]^2 \cdot dt} \tag{16.8}$$

The RMS current (I_{RMS}) is:

$$I_{RMS} = \sqrt{\frac{1}{T} \cdot \int_0^T \left[i(t)\right]^2 \cdot dt} \tag{16.9}$$

$$V_{RMS} = I_{RMS} \cdot Rz \tag{16.10}$$

where $v(t)$ is the time-varying voltage waveform and $i(t)$ is the time-varying current waveform, T is the period of the waveform, and R_Z is the real part of a complex load impedance.

In terms of voltage or current,

$$P_{RMS} = \frac{\left(V_{RMS}^2\right)}{Rz} = \left(I_{RMS}^2\right) \cdot Rz \tag{16.11}$$

16.4.2 Thermistor-Based Power Meters

Thermistor-based power meters are primarily used in instrumentation applications. These devices consist of a 50Ω resistive load thermally isolated such that small temperature changes can be attributed to power dissipated from a microwave signal.

The attributes of this power monitoring configuration are the following. The advantages of the thermistor-based power meter are that the measurement is based on power dissipated in a load causing the temperature in the load to rise (this a true

RMS power measurement independent of the signal waveform), since the microwave load is essentially a resistor, it has a very low VWSR over a very wide range of frequencies, the reflected signals are usually very small increasing the accuracy of the incident signal measurement into a characteristic 50Ω load, and the temperature of the load can be measured very accurately making this technique accurate enough to be used in laboratory instruments. The disadvantages of the thermistor-based power meter are that this technique is very slow because it is necessary for the 50Ω load to thermally stabilize to get an accurate measurement, the power head consisting of the thermally isolated 50Ω load is larger than other power measurement techniques making difficult to use in field equipment, and the accuracy falls off quickly when the signal level is low because the change in temperature of the 50Ω load is very small.

16.4.3   Peak Power Detectors

Peak power detectors, as the name implies, rise as a function of the RF signal voltage and holds the peak value. The peak power is usually calibrated in terms of RMS power (watts) by assuming the signal is sinusoidal. The relationship between the peak of a pure sine wave, V_P, and the RMS value of that sine wave, V_{RMS} [see (16.8)], is:

$$V_{\mathrm{RMS}} = \frac{V_P}{\sqrt{2}} \approx 0.707 \cdot V_P \tag{16.12}$$

Figure 16.6 is a typical peak detector. The respective wave form is shown in Figure 16.7. The input rises with a time constant (τ_{RISE}) where:

$$\tau_{\mathrm{RISE}} \approx \left(\frac{R_S}{2}\right) \cdot C \tag{16.13}$$

After the voltage peak, the diode D1 is no longer forward-biased. The waveform therefore falls with a time constant (τ_{FALL}):

$$\tau_{\mathrm{FALL}} \approx R_D \cdot C \tag{16.14}$$

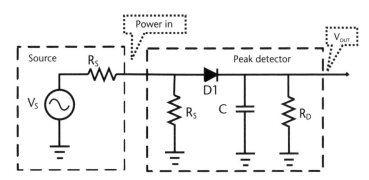

Figure 16.6   Typical peak detector used to monitor RF power.

Since by design $R_D \gg (R_S/2)$, the rise time is much smaller than the fall time. With constant-wave signals, the issue of rise and fall times is not critical, but when used in pulse RF applications the time constant (τ_{RISE} and τ_{FALL}) are critical to monitor and report the correct signal level.

The rise time, *RT*, the time that the wave takes to go from 10% to 90% of its final value, should be less than the pulse width (PW) of the pulsed RF signal. The fall time, *FT*, the time that the wave takes to go from 90% to 10% of its final value, should be less than the pulse repetition interval (PRI) of the pulsed RF signal.

The rise time (RT) and fall time (FT) for a first-order system are usually approximated to be [9]:

$$\text{Rise Time (RT)} \approx 2.2 \cdot \tau_{RISE} \tag{16.15}$$

$$\text{Fall Time (FT)} \approx 2.2 \cdot \tau_{FALL} \tag{16.16}$$

Some attributes of a peak detector are as follows. The advantages of a peak power detector are that it is a simple, low-cost design, it can capture the peak power of pulse-modulated RF signals, and it correctly converts peak voltage to power for CW, frequency modulated (FM), or phase modulated (PM) signals. The disadvantages of a peak power detector are that it measures peak voltage, not peak power (peak power is calculated correctly for sinusoidal signals that are not amplitude modulated), and it does not correctly convert the power of multiple carriers or amplitude modulated (AM) carriers.

16.4.4 RMS Power Detectors

True RMS power detectors are circuits that perform the mathematical function in (16.7) and (16.11). This type of power detector monitors the actual average power emitted for any type of complex modulated waveform [10]. The complex RMS calculation can be performed digitally for low-frequency waveforms. At higher frequencies, the RMS power of RF signals are determined using analog techniques and calibration algorithms that may be less accurate than the digital counterpart.

Some attributes of RMS power detectors are the following. The advantages of RMS power detectors are that the monitored results are a true average power independent of the signal wave shape or modulation [10] and the total power of multiple signals is correctly reported. The disadvantages of RMS power detectors are that

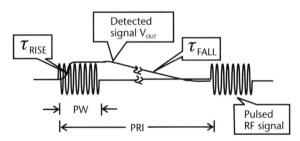

Figure 16.7 Typical waveform through the peak detector shown in Figure 16.6.

only average power is correctly reported and systems that utilize pulse modulated waveforms such as radar or navigational system usually required the peak power measurements. Average power is typically calculated from the peak power measurement and the waveform duty factor (DF).

References

[1]   Marki Microwave, "Microwave Power Dividers and Couplers Tutorial."

[2]   Jain, A., et al., "High-Power, Low-Loss, Radial RF Power Dividers/Combiners," Raja Ramanna Centre for Advanced Technology (RRCAT), Indore, India, APAC, 2007.

[3]   M/A-COM Division of AMP Incorporated, "Power Dividers/Combiners," Application Note M561.

[4]   Giga-tronics Incorporated, "Six Things about Microwave Power Amplifier," Application Note AN-GT149C, 2013.

[5]   Vectawave Technology Limited, "High Power Solid State Amplifiers Under Mismatch Conditions," Application Note 01-1214, 2014.

[6]   Rogers Corp., "RO4000® Series High Frequency Circuit Materials," 2017.

[7]   Keysight, "Fundamentals of RF and Microwave Power Measurements," Application Note, AN64-1B.

[8]   Wireless Telecom Group, "Principles of Power Measurement," Boonton, 2011.

[9]   University of Newfoundland, "Time Response Part 1: Poles and Zeros and First-Order Systems," *Control Systems*, 2010.

[10]   Hittite Microwave Corp., "Latest Advance in RF Power Detection: RMS Detector with iPAR," *Microwave Journal*, December 1, 2008.

CHAPTER 17

DC Power Interface with the RF Signal Path

17.1 Introduction

SSPAs are designed to deliver a significant amount of RF/microwave power, which is a percent of a larger amount of DC power. In field-effect transistor (FET)-based power amplifier modules, the DC power is primarily made up of a fixed drain voltage times a variable drain current. The drain voltage and current are optimized for a desired output power. Usually, the drain voltage is regulated and fixed. The drain current is set by the device's gate voltage (assuming the typical configuration of a grounded substrate and source). Typically, each MMIC module in the SSPA chain requires individual management of its primary DC power and biasing conditions necessary for optimum performance, efficiency, and reliability. Some of the primary issues considered in DC powering the devices are:

1. Maximizing the RF output power by maintaining the required drain current with a minimal drop in the drain voltage;
2. Decoupling the bias voltages to minimize the loss and interference in the RF circuit;
3. Minimizing the power dissipated to maximize the efficiency and enhance the device reliability.

17.2 Power Amplifier Circuits Using Depletion-Mode FETs

Most RF/microwave power amplifier modules (Z_1) are based on a depletion-mode FET. Figure 17.1 is a typical power amplifier module shown with a typical RF path and DC bias circuitry controlling the device drain D and gate G.

The RF gain device (Q_1) shown in Figure 17.1 is a depletion-mode FET. The drain (D) to source (S) channel width is controlled by the gate (G) to the substrate (Sub) voltage ideally electrically isolated from the channel. The source (S) is typically grounded for maximum amplifier gain. The substrate (Sub) is connected to ground such that a negative voltage on the gate (G) has the effect of the constricting the channel width and limiting the drain current (I_d) [1].

Typically, the RF signal enters (J_1) the power amplifier module through a coupling capacitor (C_1). The module (Z_1) could have an input matching network to maximize power transfer before the signal enters the FET gate (G). The amplified

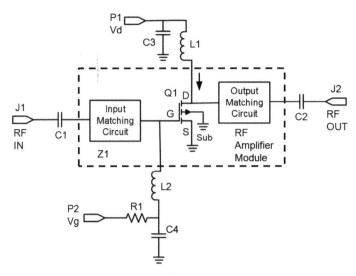

Figure 17.1 Typical RF/microwave power amplifier module.

signal out of the drain may pass through an output matching network before it is AC-coupled (C_2) to the output port (J_2).

The drain voltage (P_1) connects to the MMIC through capacitor C_3 and inductor L_1. The inductor prevents loss in the RF signal path. The capacitor is part of an L-C lowpass filter attenuating the RF signal and minimizing RF signal on the DC power lines.

The gate voltage (P_2) connects to the MMIC through resistor R_1, capacitor C_4, and inductor L_2. The inductor isolates the DC bias circuit from the RF circuit minimizing loss in the RF signal path. The capacitor is part of an L-C lowpass filter attenuating the RF signal and minimizing RF signal on the DC power lines. The series resistor limits the current in the gate bias circuitry without significantly altering the gate bias voltage.

Typically, the gate of a power FET device behaves like the Schottky diode. Under high-power conditions, the gate circuitry can detect the signal and cause the gate current to flow in the opposite direction. Therefore, depending on the RF drive level of the FET, the gate current can flow in either direction, and the gate bias generators must be able to source or sink significant amounts of both positive and negative gate current while maintaining the proper gate voltage [2, 3].

17.3 DC Bias for the Depletion Mode FETs

Each RF power amplifier module is optimized to operate at a specific drain voltage (V_d) and drain current (I_d). The gate voltage (V_g) is set to create and maintain these optimum bias conditions.

17.3.1 Gate Voltage (V_g)

The gate voltage (V_g at port P2) is usually some negative bias less than 0V. At $V_g =$ 0V, the drain to source channel is unrestricted allowing maximum drain current to flow. At $V_g \leq V_p$ where V_p is the channel pinch-off voltage, the drain channel is almost completely restricted, minimizing the drain current. The gate voltage is isolated from the RF signal path by inductor L_2, which is high impedance to the RF signal. Resistor (R_1) and bypass capacitor (C_4) minimize the low-frequency noise and interference entering the RF path and modulating the RF signal emanating from the gate voltage generator.

17.3.2   Drain Voltage (V_d) and Drain Current (I_d)

The optimum drain voltage (V_d at port P1) and optimum drain current (I_d) is usually recommended by the device manufacturer. The drain current of a power amplifier module is a relatively high value so a resistor or current limiting device in the DC drain circuit path can seriously degrade the device efficiency and therefore is rarely used. An inductor L_1 is used to isolate the RF path from the DC bias circuitry and minimize the RF path loss. A capacitor C_3 minimizes power supply noise and interference from entering the RF path and provides a short to RF signals passing from the drain through inductor L_1.

17.3.3   Gate and Drain Voltage Sequencing

To maximize the drain voltage (V_d), many times the drain current is not limited. When the gate voltage is near zero (i.e., no constriction of the drain to source channel width in the drain current can be large enough to destroy the device). An operational requirement of most depletion-mode microwave FET devices is that the gate voltage (V_g) must be set before the drain voltage is applied. This requirement is typically known as bias sequencing [4].

Sequencing is required when the device is turned on or off. A typical order of the bias sequencing shown in Figure 17.2 is as follows.

The power-on sequence is as follows:

1. At time less than T_0, the initial conditions (at time $t < 0$) are that the drain voltage is zero ($V_D = 0$ VDC), the gate voltage is zero ($V_G = 0$ VDC), and the RF signal is off.
2. At time T_0 (the start of the turn-on sequence), the drain voltage (V_D) remains at zero ($V_D = 0$ VDC), the gate voltage (V_G) goes from 0V to V_P (the pinch-off voltage of the FET ($V_G \rightarrow V_P$ VDC), and the RF signal is off.
3. At time T_1, the drain voltage (V_D) is activated, going from 0V to V_D(ON) volts, where V_D(ON) is the desired drain operating voltage. The gate voltage (V_G) remains at V_P [the pinch-off voltage of the FET ($V_G = V_P$ VDC)]. The RF signal is off.
4. At time T_2, the drain voltage (V_D) remains at V_D(ON), the desired operating voltage. The gate voltage (V_G) goes from V_P to V_G(ON), the desired gate operating voltage. The RF signal is off.
5. At time T_3, the drain voltage (V_D) remains at V_D(ON), the desired drain operating voltage. The gate voltage (V_G) remains at V_G(ON), the desired gate operating voltage. The RF signal is turned on and the module is operational.

Sequencer timing diagram

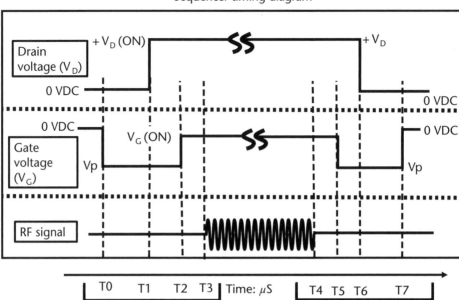

Figure 17.2 Turn-on and turn-off sequences for FET-based amplifier modules.

In the module operational mode, the power amplifier module is operational from time $t > T_3$ to time $t < T_4$.

In the power-off sequence:

1. At time T_4, the turn-off sequence begins. The drain voltage (V_D) remains at V_D(ON), the desired drain operating voltage. The gate voltage (V_G) remains at V_G(ON), the desired gate operating voltage. The RF signal is turned off.
2. At time T_5, the drain voltage (V_D) remains at V_D(ON), the desired drain operating voltage. The gate voltage (V_G) goes from V_G(ON) to V_P, the channel pinch-off voltage. The drain current is constrained. The RF signal is off.
3. At time T_6, the drain voltage (V_D) is turned off ($V_D \rightarrow 0$ VDC). The gate voltage (V_G) goes remains at V_P, the channel pinch off voltage. The RF signal is off.
4. At time T_7, the drain voltage (V_D) is turned off ($V_D \rightarrow 0$ VDC). The gate voltage (V_G) goes from V_P, to 0V ($V_G \rightarrow 0$ VDC). The RF signal is off.

Note that, on power-on and power-off sequencing, each step in the turn-on and turn-off sequence has a finite rise time (τ_R) and fall time (τ_F). All subsequent steps should be delayed in time by at four or five time constants (τ_R or τ_F as applicable) relating to the preceding step change. It is important that the drain voltage and gain voltage are stable before the signal is applied because the amplifier device may pass through unstable biasing states before it reaches its recommended quiescent operating condition.

17.4   Selection of DC Bias Capacitors

Careful selection of coupling capacitors (C_1 and C_2 in Figure 17.1) and decoupling capacitors (C_3 and C_4 in Figure 17.1) is important to achieve an optimized power amplifier design.

17.4.1   Coupling (Series) Capacitors and Decoupling (Shunt) Capacitors

Coupling capacitors are placed in the transmission path to block the DC voltage between amplifier stages or modules and pass the RF signal ideally without loss. Real capacitors have limitations that must be weighed to achieve a result close enough to the ideal coupling component such that the signal degradation is not significant.

The application of decoupling capacitors is to shunt the RF/microwave signals to ground in the DC voltage path and prevent noise and power lines spurious components in the DC power lines from entering the microwave signal path. The parameters of a real capacitor that are pertinent to the design of the SSPA are: capacitance value, capacitor resonant frequency, equivalent series resistance, and breakdown voltage. Each of these parameters is important to different aspects of the design.

17.4.2   Capacitor RF Model

Figure 17.3 is an RF chip capacitor and an RF equivalent electrical model [5] where C is the primary capacitance; ESR is the equivalent series resistance made up of two components: the lead or end-termination resistance, and the dielectric loss; L_s is the inductance of the conductor leads; and C_P is the capacitance of the chip capacitor termination pads with respect to ground. The value of capacitance C_P depends on mounting orientation (e.g., when the capacitor is vertically mounted the value of C_P decreases and the effective resonant frequency of the device increases). The calculation of resonant frequency is usually an estimate and since the effect of C_P is small it is many times neglected. R_P is the resistance of the dielectric material. This resistance is very high and usually neglected [6].

17.4.3   Capacitance Resonance and the Applicable Frequency Range

The usable frequency range of any device has an upper boundary and a lower boundary. These boundaries with respect to capacitors are very subjective and should be made in context of the entire system performance.

The upper bound of the usable frequency range of a coupling capacitor is typically up to and somewhat beyond the capacitor's resonant frequency. As the operating frequency increases and approaches the resonant frequency, the effective impedance decreases. The device impedance increases as frequency increases above resonance [7]. The usable upper limit of acceptability above resonance is determined by the system's acceptable loss.

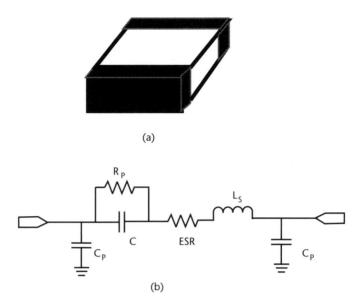

(a)

(b)

Figure 17.3 (a) The physical model of an RF chip capacitor and (b) the RF electrical model of a chip capacitor.

Note on the usability of a capacitor beyond its resonant frequency that frequency resonance is not very well defined, so there may be a risk associated with using the device up to and beyond resonance.

In the lower bound of a capacitor's usable frequency range, the low-frequency usability of a coupling capacitor is at a point where the capacitance impedance is much less than the typical 50Ω characteristic impedance such that the reflected signal and forward loss from and through the coupling capacitor is insignificant.

Notes on the capacitor model are as follows:

1. *ESR:* Dielectric loss and the skin effect make ESR a function of frequency.
2. *Capacitance changes as function of voltage across the dielectric:* The dielectric material constant decreases as a function of voltage. This causes the capacitance to decrease as a function of applied voltage. Typically, capacitor values are rated at zero applied DC voltage.
3. *Termination capacitance (C_P):* The termination capacitance has a negative second-order effect on the self-resonance of the capacitor. When the capacitor is mounted vertically (on its edge), the total capacitance with respect to ground is diminished thereby the self-resonance frequency of the device increases [8]. Typically, the self-resonant frequency is calculated assuming the termination capacitance to be insignificant.
4. *Dielectric breakdown:* The capacitor dielectric material has a breakdown voltage that when exceeded changes the insulation between the plates to a virtual short circuit. This dielectric breakdown could occur when the DC voltage plus the peak RF voltage exceeds the manufacturer's rating. This condition by itself is not destructive but excessive amount of current could cause overheating which could result in a failure of the device.

An example of capacitance change as a function of applied voltage [9] is:

Capacitance: 1,000 pF at 0 VDC;

Dielectric material: X7R;

Rated voltage: 100V;

Capacitance at 50V (50% of rated voltage): 910 pF at 50 VDC.

An example of capacitance impedance and transmission loss as a function of frequency before and through resonance is that the parallel capacitance C_P and parallel resistance R_P are assumed to be negligible and therefore omitted from this example [10]. Figure 17.4 is a model of the 1,000-pF capacitor analyzed as a series coupling capacitor and as a shunt decoupling capacitor. The capacitor characteristics are:

Capacitance: 1,000 pF at 0V;

Chip size: 0603 (60 mils × 30 mils);

Dielectric material: X7R;

Rated voltage: 100 VDC;

ESR: 0.272Ω;

LS: 0.5 nH (nanohenry).

Figure 17.5 is a graph of the 1,000-pF capacitor used as a series coupling element. The resonant frequency is approximately 210 MHz but the usable frequency without a significant amount of added loss is much higher than the resonant frequency.

Figure 17.6 is a graph of the magnitude of the impedance of the 1,000-pF capacitor used as a shunt decoupling element. The resonant frequency is similar to the series coupling model, but the usable frequency for series and shunt devices is not necessarily the same because the criteria of shunting an RF signal is not the same as a passing an RF signal.

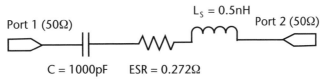

1000pF Chip Capacitor used as series coupling device

(a)

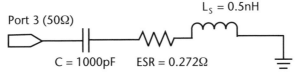

1000pF Chip Capacitor used as shunt decoupling device

(b)

Figure 17.4   Simulation model of a 1,000-pF chip capacitor: (a) used in series and (b) used in shunt.

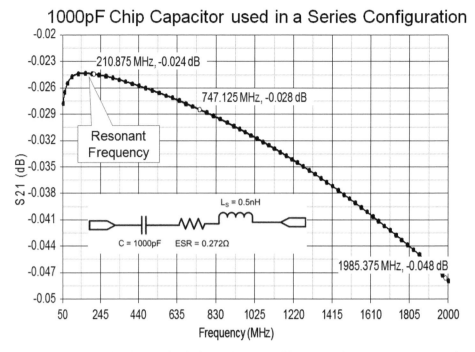

Figure 17.5 Magnitude of S_{21} in decibels as a function of frequency for a typical 1,000-pF chip capacitor used in series.

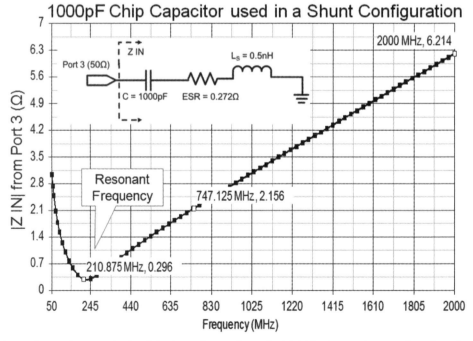

Figure 17.6 The magnitude of the impedance of a typical 1,000-pF capacitor as a function of frequency.

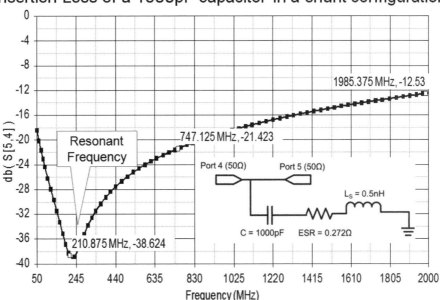

Figure 17.7   The magnitude of the loss (S[5,4] in dB) as a function of frequency when the input (Port 4) and output (Port 5) impedances are 50Ω.

Figure 17.7 shows the available loss with respect to a 50Ω transmission line as a function of frequency when the capacitor is used to shunt the RF signal (e.g., when used in isolated DC power lines). Maximum loss occurs at the resonant frequency. In this mode, the usable frequency beyond resonance is limited because the required loss decreases rapidly. The usable decoupling range can be extended by using small capacitors with higher resonant frequencies in parallel with the 1,000-pF capacitor shown. Increase loss (signal isolation) can also be improved by using a series inductance on either side of the shunt capacitance.

17.5   Selection of DC Bias Inductors

Inductors in the bias circuit isolate the drain power supply (L_1) and the gate power supply (L_2) from the RF signal path, as shown in Figure 17.1. The RF functionality in the two inductors is similar (i.e., they present a high enough impedance such that the power amplifier input and output signals are unaffected by the input and output DC bias circuit). The inductor on the FET drain (at the amplifier output) in addition to providing RF isolation must pass a significant amount drain bias current.

The parameters of real inductors that are pertinent to the design of the SSPA are: inductance value, inductor self-resonant frequency, ESR, and maximum current rating. Each of these parameters is important to different aspects of the design (i.e., DC bias paths and RF signal paths).

17.5.1   Inductor RF Model

Figure 17.8 is an RF inductor model showing the primary parasitic effects [11]. L is the primary inductance, and ESR is the equivalent series resistance made up of two components: the resistance of the coil windings, and the skin effect of the inductor wire, which decreases the wire penetration as frequency increases. This effectively decreases the area of the inductor, increasing the ESR. C_P is the parallel capacitance due to the proximity of the coil windings.

17.5.2   Inductor Resonance and the Applicable Frequency Range

The usable upper-frequency range of an inductor in the DC bias path is the typically up to and a little beyond the inductor's resonant frequency. As the operating frequency approaches the inductors resonant frequency, the effective impedance increases, becomes a maximum at the resonant frequency, and decreases after resonance [12].

The low-frequency and the upper-frequency usability of an inductor is at a point where the inductance impedance is a significant load on a 50Ω transmission line such that the transmission loss is noticeably degraded. In the example of the inductance impedance and transmission loss as a function of frequency through resonance, the inductance is a 47-nH coil analyzed as a shunt decoupling device with the following primary and secondary parasitic parameters:

Inductance (L): 47 nH;

DCR: 0.052Ω;

Self-resonance frequency (F_{SR}): 2,250 MHz;

Current IRMS: 1.7A.

For this example, the skin effect of the coil wires is neglected, that is, the DC resistance (DCR) is assumed approximately equal to the ESR and the inductor capacitance to ground is neglected.

The self-resonant frequency (F_{SR}) is given by (17.1) and (17.2):

$$F_{SR} = \frac{1}{\sqrt{L \cdot C_P}} \qquad (17.1)$$

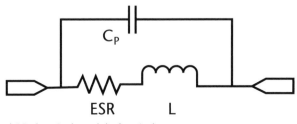

Figure 17.8   Typical RF electrical model of an inductor.

Solving for C_P:

$$C_P = F_{SR}^2 \cdot LC_P = 0.106 \text{ pF} \qquad (17.2)$$

Figure 17.9 is a graph of the magnitude of the impedance of the 47-nH inductor used as a shunt decoupling element.

Figure 17.10 shows the loss with respect to a 50Ω transmission line as a function of frequency. Minimum loss occurs at the resonant frequency. In this mode, the usable frequency beyond resonance is limited because the loss tends rapidly increase. The upper usable frequency range can be extended by using smaller inductors with higher resonant frequencies in series with the 47-nH coil shown. Note that for multiple series inductors it is normally a good practice to put the lowest inductance coil closest to the signal path and smaller inductors tend to have lower parasitic effects.

17.5.3   Rated Current of an Inductor

Inductors used to couple DC bias to the output of a power amplifier module can be required to carry a large amount of RMS or peak current. The rated current of an inductor is generally a function of the RMS current, which produces an acceptable temperature rise in the device [13]. The temperature rise is usually related to the published maximum current through the inductor and the maximum operating ambient temperature. An example of a rated current in an inductor is the typical manufacturers' specifications pertinent to the rated current of an inductor, given in Table 17.1.

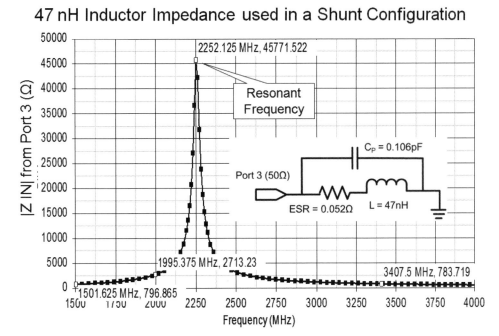

Figure 17.9   The magnitude of the impedance of the 47-nH inductor as a function of frequency.

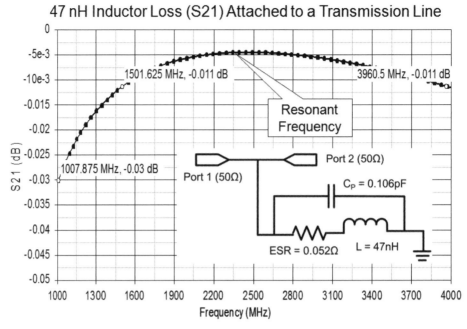

Figure 17.10 The magnitude of the loss S_{21} in decibels as a function of frequency when the input (Port 1) and output (Port 2) impedances are 50Ω.

Items in rows 1 through 4 in Table 17.1 are usually given by the inductor manufacturer. Row 4 is the RMS current that will produce the differential temperature rise from the maximum ambient temperature, row 1, to the maximum part temperature, row 2. The differential temperature rise delta temperature, row 5, is 20°C in this example. From this information, the thermal resistance (Rth) can be calculated:

$$Rth = \frac{\Delta T}{RDC \cdot I_{RMS}^2} \quad (°C/watt) \tag{17.3}$$

where the DC power dissipated in the inductor ($P_{INDUCTOR}$) is:

$$P_{INDUCTOR} = RDC \cdot I_{RMS}^2 \tag{17.4}$$

Table 17.1 Typical DC Characteristics of an Inductor

Row	Description	Value	Units
1	Maximum ambient temperature:	125	°C
2	Maximum part temperature:	145	°C
3	DC persistence (RDC)	0.0113	Ω
4	I_{RMS}	4.4	Amps
5	Delta temperature (ΔT)	20	°C
6	Thermal resistance (Rth)	91.42	°C/W

Note that for inductor DC ratings the maximum operating temperature is a related to the inductor construction and not the temperature rating of the copper wire windings, which is usually much higher, and inductor thermal resistance is a function of the body style, insulation coating, and thermal flow paths. This parameter is often not stated by the device manufacturer.

17.6   Quarter-Wave Transmission Lines to Bias Power Amplifier Modules

Quarter-wave transmission lines with an RF short at one end theoretically presents an open circuit at the other end [14]. With the open circuit end connected to a signal transmission line and the other end RF shorted at a DC drain or gate supply voltage, the quarter-wave transmission line can be used effectively with a minimum of loss to the RF signal. The quarter-wave line also provides some distance between the RF and DC circuitry, which can be very useful in helping to isolate the two functions. Figure 17.11 is a typical circuit using quarter-wave transmission lines to separate the RF signal path from the DC power supplies driving the gate and the drain.

Figure 17.12 shows the line impedance as seen by a 50Ω transmission line passing the RF signal. The impedance of the quarter-wave transmission line providing DC bias voltages is designed to resonate at approximately 1,521 MHz.

Figure 17.13 shows the added insertion loss of a 50Ω transmission line when a 100Ω shorted quarter-wave line is connected perpendicular to the signal path. As shown the quarter-wave transmission line is bandwidth-limited to about 30% of the centered frequency for 0.1-dB loss. The loss is more significant at the module output (drain) because it directly degrades the output power decibel for decibel.

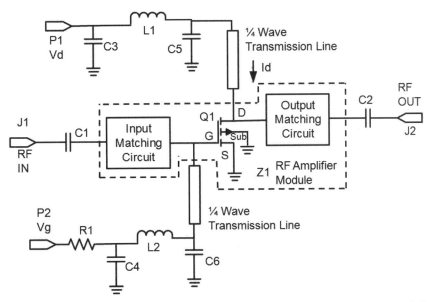

Figure 17.11 Typical RF/microwave power amplifier module using quarter-wave transmission lines to isolate the DC bias from the RF signal path.

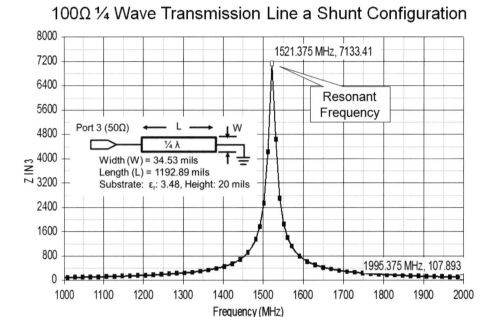

Figure 17.12   The magnitude of the impedance of the quarter-wave transmission line as a function of frequency.

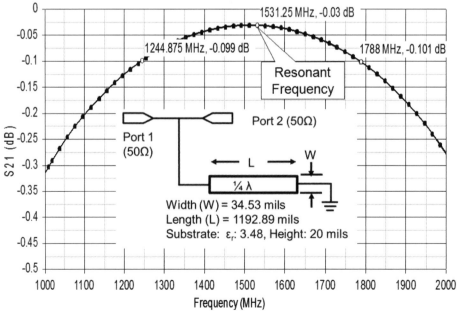

Figure 17.13   The magnitude of the quarter-wave transmission line loss (S_{21}) in decibels as a function of frequency when the input (Port 1) and output (Port 2) impedances are 50Ω.

Higher-impedance shunt quarter-wave transmission line (>100Ω) will result in a lower loss but because the line width is smaller, they tend to have lower current-carrying capability.

Note that on quarter-wave transmission lines used for drain and gate biasing, as shown in Figure 17.10, the quarter-wave transmission line in the DC bias path is usually an L-C filter network. The additional L-C filtering is usually necessary to improve the RF signal to DC isolation in both direction (i.e., RF signal entering the DC power path and spurious signal on the DC power path interfering with the RF signals). More complex microwave isolation circuits can be employed to widen the bandwidth of the simple single quarter-wave transmission line [15]. A small resistor in series with the gate bias quarter-wave transmission line between the bias line and the signal path often improves the isolation over a wide frequency band [15].

References

[1]   Lidow, A., et al., *GaN Transistors for Efficient Power Conversion*, New York: Wiley, 2014.

[2]   Ampleon, The Netherlands B.V. "Bias Module for 50 V GaN Demonstration Boards," Application Note AN11130, September 2015.

[3]   Component Distributors Inc., "CDI GaN Bias Board User's Guide," Application Note CDIAN003, February 2015.

[4]   Hittite Microwave Corporation, "Design and Implementation of a Bias Sequencing Circuit for the HMC463LP5 Low Noise Amplifier," 2004.

[5]   Niknejad, A., and J. Dunsmore, "High Frequency Passive Components," Laboratory Manual, University of California Berkeley, 2013.

[6]   Blewett, M., "Capacitor Self-Resonance" Educator's Corner Experiments, Agilent Technologies.

[7]   Fiore, R., "Capacitors in Coupling and DC Blocking Applications," American Technical Ceramics Corp., Circuit Designers' Notebook, January 2005.

[8]   Busse, M., "Capacitors for RF Applications," Dielectric Laboratories, Inc.

[9]   AVX Corp, "SpiCap 3.0," 2003.

[10] Cain, J., "Parasitic Inductance of Multilayer Ceramic Capacitors," AVX Corporation.

[11] O'Hara, M., "Modeling Non-Ideal Inductors in SPICE," Newport Components, November 1993.

[12] Coilcraft, "Key Parameters for Selecting RF Inductors," Document 671-2 Revised, February 2015.

[13] Coilcraft, "How Current and Power Relates to Losses and Temperature Rise," Document 1055-1 Revised, September 2012.

[14] Gabriela, G., and G. Gonzalez, "Measurements for Modelling of Wideband Nonlinear Power Amplifiers for Wireless Communications," Helsinki University of Technology, September 2004.

[15] Basset, R., "High-Power GaAs FET Device Bias Considerations," Fujitsu Compound Semiconductor, Inc., Application Note Number 010.

CHAPTER 18

SSPA DC Voltage and Current

18.1 Introduction

The typical SSPA is made up of a series of amplifier modules, each handling successively higher-power levels from the input to the output. As the individual-module RF output power requirements increase, the required DC power increases with each module designed to be most efficient at a specific DC voltage and DC. Since efficiency is a key factor in the SSPA design, forcing the system to use a common voltage is usually not optimal.

Efficiently satisfying the various DC power requirements is a challenge that, when met, can significantly improve the SSPA design.

Note that, on the drain voltages shown in Figure 18.1, the power amplifiers shown in Figure 18.1 utilize GaN semiconductor FETs, which require higher drain voltages than amplifiers designed around GaAs semiconductor FETs.

18.2 DC Power Requirements for a Typical SSPA

A typical SSPA chain of devices and the DC voltage requirements of each module are shown in Figure 18.1. The different voltage, current, and bias condition with

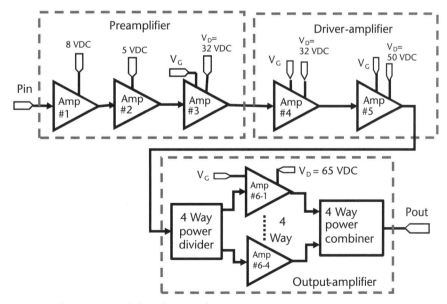

Figure 18.1 The active modules of a typical SSPA.

285

successively increasing DC power as the signal level increases from the input to the output is not uncommon. If the modules in the power amplifier chain are FET-based, they typically have a specified drain voltage and a range of gate voltages, where the gate voltage is tuned to obtain a specified optimum drain current. Since the amplifier modules usually operate in a class AB mode (for a quasilinear operation) or a class C mode (for saturated operation) [1], there are usually two currents specified: a quiescent current (no signal) and an operational current (the current when the signal present at its maximum rated power).

18.2.1   DC in a Linear (Class AB) Mode SSPA

Amplifiers operating in a quasilinear (class AB) [2] have a smaller difference in quiescent current and operational current than amplifiers operating in a class C mode. The difference is related to the class AB bias conditions and the respective amount of acceptable operational distortion. Most linear (class AB) SSPAs have a continuous signal presence and are operating at their maximum rated power. Under these conditions, the primary DC power and thermal design factors for the same peak power level of saturated amplifiers are more of a design challenge.

18.2.2   DC in a Saturated Mode SSPA

Saturated-mode amplifiers are usually biased in the class C mode. The no signal quiescent current of these amplifiers is usually much less than the operational (signal present) current. Most saturated-mode SSPAs operate on pulse modulated microwave signals with a low duty cycle resulting in a low average current and therefore a low average power dissipated in the device.

18.2.3   Power Requirements of Depletion-Mode FET

Many high-power MMIC modules use FETs as their primary output power element. The FET voltage sequencing requirements and the different current drawn during the no signal condition and the maximum output power condition require that the DC voltage and currents be analyzed under the various modes of operation

18.2.3.1   The Depletion-Mode FET Drain Current

The drain current in a depletion-mode FET used in a power amplifier has three primary modes of concern: pinch-off mode, quiescent current mode, and peak power output current mode. During each of these modes, the drain voltage is constant at its rated value.

1. *Pinch-off mode:* Pinch-off occurs when the gate voltage constricts the flow of drain current. The drain current during this mode is small and often neglected (considered to be zero).
2. *Quiescent current mode:* The quiescent current is the value set by the gate bias voltage under no signal conditions.

3. *Peak power output current mode:* The peak power output current is the value that is expected when the signal power is at its rated level for the respective device. The gate voltage can be adjusted to obtain the required output signal power. This adjustment may cause a change in the drain current. Note that, on the peak drain current, the peak drain current at the rated output power is an approximate value that varies under various environmental conditions [3].

4. *Drain current temperature compensation:* The gate voltage is often varied with variations in ambient temperature to maintain the drain current and the signal at the rated output power.

18.2.3.2   The Depletion-Mode FET Gate Requirements

The gate voltage of the depletion-mode FET used in a power amplifier has three modes of primary concern: pinch-off voltage, zero volts mode (maximum drain current), and operational gate voltage. The gate voltage is actually the gate to substrate voltage (or gate to source voltage when the source and the substrate are shorted) when the substrate is grounded.

1. *Pinch-off voltage mode:* The pinch-off voltage is typically a negative voltage gate to substrate (or gate to source when the substrate and gate are shorted) that constricts the drain-to-source channel such that the drain current is small and often considered zero. For optimum FET device sequencing, the drain voltage is not applied until the gate voltage is set to pinch-off.

2. *Zero volts mode:* When the gate voltage is zero, the drain-to-source channel is unrestricted and the drain current can reach a high enough level to destroy the device. Zero gate voltage is an initial condition that should only exist when the drain voltage is zero and there is no signal flowing into the devices.

3. *Operational gate voltage mode:* The operational gate voltage is the initial set level that allows the drain current to reach its specified quiescent value. It is sometimes changed such that the output signal reaches its rated output power level. The operational gate voltage is often varied as a function of ambient temperature to maintain the optimum drain current and output signal level under a wide variety of operating temperatures.

Notes that, on gate voltage, there are devices configured with the substrates not grounded [4]. If the substrate is tied to a high voltage typically related to the drain voltage, the gate voltage will be at a high positive voltage, but negative with respect to the substrate voltage. The gate current can flow in either direction due to signal detection at the gate junction [5]. The gate voltage must be designed to sink or source the gate current.

18.3   Example: Power Requirements of Each Module in the SSPA Chain

The amplifier chain in Figure 18.1 shows the active elements of a typical SSPA. In the example shown, the SSPA is divided into three functional sections: the preamplifier, the driver amplifier, and the output amplifier. The output amplifier is four theoretically identical power amplifier modules in parallel.

Functionally, the SSPA in this example is operating in the saturation mode with an input pulse signal having a duty cycle of approximately 1% (1% on and 99% off). In this mode, the amplifier modules that are operating in a saturated mode have a significant difference in the quiescent current and the operational current.

18.3.1 DC Power Requirements of the Preamplifier

The preamplifier in this example is designed to have three amplifier modules increasing the signal level from around 1 mW (P_{in}) to 10W (P_{out}) (see Figure 18.2).

18.3.1.1 Preamplifier Bipolar Amplifiers (Amp #1 and Amp #2)

Amplifiers Amp #1 and Amp #2 are bipolar devices operating in the linear mode (class A) with a DC power of +8 VDC at 65 mA and +5 VDC at 147 mA, respectively, listed in Table 18.1. The last column is the DC power consumed by each amplifier.

18.3.1.2 Preamplifier Output FET Amplifiers (Amp #3)

The output of Amp #2 feeds a class C depletion-mode FET amplifier module. The specified drain voltage (V_D) is 32 VDC with a no signal quiescent drain current (I_{DQ}) of 50 mA and a peak drain current (I_{DP}) of 557 mA (signal present at the rated output power approximately 10W). The gate voltage is negative with a maximum value of 0V (the drain current has a potential to be at its maximum when $V_G = 0V$)

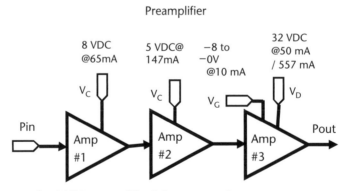

Figure 18.2 Example of SSPA preamplifier DC power requirements.

Table 18.1 DC Voltage and Estimated Currents for Bipolar Devices Amp #1 and Amp #2 at the Input of the Preamplifier Chain

Amplifier Device No.	Device Type	Output Voltage (V)	Output Current (mA)	DC Power (W)
1	Bipolar	8	65	0.52
2	Bipolar	5	147	0.74

and a minimum value of −8VDC (drain-to-source channel at or near pinch-off, that is, minimum drain current). The no-signal quiescent drain current (50 mA) is set by adjusting the gate voltage (typically in the −2 VDC to −4 VDC range). The peak drain current at the rated output power (557 mA) is an approximate value set by the gate voltage for the expected signal output power level. Table 18.2 is a list of the DC voltages and currents for Amp #3.

18.3.1.3   Preamplifier Output Power and Efficiency

Output power and efficiency are key parameters in any SSPA design. Although the preamplifier is a minor contributor to the overall SSPA, the DC consumed power is important with respect to its voltage regulator design, mechanical design, thermal design, and reliability.

Table 18.3 lists the pertinent power and efficiency characteristics of the pre-amplifier. The information listed comes from a description of the SSPA example, Table 18.1, and Table 18.2.

- *Row 1, Peak RF output power (P_{RF}):* The peak RF output power (P_{RF}) is 10W listed earlier. The peak output power is that of a CW signal or the peak power of a pulsed RF signal.
- *Row 2, Peak DC power (P_{DC}):* Peak DC power (P_{DC}) is the peak DC power in Table 18.2 column 8 plus the DC power for Amp #1 plus the DC power for Amp #2 both in Table 18.1.

Table 18.2   DC Voltage and Estimated Currents for Amp #3 in the Preamplifier Chain

1 Amp No.	2 V_D (V)	3 I_{DQ} (mA)	4 I_{DP} (mA)	5 V_G Max. (V)	6 V_G Min. (V)	7 I_G Max. (mA)	8 ≈ Peak DC Power (P_D) (W)
3	32	50	557	0	−8	10	17.82

Column 1: amplifier number, column 2: nominal drain voltage, column 3: quiescent drain current (no RF signal), column 4: peak drain current (signal output is at rated RF power), column 5: maximum gate voltage (there is no drain-to-source channel constriction), column 6: minimum gate voltage (the drain-to-source channel is constricted such that there is only a small drain to source leakage current), column 7: maximum gate current (due to signal detection at the gate this current could be sinking or sourcing), and column 8: the approximate calculated DC peak power ($P_D = V_D \cdot I_{DP}$)

Table 18.3   Preamplifier Power and Efficiency Calculations

1	Peak RF Output Power (P_{RF})	10W
2	Peak DC power (P_{DC})	19.08W
3	Quiescent DC Power (P_{QDC})	2.86W
4	CW and Peak Power Efficiency (Eff_{CW})	52.4%
5	Pulse Duty Factor (DF)	1.0%
6	Pulsed Mode Power Dissipated ($DCPower_{Pulse}$)	3.017W

- *Row 3, Quiescent DC power (P_{QDC}):* The quiescent DC power (P_{QDC}) is I_{DQ} (column 3) times V_D (column 2) in Table 18.2.

$$P_{QDC} = I_{DQ} \cdot V_D \tag{18.1}$$

- *Row 4, CW and peak power efficiency (Eff_{CW}):* The CW and peak power efficiency (Eff_{CW}) is the peak RF output power (row 1) divided by the peak DC power (row 2), both in Table 18.3.

$$Eff_{CW} = \frac{P_{RF}}{P_{DC}} \tag{18.2}$$

- *Row 5, Pulse duty factor (DF):* The pulse duty cycle is given in (18.3).
- *Row 6, Pulsed mode power dissipated ($DCPower_{Pulse}$):* The pulsed mode power dissipated is the peak DC power (row 2) times the pulse duty factor (row 5) plus one minus the duty factor times the quiescent DC power.

$$DCPower_{Pulse} = P_{DC} \cdot DF + P_{QDC} \cdot (1 - DF) \tag{18.3}$$

The DC power dissipated in the CW mode (19.08 W) is significantly greater than the power dissipated in the pulse mode (3.017 W). Operating in the CW mode instead of the pulsed mode could have a profound effect on the mechanical and thermal design of the preamplifier.

18.3.2 DC Power Requirements of the Driver Amplifier

The driver amplifier in this example has two amplifier module stages increasing the signal level from around 10 W to 280 W. Both amplifiers (Amp #4 and Amp #5) are class C depletion-mode FET amplifier modules operating in the saturated mode (see Figure 18.3).

Amplifier number 4 (Amp #4) has a specified drain voltage of 32 VDC with a quiescent drain current (no signal) of 70 mA and a peak drain current of 1.5A.

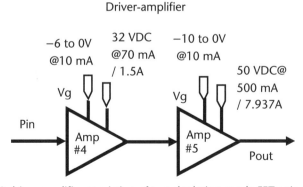

Figure 18.3 SSPA driver amplifier consisting of two depletion-mode FET gain stages.

The peak drain current occurs when signal is present at the rated output power of approximately 25 W. The gate voltage is negative with a maximum value of 0 V (maximum potential drain current) and a minimum value of −6 VDC (drain to source channel at or near pinch off, that is, minimum drain current). The no-signal operational drain current (70 mA) is set by adjusting the gate voltage, which for this device is in the approximate range of −2.5 VDC. The peak drain current at the rated output power (1.5A) is also approximate value [3].

Amplifier number 5 (Amp #5) has a specified drain voltage of 50VDC with a quiescent drain current (no signal) of 70 mA and a peak drain current of 7.937A (signal present at the rated output power approximately 280W). The gate voltage is generally negative with a maximum value of 0V (maximum potential drain current) and a minimum value of −10VDC (drain to source channel at or near pinch-off, that is, minimum drain current). The gate voltage must be designed to sink or source 10 mA of current. Table 18.4 is a list of the DC voltages and currents for Amp #4 and Amp #5.

18.3.2.1   Driver-Amplifier Output Power and Efficiency

The driver amplifier is a significant contributor to the overall SSPA DC power consumed. This assembly is critical, supplying sufficient RF power to drive the output stage into saturation. The power consumed in the driver stage can have a significant impact on the overall SSPA efficiency. The design requires an efficient voltage regulator design, an electrically stable mechanical design, and a reliable thermal design.

Table 18.4 DC Voltage and Estimated Currents for Amp #4 and Amp #5 in the Driver-Amplifier Chain

1 Amp Device No.	2 V_D (V)	3 I_{DQ} (mA)	4 I_{DP} (mA)	5 V_G Max. (V)	6 V_G Min. (V)	7 I_G Max. (mA)	8 $P_D \approx$ Peak DC Power (W)
4	32	70	1,500	0	−6	10	48.00
5	50	500	7,937	0	−10	10	396.85

Column 1: amplifier number, column 2: nominal drain voltage, column 3: quiescent drain current (no RF signal), column 4: peak drain current (signal output is at rated RF power), column 5: maximum gate voltage (there is no drain-to-source channel constriction), column 6: minimum gate voltage (the drain-to-source channel is constricted such that there is only a small drain to source leakage current), column 7: maximum gate current (due to signal detection at the gate, this current could be sinking or sourcing), and column 8: the approximate calculated DC peak power ($P_D = V_D \cdot I_{DP}$).

Table 18.5 Driver-Amplifier Power and Efficiency Calculations

1	Peak RF output power	280W
2	Peak DC power	444.85W
3	Quiescent DC power	27.24W
4	CW and peak power efficiency	62.9%
5	Pulse duty factor	1.0%
6	Pulsed mode power dissipated	31.4161W

Table 18.5 listed the pertinent power and efficiency characteristics of the driver-amplifier. The information listed comes from a description of the SSPA and Table 18.4.

- *Row 1, Peak RF output power (P_{RF}):* The peak RF output power (P_{RF}) is 280W.
- *Row 2, Peak DC power (P_{DC}):* Peak DC power when the driver amplifier is delivering 280W.
- *Row 3, Quiescent DC power (P_{QDC}):* The quiescent DC power (P_{QDC}) is the sum total quiescent power [I_{DQ} (column 3) times V_D (column 2) in Table 18.4] for each amplifier (Amp #4 and Amp #5).
- *Row 4, CW and peak power efficiency (Eff_{CW}):* The CW and peak power efficiency (Eff_{CW}) is the peak RF output power (row 1) divided by the peak DC power (row 2), both in Table 18.5.
- *Row 5, Pulse Duty Factor (DF):* The pulse duty factor is given in (18.3).
- *Row 6, Pulsed mode power dissipated ($DCPower_{Pulse}$):* The pulsed mode power dissipated is the peak DC power (row 2) times the pulse duty factor (row 5) plus one minus the duty factor times the quiescent DC power.

The DC power dissipated in the CW mode (444.85W) is significantly greater than the power dissipated in the pulse mode (31.4161W). Operating the driver amplifier in the CW mode will have a profound effect on the mechanical and thermal design.

18.3.3   DC Power Requirements of the Output Parallel Amplifier

The output amplifier in this example has four ideally identical single amplifier modules in parallel increasing the signal level from around 280W to approximately 4 kW. The four parallel amplifiers (Amp #6-1 through Amp #6-4) in Figure 18.4(a) are class C depletion-mode FET amplifier modules operating in a saturated mode,

Figure 18.4(b) shows the DC power requirements for each of the four parallel amplifiers. Each amplifier requires a drain voltage of 65 VDC with a quiescent drain current (no signal) of 130 mA (I_{DQ}) and a peak drain current of 27.3A (signal present at the rated output power approximately 1,000W). The gate voltage is negative with a maximum value of 0V (maximum potential drain current) and a minimum value of −8 VDC (drain-to-source channel at or near pinch-off, that is, minimum drain current). The no-signal operational drain current (130 mA) is set by adjusting the gate voltage, which for this device is in the approximate range of −2.8 VDC. The peak drain current at the rated output power (27.3A) is also the approximate value.

Each of the four parallel amplifiers are theoretically identical with the output powers from each adding up to 4 kW minus the loss in the output combiner and the channel mismatch tracking loss. Table 18.6 listed the pertinent power and efficiency characteristics of the output amplifier.

The following rows appear in Table 18.6:

- *Row 1, Output combiner loss (P_{LOSS}):* Loss of the output four-way combiner plus the loss due to the channel gain and phase mismatch.

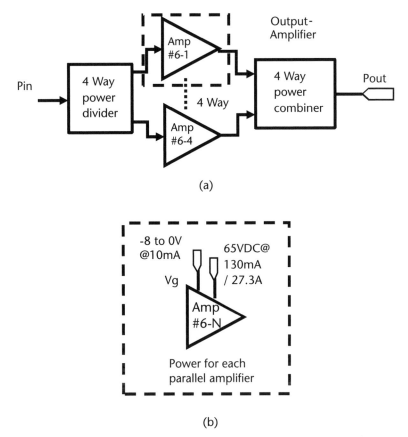

Figure 18.4   (a) SSPA output amplifier consisting of four parallel depletion-mode FET gain stages and (b) the voltage and current of each stage.

- *Row 2, Number of output parallel amplifiers (N):* There are two columns listing the results of a single amplifier and the four parallel amplifiers through the four-way combiner. The single amplifier results ($N = 1$) does not consider the combiner loss (row 1). The four parallel amplifier results ($N = 4$) considers the combiner loss (row 1).

Table 18.6   Output Parallel Amplifier Power and Efficiency Calculations

		dB	Linear
1	Output combiner loss (P_{LOSS})	−0.9	0.813
2	Number of output parallel amplifiers (N)	1	4
3	Peak RF output power (P_{RF})	1,000	3,251W
4	Peak DC power (P_{DC})	1,774.50	7,098W
5	Quiescent DC power (P_{QDC})	8.45	33.8W
6	CW and peak power efficiency (Eff_{CW})	56.4%	45.8%
7	Pulse duty factor (DF)	1.0%	1.0%
8	Pulsed mode power dissipated ($DCPower_{Pulse}$)	26.1105	104.442W

- *Row 3, Peak RF output power (P_{RF}):* The peak RF output power (P_{RF}) is 1,000W for each amplifier but the combined RF output power is 3,251W after accounting for the combiner and mismatch loss (row 1).
- *Row 4, Peak DC power (P_{DC}):* Peak DC power when each output amplifier is delivering 1,000W. The DC power for the combined parallel amplifiers is four times the single amplifier.
- *Row 5, Quiescent DC power (P_{QDC}):* The quiescent DC power (P_{QDC}) is I_{DQ} times (V_D) for each amplifier and multiplied by 4 when considering the total of the four parallel amplifiers.
- *Row 6, CW and peak power efficiency (Eff_{CW}):* The CW and peak power efficiency (Eff_{CW}) is the peak RF output power (row 3) divided by the peak DC power (row 4), The efficiency of the single power amplifier is better than the efficiency of the four parallel combined amplifiers because the combiner loss (row 1) is considered in combined amplifier calculation.
- *Row 7, Pulse duty factor (DF):* The pulse duty cycle is given in Section 18.3.
- *Row 8, Pulsed mode power dissipated ($DCPower_{Pulse}$):* The pulsed mode power dissipated is the peak DC power (row 4) times the pulse duty cycle (row 7) plus one minus the duty factor times the quiescent DC power (P_{QDC}) given in row 5.

The DC power dissipated in the CW mode (7,098W) is significantly greater than the power dissipated in the pulse mode (104.442W). Handling the power dissipated in the CW mode requires a major mechanical and thermal design effort.

18.3.4   DC Power Requirements and Efficiency of the SSPA (Preamplifier, Drive Amplifier, and Output Parallel Amplifier Combined)

The SSPA separated into three functional segments: the preamplifier (3 stages), the driver amplifier (2 stages), and output amplifier (4 parallel amplifiers); these are shown in Figure 18.5. Each amplifier stage is labeled with its specified voltage and current.

The total power and efficiency of the SSPA in this example is shown in Table 18.7. The total DC power dissipated is the sum of the power dissipated from each segment of the design, the preamplifier, the drive amplifier, and the output parallel amplifier. The RF output power is the combined output power of the last stage (output parallel amplifier).

Each of the segments (preamplifier, drive amplifier, and output parallel amplifier) that make up the SSPA chain contribute to the available RF power and consumed DC power and affect the amplifier efficiency. As shown, the output stage is the primary contributor to the results, but a nonoptimum design could change the secondary contributions and significantly degrade the systems performance.

The CW and peak power efficiency (Eff_{CW}) (row 4) of the entire SSPA in the CW mode is the peak RF output power (P_{RF}) (row 1) divided by the total peak DC power (P_{DC}) (row 2). The efficiency is 43.0%. Under pulse conditions with a pulse duty factor (DF) of 1% (row 5), the pulsed mode power dissipated ($DCPower_{Pulse}$) (row 6) is 138.88W, with almost half of that power, the quiescent DC power (P_{QDC}) (row 3), 63.9W, consumed during the idle state of the amplifier (100% − DF = 99%).

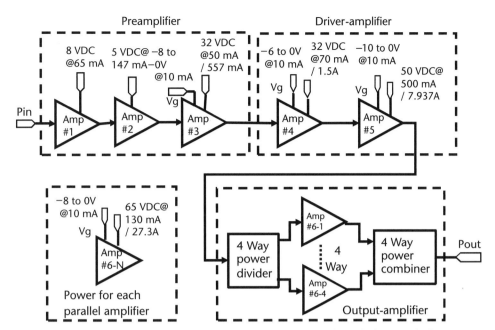

Figure 18.5   Three segments of the SSPA amplifier in this example with the specified respective voltages and currents.

Table 18.7   The Final Expected RF Output Power and Respective DC Power Consumed for the CW and Pulse Modes

	Total SSPA Power	
1	Peak RF output power (P_{RF})	3,251W
2	Peak DC power (P_{DC})	7,561.93W
3	Quiescent DC power (P_{QDC})	63.90W
4	CW and peak power efficiency (Eff_{CW})	43.0%
5	Pulse duty cycle (DF)	1.0%
6	Pulsed mode power dissipated ($DCPower_{Pulse}$)	138.88W

This suggests that there will still be a significant amount of power consumed even if the signal duty factor were lowered.

18.4 Meeting the Drain Current Requirements for a Pulsed Saturated Mode SSPA

In the example given in Section 18.3, the drain current for each output amplifier stage was 27.3A. This substantial amount of current, when required by a low duty factor pulsed signal driving a saturated amplifier, does not have to be derived directly from the DC voltage source. The peak current required by a pulse mode saturated power amplifier can be stored in a reservoir capacitor close to the FET drain. The reservoir capacitor is charged during the off time between pulses and discharge during the RF pulse width. The required charging current between RF pulses is a small fraction of the discharge current. The reservoir capacitor must hold enough charge such that the drain voltage is not significantly lowered during the peak drain current discharge time (i.e., the pulse width of the RF signal).

18.4.1 A Drain Voltage and Current Charging and Discharging Model

The simulation model of the reservoir capacitor charging and discharging function is shown in Figure 18.6.

The major functions of the model are:

1. *Drain voltage regulator:* The source voltage (Vs) of the drain voltage regulator has a short-circuit current $Is = Vs/Rs$, where Rs is the equivalent output impedance under short-circuit conditions.
2. *Reservoir capacitor equivalent circuit:* The reservoir capacitor equivalent circuit is a simplified model of the primary capacitor Cc. The capacitor's equivalent series resistance is resistance ESR. In this model, the series inductance of the capacitor is not shown because it usually can be neglected without significantly affecting the results provided that the capacitor is located close to the FET drain.
3. *FET drain and gate circuit:* The FET drain and gate circuit is modeled as a drain resistance RD being switched on and off. The switch is shorted when the RF gate pulse is active and open when the RF gate pulse is not present. The source S and the substrate Sub are grounded.

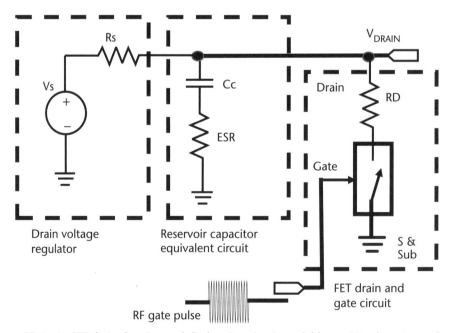

Figure 18.6 An FET drain charging and discharging circuit model for an RF pulse-saturated high-power amplifier.

18.4.2   Charging and Discharging the Reservoir Capacitor (Cc)

When the RF gate pulse is not active and the drain switch is open, the capacitor (Cc) charges to approximately *Vs* through time constant τ_C:

$$\tau_C \approx (Rs + ESR) \cdot Cc \qquad (18.4)$$

After the initial power turn-on, which is typically a much longer time than the pulse repetition interval, the capacitor recharges between RF gate pulses.

18.4.2.1   Capacitor (Cc) Discharging

When the RF gate pulse is present the drain switch closes and the capacitor (Cc) discharges through time constant τ_D for the time interval of the pulse width

$$\tau_D \approx \left(\frac{Rs \cdot RD}{Rs + RD} + ESR \right) \cdot Cc \qquad (18.5)$$

While the capacitor is discharging through the drain, there is a relatively small current (less than the reservoir discharge current) from the DC regulator continuing to charge the capacitor (Cc). The drain resistance (RD) is small so the voltage regulator may sense a short-circuit and go into a constant current mode. The short-circuit current (*Is*) is:

$$Is = \frac{Vs}{Rs} \qquad (18.6)$$

Caution: Some voltage regulators go into a fold-back, current-limiting mode when they sense a short-circuit on the output [6]. These types of regulators may hang up and take a long time to return to their normal output voltage (*Vs*). Regulator that do not return to their normal output voltage within a short time (<< PRI) should not be used in this function.

The reservoir capacitor (Cc) should be large enough such that the drain voltage discharge will be small such that the RF signal output power remains relatively constant through the time of the RF pulse width.

18.5   Example of Changes in Drain Voltage When the Reservoir Capacitor Charges and Discharges

Figure 18.7 is a model used to simulate a 1,000-W high-power amplifier module driven from a 65 VDC regulated voltage source (*Vs*) with a 32.5Ω source impedance (*Rs*). The short-circuit current (*Is*) is 2A. When the RF pulse is active, the gate bias is set to allow the drain current (*Id*) to be approximately 27.3A.

The simulation in this example is designed to show a typical response expected when an RF pulse drives the gate to close the drain-to-source switch producing the high current necessary to realize the expected high-power RF signal at the output

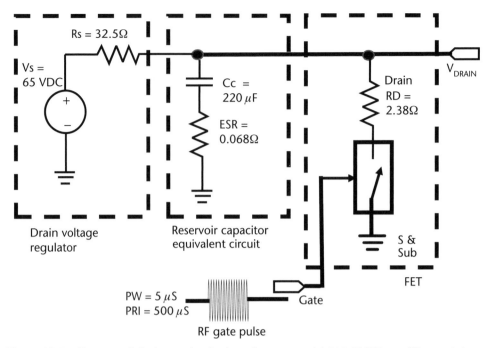

Figure 18.7 Charge and discharge circuit of a pulse saturated 1,000-W FET amplifier module.

of the module. This example illustrates the issues encountered when considering real capacitors with their associated ESR and voltage sources (*Vs*) with finite current capabilities.

The major functions of concern are:

1. *Drain voltage regulator:* 65 VDC with a short circuit current ≈ 2A (*Rs* = 32.5Ω).
2. *Reservoir capacitor equivalent circuit:* In this example the reservoir capacitance (*Cc* = 220 µF) has an ESR equal to 0.068Ω.
3. *FET drain circuit:* The FET drain circuit is coarsely modeled as an open circuit between RF pulses and short-circuit to a load resistor (RD) during the RF pulse width. The pulse repetition interval (PRI) is 5,000 µS. The pulse width is 5 µS. The drain (I_D) current when the RF pulse is present is ≈ 27.3A (RD ≈ 2.38Ω) and approximately zero between RF pulses.

18.5.1 Initial Turn-On

The reservoir capacitor (*Cc*) charges from the 65-VDC, 2-A DC power supply at start-up (see Figure 18.8) and between RF pulses. It discharges during the RF signal pulse (pulse width is 5 µS).

The charging time constant when the drain switch is open is

$$\tau_C = (Rs + ESR) \cdot Cc = 7.165 \text{ mS} \tag{18.7}$$

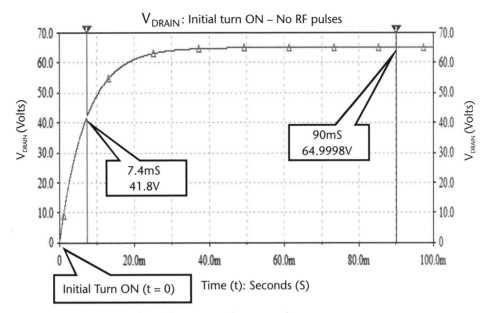

Figure 18.8 Simulation of initial turn-on without RF pulses.

18.5.2 The Drain Voltage Transient Before, During, and After the RF Pulse

The simulation of the transient response of the drain voltage is shown in Figure 18.9. The RF pulse triggers the gate circuit at time t_1. At time t_2 the RF pulse ends. The analysis breaks the transient response of the drain voltage into four moments or periods in time:

1. t_1-: The time just before the RF pulse starts;
2. t_1+: The time just after the RF pulse starts;
3. t_1+ to t_2-: The time during the RF pulse width (PW);
4. t_2+: The time just after the RF pulse ends.

18.5.2.1   Time Between RF Pulses Leading up Time t_1

Since the pulse repetition interval is 5 mS and charging time constant is 7.165 mS, the reservoir capacitor (Cc) does not fully charge between RF pulse discharges. As shown in Figure 18.9, the reservoir capacitor charges to voltage V1 (64.4V).

18.5.2.2   Voltage Drop When RF Pulse Turns On (Time Period: t_1- to t_1+)

At time t_1- (the small interval just before time t_1), the drain voltage is 64.4V. At time (t_1), an RF pulse into the gate causes the drain switch to close and discharge the reservoir capacitor (Cc). The discharge starts time t_1+ (the small interval just after time t_1) when the drain is near 65 VDC and drain resistance (RD) is near 2.38Ω. At that point, approximately 27.1A flows from the capacitor into the FET drain. The voltage across the capacitor's ESR resistor becomes negative. The FET drain voltage at the start of the discharge cycle (time t_1+) is V2 (62.6V) a drop of approximately

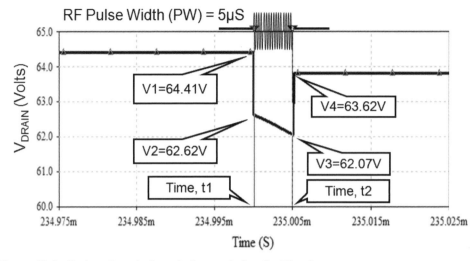

Figure 18.9 Drain voltage before, during, and after the RF pulse.

1.8V from the drain voltage at time t_1-. The voltage V2 is approximated by (18.8), where the first factor $(Vi \cdot Rp/(Rp + ERS))$ is derived from the charged capacitor Cc (voltage V1 at time $t_1- = 64.4\text{V}$) and the second factor $(Is \cdot (ESR \cdot Rp)/(ESR + Rp))$ is due to the short-circuit current (Is) from the voltage source (Vs).

$$V2 = V1 \cdot \frac{Rp}{Rp + ERS} + Is \cdot \frac{ESR \cdot Rp}{ESR + Rp} = 62.62\text{V} \tag{18.8}$$

where V1 is the drain voltage (64.41V) just before time t_1 (t_1-), V2 is the drain voltage (62.62V) just after time t_1 (t_1+), t_1 is the time that RF pulse is active at the gate, Rp is the parallel combination of resistors RS and RD $(Rp = [Rs \cdot RD]/[Rs + RD])$, ESR is the equivalent series resistance of reservoir capacitor Cc, and current Is is the short-circuit current from the voltage regulator.

18.5.2.3 Drain Voltage Discharge During RF Pulse

The discharge time constant during the RF pulse width is:

$$\tau_D = \left(\frac{Rs \cdot RD}{Rs + RD} + ESR \right) \cdot Cc = 503 \ \mu S \tag{18.9}$$

Since the pulse width (PW) of the RF signal is much less than the discharge time constant (τ_D) the decrease in voltage is relatively linear as shown in Figure 18.9. The actual voltage at the end of the 5-μS pulse is given by:

$$V3 = V2 \cdot e^{\frac{-PW}{\tau_D}} + \frac{Is \cdot PW}{C_C} \approx 62.07\text{V} \tag{18.10}$$

The change in voltage:

$$\Delta Vd = V3 - V2 = (62.62 - 62.07) = -0.55\text{V} \qquad (18.11)$$

where $V2$ is the drain voltage at the start of the RF pulse (volts), $V3$ is the drain voltage at the end of the RF pulse (volts), τ_D is the capacitance discharge time constant in second [see (18.9)], PW is the pulse width of the RF signal in seconds, Is is the short-circuit current of the voltage source in amps, and C_C is the capacitance of the reservoir capacitor in farads.

The first part of (18.10) ($V2 \cdot e^{-PW/\tau_D}$) accounts for the discharge of capacitor Cc through time constant τ_D and second part of (18.10) (($Is \cdot PW)/C_C$) accounts for the charging of capacitor C_C by the short circuit current from the drain voltage regulator.

18.5.2.4   Voltage Increase When RF Pulse Turns Off (Time Period: t_1- to t_2+)

When the RF pulse turns off the drain switch opens and the voltage goes from 62.07V at time t_2- to 63.62V at time t_2+. The magnitude of the delta voltage increase is less than the magnitude of the delta voltage decrease because the initial voltage is less at time t_2- than the initial voltage at time t_1-.

18.5.3   Voltage Increase When Between RF Pulses

The voltage increases from time t_4+ (63.62V) to the time just before the next RF pulse (t_1-) exponentially approximately to 64.41V (t_1-) through time constant $\tau_C = (RS + ESR) \cdot C_C = 7.165$ mS (18.7).

18.5.4   Notes on the Model Used in this Example

The model was selected to illustrate the issues concerning the voltage regulator and reservoir capacitor used to develop the high drain current necessary to produce the required high RF power. Second-order and third-order effects that would have a minor effect on the results were not included in the model.

The impedance (Rs) is typically lower than that shown when the regulator is operating in the linear mode, that is, below its maximum short circuit current (2A). The resonant frequency of the reservoir capacitor (Cc) was considered much higher than the frequency of interest in this model. This may not be a valid assumption if the inverse of the RF pulse width was near the capacitor's resonant frequency. Adding an effective series inductance to the model would account for the resonant frequency of the capacitor (Cc). The model of the drain circuit is a drain resistor (RD) in series with a near-ideal switch. The drain-to-source model is actually a linear approximation of the nonlinear drain circuit. The RF switch is modeled as an open circuit when the RF pulse is not present. Actually, there is a small leakage quiescent current present when the switch is open that is assumed to have a negligible effect on the result. The equations shown illustrate components that have a significant cause and effect on the dependent parameters. Second-order and minor parasitic effects were omitted, which may account for minor discrepancies in the simulated versus calculated results.

18.6   Notes on SSPA DC Voltage, Current, and Efficiency

The device manufacturer usually recommends the DC voltage and current for the rated RF output power of each individual amplifier module. From these parameters, the efficiency of the device (RF power divided by DC power) is often computed. Variations to the recommended values are common but, when implemented, a careful analysis is usually required to make sure that the device remains reliable and stable.

Device efficiency typically expressed by the device manufacturer is at the maximum power delivered to a matched load impedance. Actual efficiencies are related to the actual power delivered, which may not be the same as the maximum power.

18.6.1   General Comments on SSPA Efficiency

There are many factors not included in the in the device manufacturers efficiency specification. Some of these factors are the efficiency of the voltage regulators, VSWR mismatch loss, and power consumed by the gate circuitry (usually small compared to the overall power dissipated).

Efficiency specifications are almost always given as typical because of the number of manufacturing and implementation variables related to the implementation of the designed device.

18.6.2   Notes on the Efficiency of SSPAs Operated in a Saturated Pulse Mode

The power dissipated when the pulse signal is a low duty factor can be largely due to the quiescent power (no signal present). The SSPA efficiency factor could change drastically as a function of the duty factor. The power dissipated in the pulse mode and the power dissipated in the CW mode can be drastically different requiring a substantial difference in the mechanical and thermal design of the amplifier. Long pulses, even when the duty factor is low, may have to be treated thermally as a hybrid between short-pulse, low duty factor signals and CW signals. The distinguishing factor of when to thermally analyze the circuit in terms of peak power instead of its average power is the length of the pulse and the thermal transient time constant of the devices channel or junction [7].

18.6.3   Efficiency of CW Communications Amplifiers

CW communication systems typically require power amplifiers to operate in a linear (class A) or quasilinear (class AB) mode. The efficiency in these modes is related to the amount of acceptable distortion and usually much less than SSPAs operating in a saturated mode (maximum distortion).

18.6.3.1   Example of Saturated Mode and Linear Mode Efficiency Differences

The efficiency of a pulse saturated amplifier in the previous example (Section 18.3) was given as 43% during the time the RF pulse was active or if a CW signal was present under the same operating conditions. To satisfy the linearity requirements of a typical communications system (third-order intermodulation interference, adjacent channel rejection ratio, noise power ratio, and so forth), the output power is typically backed off at least 3 dB (0.5 in linear units) from the 1-dB gain compression point and more from the saturated power level [8]. Since the DC power is unchanged and the RF power is reduced by a factor of 0.5 or less, the efficiency of the amplifier degrades proportionally. A 43% efficiency factor can easily decrease to less than 20%.

References

[1]   Gabriela, G., and G. Gonzalez, "Measurements for Modelling of Wideband Nonlinear Power Amplifiers For Wireless Communications," Master's Thesis, Helsinki University of Technology, September 2004.

[2]   Liang, C., J. Jong, and W. Stark, "Nonlinear Amplifier Effects in Communications Systems," *IEEE Transactions on Microwave Theory and Techniques*, August 1999.

[3]   Triquint Semiconductor, "Biasing GaN on SiC HEMT Devices."

[4]   Qorvo Semiconductor, "TGA2216 GaN Power Amplifier."

[5]   Cree, Inc., "GaN HEMT Biasing Circuit with Temperature Compensation," Application Note-011 Rev. B, 2012.

[6]   Simpson, C., "Linear and Switching Voltage Regulator Fundamentals Part 1," Texas Instruments, 2011.

[7]   Freescale Semiconductor, Inc., "Thermal Analysis of Semiconductor Systems," Document Number: Basicthermalwp/ Rev 0, 2008.

[6]   Gouba, O., and Y. Louët, "Theoretical Analysis of the Trade-Off Between Efficiency and Linearity of the High Power Amplifier in OFDM," IETR/SUPELEC, Campus de Rennes, France, 2012.

CHAPTER 19

Thermal Design and Reliability

19.1 Introduction

SSPAs dissipate a significant amount of primary DC power above the RF power delivered. The thermal management of the power dissipated is crucial to a reliable design. Overdesigning the thermal paths adds cost, size, and weight while under designing the thermal paths leads to excessive heat, a less reliable product, and ultimately higher failure rates. An optimum solution requires an understanding of the heat sources, the thermal paths, and the heating effects on reliability.

19.2 Device Reliability as a Function of Temperature

Theoretically, the reliability of a device, typically measured by its mean time to failure (MTTF) or mean time between failures (MTBF), can be found by analyzing data from a large the number of devices operated in a controlled environment. Practically, the MTTF could and should be a very long time making this technique neither practical nor cost-effective, given the relatively short time period between the design and the release of most products. This problem is resolved by stressing the product, thereby accelerating the time between failures. The results are extrapolated to normal failure rates under normal operating conditions. This process is commonly known as accelerated life testing. Note that MTTF usually refers to nonrepairable failures, and MTBF usually refers to repairable failures, and the terms are commonly used interchangeably without a distinction to their actual definition.

19.2.1 Accelerated Life Test

The purpose of an accelerated life test is to measure the MTTF of a device, combination of devices, or systems under a controlled stressed environment such that the acquired data can used to estimate the MTTF under normal operating conditions. Accelerated life testing is the process of testing a device under conditions in excess of its normal operating parameters. The objective is to discover the faults and modes of failure in a reasonable time period much less than the actual usable life of a device. The parameters usually tested individually could be temperatures, applied voltages, vibration, shock, and acceleration. The environmental parameter that primarily affects the reliability of the electronics and SSPAs is temperature and therefore of most concern to the design engineer [1].

Accelerated life testing at a higher temperature than the normal operating environment [2] is a common method used to predict SSPA MTTF. The accelerated life test is usually performed on a random sample of devices selected from a larger manufacturing lot. The results are more meaningful to that particular lot but, given a certain consistency in the manufacturing process, the data is assuming meaning over similar production lots. Accelerated life tests are often periodically repeated to confirm the assumed reliability numbers and prove the manufacturing consistency. The temperature used in the accelerated life tests should not be so high such that unusual failure mechanisms are induced.

19.2.2   Determining MTTF from Accelerated Life Test Data

The failure rate during accelerated life tests is extrapolated to normal expected operating conditions by applying a statistical acceleration factor [3]. A commonly used failure acceleration factor (AF) is the Arrhenius equation [4].

$$AF = e^{\left(\frac{Ea}{k}\cdot\left(\frac{1}{T_{\text{use}}}-\frac{1}{T_{\text{Test}}}\right)\right)} \tag{19.1}$$

where AF is the acceleration factor due to changes in temperature; Ea is the apparent activation energy (eV) of the failure mode; k is the Boltzmann's constant = 8.617×10^{-5} eV/K; T_{use} is the absolute temperature that the device will be used (K), T_{test} is the absolute accelerated life test temperature (K), K is the degrees kelvin where K = $273.15° + °C$, and eV (electron volts) is approximately 1.602×10^{-19} joules, the energy increase of an electron moving across an electric potential of 1V.

Activation energy (Ea) is an empirical value that is the minimum energy required to initiate a specific of failure mode. The higher the activation energy, the lower the failure rate. Activation energy is usually the mean value (50% probability of a failure). In typical activation energies, silicon devices are usually given as 0.7 eV for silicon [1], and GaN devices generally have a higher activation energy, which could be in the 2.15-eV range for a 50% probability and 1.94 eV for a 90% probability [5]. The lower the activation energy level, the higher the probability of a failure at an elevated temperature.

19.2.3   Example: Using Accelerated Life Test Data to Determine MTTF

The following example illustrates how MTTF data is extrapolated from an accelerated life test data. In the test condition, assume that an accelerated life test was run on a GaN FET device with the following parameters and results: channel temperature of 250°C (T_{TEST}) and device exhibited 1 failure in 1,000 hours ($MTTF_{\text{TEST}}$).

Problem: What is the expected failure rate ($MTTF_{\text{ACTUAL}}$) when the channel is operating at a temperature, that is, the used temperature, T_{USE}, of 125°C? Assume an activation energy (Ea) of 2.15 eV.

Solution: The AF is

$$AF = e^{\left(\frac{Ea}{k}\left[\frac{1}{T_{\text{use}}}-\frac{1}{T_{\text{Test}}}\right]\right)} = 3.176 \cdot 10^6 \tag{19.2}$$

$$MTTF_{\text{ACTUAL}} = MTTF_{\text{TEST}} \cdot AF = 1{,}000 \cdot 3.176 \cdot 10^6 = 3{,}176 \text{ million hours} \quad (19.3)$$

19.2.4   Notes on Device Reliability as a Function of Temperature

Reliability is a statistical function where the reliability number (MTTF or MTBF) is the mean value. Note that with regard to the effect of increasing the used temperature (T_{USE}), if the use temperature (T_{USE}) increased by 10°C to 135°C, the actual MTTF ($MTTF_{\text{ACTUAL}}$) results would decrease to:

$$AF = e^{\left(\frac{Ea}{k}\left(\frac{1}{T_{\text{use}}} - \frac{1}{T_{\text{Test}}}\right)\right)} = 6.852 \cdot 10^5 \quad (19.4)$$

$$MTTF_{\text{ACTUAL}} = MTTF_{\text{TEST}} \cdot AF = 1{,}000 \cdot 6.582 \cdot 10^5 = 685.2 \text{ million hours}$$
$$(19.5)$$

In this example, a 10°C increase in channel temperature degraded the reliability of the device 4.636 times for the same device with all other conditions the same.

In the example of operating temperature rising, $\Delta T = +10°C \rightarrow$ reliability drops 4.636 times, this example shows the importance of operating the power amplifier to be as cool as possible. Note that the Arrhenius equation describes chemical reaction rates and it is an approximate model for the thermal acceleration of semiconductor device failures [2].

Note that a failure is defined by the user typically when a parameter degrades such that the device is not useable [1]. Failures are not necessarily catastrophic (e.g., no output signal). For example, a failure criterion can be defined as the maximum output signal level degrading by more than 1 dB. The commonly used acceleration factor equation (the Arrhenius equation) is an approximation. The used temperature (T_{USE}) is not necessarily the device's highest operating temperature. Systems reliabilities are typically specified near or a little above the average operating temperature, not necessarily the worst-case highest temperature.

19.3   Calculating the Power Dissipation and Temperature Rise in an FET Amplifier Module

Temperature rise in a device is caused internally by the power dissipated in the device and externally by power dissipated in surrounding devices. The power dissipated in surrounding devices raises the ambient temperature of all the devices in the system.

Thermal analysis has certain similarities to circuit analysis. One similarity is the concept of transient response and steady-state response, where transient response relates to the time that it takes the power dissipated to convert to an increase in temperature and the other considers the temperature rise over a long time (i.e., after the system is thermally stabilized). An efficient design requires that most of the DC power is concentrated in the power amplifier module. Power dissipated in peripheral devices are necessary but should be limited as much as possible because they degrade the efficiency of the of the overall SSPA system.

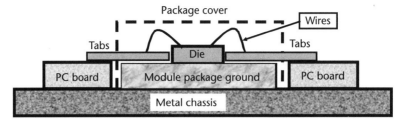

Figure 19.1 Power amplifier module mounted on a chassis and connected to a printed circuit board.

Figure 19.1 shows a power amplifier module mounted on a metal chassis and its connections to a printed circuit board. A power device is rarely attached to a printed circuit board because thermal resistance of a printed circuit board is usually high (about 20°C/W to 60°C/W) [6] depending on the thickness of the board and the quantity of copper traces to a highly conductive thermal heat sink (e.g., a metal chassis).

The term junction temperature (T_J) originally referred to bipolar transistor junctions but became widely used to indicate the temperature of the MMIC chip or die or, in the case of an FET MMIC, the temperature of the conduction channel. It is generally assumed that the die's temperature is uniform and the whole surface of the die is at the same temperature [7]. This approximation, generally considered valid, ignores thermal gradients and multiple heat sources within the die.

19.3.1 FET Channel to Case Temperature Differential

The key factor in determining the reliability of a device is the device's junction temperature for bipolar devices or the channel temperature for FETs (T_J). The FET's channel temperature is a function of the case temperature (T_C), the power dissipated in the channel, and the thermal resistance between the channel and the case (θ_{J-C} in degrees C/W) (see Figure 19.2).

The channel temperature is calculated from the thermal resistance [8] equation:

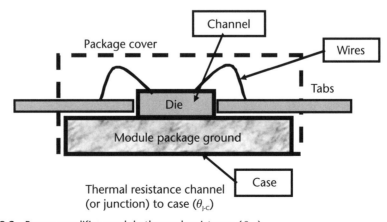

Figure 19.2 Power amplifier module thermal resistance (θ_{J-C}).

$$T_J = P_{\text{DIS}} \cdot \theta_{J-C} + T_C \tag{19.6}$$

where T_J is the FET channel temperature, P_{DIS} is the power dissipated in the channel, θ_{J-C} is the thermal resistance between the FET channel and the case (°C/W), and T_C is the case temperature.

19.3.2   FET Channel to Ambient Temperature Differential

The change in temperature (ΔT_{C-A}) from the module package ground (T_C) to the ambient temperature of the chassis (T_A) is calculated from the thermal resistance of the metal-to-metal interface (θ_I) plus the thermal resistance of the metal chassis to the ambient temperature ($\theta_{\text{Chassis-}A}$) times the power dissipated in the power amplifier module (P_{DIS}).

$$\Delta T_{C-A} = P_{\text{DIS}} \cdot \left(\theta_I + \theta_{\text{Chassis-}A} \right) = T_C - T_A \tag{19.7}$$

where ΔT_{C-A} is the temperature differential (°C) between the case of the power amplifier module and the ambient temperature around the chassis

$$\Delta T_{C-A} = T_C - T_A \tag{19.8}$$

where T_C is the case temperature of the power amplifier module (°C), T_A is the ambient temperature surrounding the chassis (°C), P_{DIS} is the power dissipated in the power amplifier module (watts), (θ_I) is the thermal resistance of the metal to metal interface between the bottom of the power amplifier module and the chassis (°C/W), and ($\theta_{\text{Chassis-}A}$) is the thermal resistance of the metal chassis to the ambient temperature (T_A) (°C/W).

Solving for the power amplifier case temperature (T_C) in (19.7):

$$T_C = P_{\text{DIS}} \cdot \left(\theta_I + \theta_{\text{Chassis-}A} \right) + T_A \tag{19.9}$$

Substituting (19.9) into (19.6),

$$T_J = P_{\text{DIS}} \cdot \theta_{J-C} + P_{\text{DIS}} \cdot \left(\theta_I + \theta_{\text{Chassis-}A} \right) + T_A \tag{19.10}$$

Factoring out the power dissipation (P_{DIS}) gives a direct relationship between the channel temperature (T_J) and the ambient temperature (T_A).

$$T_J = P_{\text{DIS}} \cdot \left(\theta_{J-C} + \theta_I + \theta_{\text{Chassis-}A} \right) + T_A \tag{19.11}$$

19.3.3   Thermal Model of an FET Amplifier Module

From (19.11), it can be seen that if the power dissipated in the amplifier module (P_{DIS}) is fixed and the ambient temperature (T_A) is fixed by the SSPA environment,

Figure 19.3 Schematic representation of the thermal resistance path.

the only way to reduce the FET channel temperature (T_J) is to minimize as much as possible the thermal resistance of each element between the FET channel and the outside environment. Figure 19.3 is a schematic representation of the direct thermal path described by (19.11).

The thermal path from the channel to the module package ground is usually the primary path but not the only thermal path in the power amplifier module. The thermal resistance diagram is more complex. Figure 19.4 shows additional but not all thermal paths from the channel [8]. Parallel thermal paths reduce the thermal resistance similar to parallel electrical paths.

19.3.4 Techniques to Reduce the Thermal Resistance Between the FET Channel and the Ambient Temperature

Reducing the thermal resistance between the channel temperature and the ambient temperature is critical to improving the reliability of the SSPA. A small reduction in the FET channel temperature results in a significant reliability improvement.

19.3.4.1 Using a Thermal Pad Between Metal-to-Metal Surfaces

The power amplifier module is typically attached directly to a metal chassis for maximum thermal efficiency. The temperature differential and the temperature of

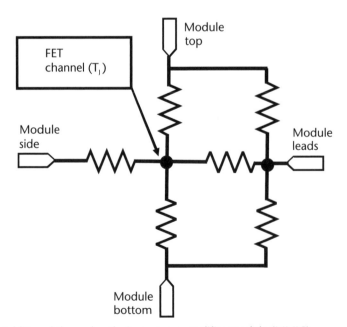

Figure 19.4 Additional thermal paths in a power amplifier module (MMIC).

the power amplifier case is determined by noting the thermal resistance of the chassis material and the thermal interface resistance [9] between the base of the power amplifier module (which is typically a metal) and the metal chassis as shown in Figure 19.5. The figure shows typical metal-to-metal surface irregularities or roughness such that contact occurs only at the high points, leaving air filling the voids. Air has a high thermal resistance forcing most of the heat to flow through the contact points. Depending on the level of roughness the thermal resistance at the interface (θ_I) could be high enough to seriously degrade the thermal performance of the power amplifier. This issue occurs at all metal to metal interfaces [10].

The thermal resistance can be lowered by use a highly conductive thermal pad or thermal grease to fill the air gaps between the metal-to-metal interfaces [9]. The thermal resistance of the thermal pad or thermal grease should be much less than the thermal resistance of the metal-to-metal interface.

Note that, with regard to thermal pads, the interface between the power amplifier and the chassis usually requires a very good electrical ground, and any substance between the module package ground and the chassis should have a very low electrical resistance as well as a very low thermal resistance.

19.3.4.2   Other Common Techniques to the Reduce the Path Thermal Resistance

Some other techniques used to lower the thermal resistance between the FET channel (T_J) and the ambient temperature (T_A) are:

1. *Chassis metals:* Use a chassis metal that has a low coefficient of thermal resistance [11].
2. *Heat sinks:* The chassis to ambient thermal resistance can many times be reduced with a heat sink that captures the air flow around the chassis [12]. The thermal resistance changes as a function of the airflow available. Airflow

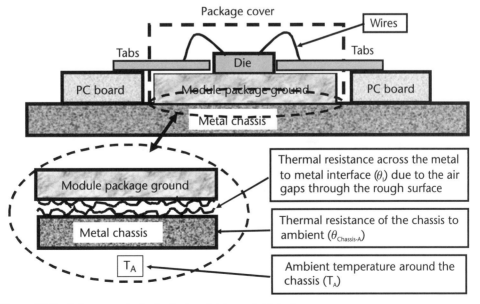

Figure 19.5 Enlargement of a typical metal-to-metal interface.

can either be forced air from a fan or natural convection. Natural convection occurs with no fan. Heat transfer depends on the air around the heat sink surface.

3. *Other common cooling techniques:* These are thermoelectric coolers, forced air cooling systems, and heat pipes.

19.4 Examples of FET Power Amplifier Module Thermal Calculations

The following examples illustrate the procedures normally followed to determine the channel temperature of an FET amplifier module.

19.4.1 Example: Calculating FET Channel Temperature

Determine the channel temperature of the power amplifier module shown in Figure 19.6. Assume that the active device in the module is a single FET amplifier stage with the input and output parameters are that described in Table 19.1.

Figure 19.6(a) shows the power amplifier FET module with the pertinent electrical and thermal analysis input and outputs. Figure 19.6(b) is the thermal model of the device in terms of its electrical equivalent circuit.

Solution:

1. Calculate the power dissipated in the power amplifier module. The power dissipated in the device (P_{DIS}) is the power going in minus the power going out:

$$P_{DIS}(W) = RF_{IN} + P_{DC} - RF_{OUT} = 21W \tag{19.12}$$

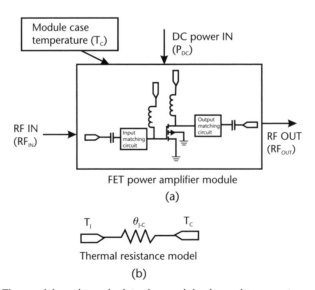

Figure 19.6 (a) The model used to calculate the module channel temperature and (b) the thermal resistance model from the module case to the FET channel.

Table 19.1   Parameters for Calculating Channel Temperature

1	RF in (RF_{IN})	1	Watts
2	RF out (RF_{OUT})	10	Watts
3	DC power in (P_{DC})	30+1p	Watts
4	Case temperature (T_C)	50	°C
5	Thermal resistance ($\theta_{J\text{-}C}$)	3	°C/W

Row 1, RF in (RF_{IN}): RF input power level (RF_{IN}) in watts; row 2, RF out (RF_{OUT}): RF output power level (RF_{OUT}) in watts; row 3, DC power in (P_{DC}): DC input power level (P_{DC}) in watts; row 4, module case temperature (T_C): case temperature of the module (T_C) in degrees centigrade (°C); and row 5, thermal resistance ($\theta_{J\text{-}C}$): thermal resistance of the module from the case to the channel ($\theta_{J\text{-}C}$) in degrees centigrade per watt (°C/W).

2. Calculate the channel temperature (T_J) using (19.6):

$$T_J = T_C + P_{\mathrm{DIS}} \cdot \theta_{J-C} = 113°C \qquad (19.13)$$

19.4.2   Example: Calculating FET Channel Temperatures Through Multiple Thermal Paths

Determine the channel temperature of the power amplifier module shown in Figure 19.7. Assume that the active device in the module is a single FET amplifier stage with the input and output parameters described in Table 19.2.

Figure 19.8(a) is a rough sketch of the physical construction inside the power amplifier while Figure 19.8(b) shows the thermal model of the two paths from the module (die) channel to the outside ambient temperature. The two thermal paths are in parallel and combine the same way that two parallel resistors combine. The additional thermal path lowers the junction temperature (T_J), assuming no change in the ambient temperature (T_A).

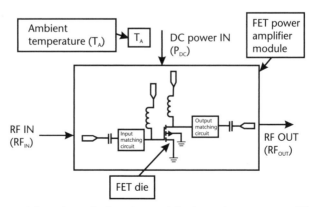

Figure 19.7   The model used to calculate the module channel temperature (T_J) with respect to the ambient air temperature (T_A).

Table 19.2 Parameters for Calculating Channel Temperatures
Through Multiple Thermal Paths

1	RF in (RF_{IN})	0.5	Watts
2	RF out (RF_{OUT})	12	Watts
3	DC power in (P_{DC})	20	Watts
4	Ambient temperature (T_A)	55	°C
5	Thermal resistance $(\theta_{J\text{-}C})$	2	°C/W
6	Thermal resistance $(\theta_{C\text{-}A})$	8	°C/W
7	Thermal resistance $(\theta_{Cover\text{-}J})$	20	°C/W
8	Thermal resistance $(\theta_{Cover\text{-}A})$	7	°C/W

Row 1, RF in (RF_{IN}): RF input power level (RF_{IN}) in watts; row 2, RF out
(RF_{OUT}): RF output power level (RF_{OUT}) in watts; row 3, DC power in (P_{DC}):
DC input power level (P_{DC}) in watts; row 4, module case temperature (T_C):
case temperature of the module (T_C) in degrees centigrade (°C); row 5, thermal
resistance $(\theta_{J\text{-}C})$: thermal resistance of the module from the case to the channel
$(\theta_{J\text{-}C})$ in degrees centigrade per watt (°C/W); row 6, thermal resistance $(\theta_{C\text{-}A})$:
thermal resistance of the module from the case to the ambient temperature
$(\theta_{C\text{-}A})$ in degrees centigrade per watt (°C/W); row 7, thermal resistance $(\theta_{Cover\text{-}J})$:
thermal resistance of the module from the cover to the channel $(\theta_{J\text{-}C})$ in degrees
centigrade per watt (°C/W); and row 8, thermal resistance $(\theta_{Cover\text{-}A})$: thermal
resistance of the module from the cover to the ambient temperature $(\theta_{Cover\text{-}A})$ in
degrees centigrade per watt (°C/W).

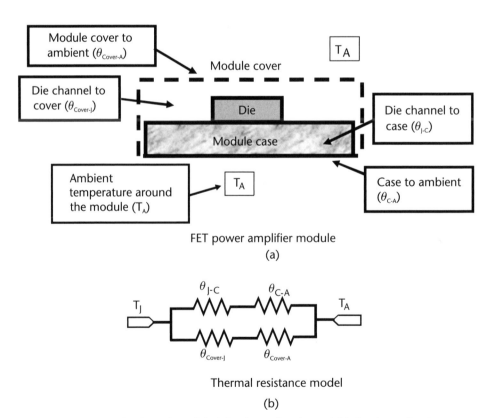

FET power amplifier module

(a)

Thermal resistance model

(b)

Figure 19.8   (a) A rough physical model of the thermal paths and (b) the thermal resistance
model.

Solution, Part 1: Channel to ambient only through the module case:

1. Calculate the power dissipated in the power amplifier module. The power dissipated in the device (P_{DIS}) is the power going in minus the power going out:

$$P_{DIS}(W) = RF_{IN} + P_{DC} - RF_{OUT} = 8.5\,W \tag{19.14}$$

2. Calculate the total path thermal resistance ($\theta 1_{J\text{-}A}$) through the case

$$\theta 1_{J-A} = \theta_{J-C} + \theta_{C-A} = 10°C/W \tag{19.15}$$

3. Calculate channel temperature (T_J) through the path thermal resistance from the channel to the outside ambient temperature through the case (neglecting the path through the cover) module.

$$T_J = T_A + P_{DIS} \cdot \theta 1_{J-A} = 140°C \tag{19.16}$$

Solution Part 2: Channel to ambient only through the module cover:

1. Calculate the power dissipated in the power amplifier module. The power dissipated in the device (P_{DIS}) is the power going in minus the power going out:

$$P_{DIS}(W) = RF_{IN} + P_{DC} - RF_{OUT} = 8.5\,W \tag{19.17}$$

1. Calculate the total path thermal resistance ($\theta 2_{J\text{-}A}$) through the cover

$$\theta 2_{J-A} = \theta_{Cover\text{-}J} + \theta_{Cover\text{-}A} = 27°C/W \tag{19.18}$$

2. Calculate channel temperature (T_J) through the path thermal resistance from the channel to the outside ambient temperature through the module cover (neglecting the path through the case).

$$T_J = T_A + P_{DIS} \cdot \theta 1_{J-A} = 284.5°C \tag{19.19}$$

The thermal path through the cover has a much higher resistance than the thermal path through the module case. If the cover were the only thermal path available, the reliability of the device would be severely impacted.

Solution Part 3: Channel to ambient through the module cover and case paths in parallel:

1. Calculate the power dissipated in the power amplifier module. The power dissipated in the device (P_{DIS}) is the power going in minus the power going out:

$$P_{DIS}(W) = RF_{IN} + P_{DC} - RF_{OUT} = 8.5\,W \qquad (19.20)$$

2. Calculate the total path thermal resistance ($\theta T_{J\text{-}A}$) which is the thermal path through the cover in parallel with the thermal path through the module case ($\theta 1_{J\text{-}A} // \theta 2_{J\text{-}A}$).

$$\theta T_{J-A} = \theta 1_{J-A} // \theta 2_{J-A} = 7.3°C/W \qquad (19.21)$$

3. Calculate channel temperature (T_J) using the parallel thermal paths through the cover and the case.

$$T_J = T_A + P_{DIS} \cdot \theta 1_{J-A} = 117.03°C \qquad (19.22)$$

The thermal path through cover although not very efficient by itself lowered the FET channel temperature by approximately 23°C over the results obtain using only the thermal path going through the module case. There are other parallel thermal paths from the channel to the outside ambient temperature that could provide additional FET device channel cooling. It is left up the design engineer to decide which if any thermal paths warrant a significant enough improvement to warrant the calculation or measurement of the respective thermal resistance.

19.5 Thermal Transients

Materials have a thermal inertia (i.e., the temperature of a mass does not change instantaneously). When an electronic device is in thermal equilibrium and a power surge dissipates in the active device, the temperature does not immediately change, but starts rising due to the heat generated. The temperature of the device eventually stabilizes at a point of thermal equilibrium where the heat generated equals the heat dissipated. The heating time prior to stabilization is called the thermal transient period [7].

The thermal mass can be modeled as a capacitor that stores heat and a thermal resistance that dissipated heat. This has an analogy to an electrical R-C network where the voltage across the capacitor does not change instantly but charges to a new steady-state value exponentially with a time constant Tau (τ), where $\tau = R \cdot C$. The thermal mass similarly stabilizes to an equilibrium temperature exponentially with a thermal time constant (τ_{th}) [13]:

$$\tau_{th} = \theta_{th} \cdot C_{th} \qquad (19.23)$$

where τ_{th} = the thermal time constant in seconds, θ_{th} = the thermal resistance in units of degrees Centigrade divided by the power dissipated (°C/W), and C_{th} = the equivalent thermal capacitance in units of energy (joules or J) divided by temperature in degrees centigrade (J/°C).

Shown in Figure 19.9(a) is an MMIC with its power amplifier die attached to the module case. The MMIC is mounted to a metal chassis. The thermal transient circuit is modeled as an equivalent electrical circuit shown in Figure 19.9(b) [7]. Using the electrical analogy, a surge in power in the die is not immediately noticed by the module case or the metal chassis since these elements usually have a much higher thermal capacitance than a small die. The thermal resistance and thermal capacitance of each device is different and therefore have separate thermal time constants. Since the die is the typically the smallest mass, its thermal time constant is shortest [14].

Each segment in Figure 19.9(b) has an effective impedance Zth1, Zth2, and Zth3, which is the respective thermal resistance in parallel with the impedance of the thermal capacitance. The effective impedance of the thermal capacitance is a function of time [15].

Note that power dissipation and the resultant temperature distribution shown in Figure 19.9 are a function of time. Most of the power is dissipated is in the power amplifier die. The rise in temperature resulting from that power dissipation must be distributed before the reliability of the die is degraded.

19.5.1 DC Power Turn-On Thermal Transients in Linear and CW-Mode SSPAs

When operating in a linear or CW mode, the power amplifier exhibits a power surge and respective temperature rise at initial turn-on. A controlled DC power turn-on consistent with the thermal time constants of all the elements in the thermal path

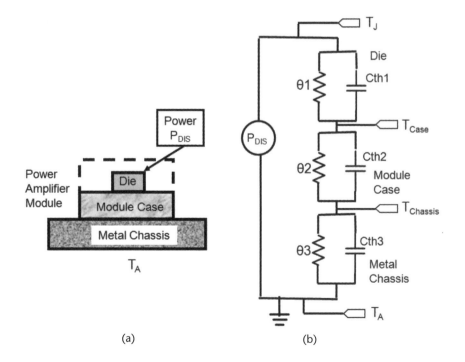

(a)   (b)

Figure 19.9 (a) An MMIC mounted on a chassis and (b) its simplified thermal transient model.

limits the temperature rise in the die to a point where the reliability of the device is unaffected.

19.5.2   Implications of Thermal Transients in the Design of a Pulse Saturated Mode SSPA

Saturated pulse mode SSPAs often operate at a much higher level than linear (CW) mode SSPAs. The saturated-mode SSPA relies on the duty factor of the pulse signal to average the total power dissipation to similar levels seen by linear (CW) SSPAs. Theoretically, the reliability of the two respective amplifiers configurations will be the same if the average power dissipated is the same given that all other parameters effecting reliability remain the same.

Note that, with regard to the reliability of devices with long pulses, this premise of relying on average power instead of peak power when determining device reliability tends to fall apart if the pulse width of the RF signal is long. Long pulses are RF pulses that are long enough to raise the temperature of the die but too short to change to temperature of module case, which have a much longer thermal time constant. The temperature rise in a die increases for the duration of the pulse assuming that the time constant of the module case is much longer than the pulse such that the module case temperature does not change. Instantaneous temperature rises cause an exponential degradation in reliability. Peak temperatures have a more detrimental effect on device reliability than the calculation of average temperature suggests. In the limit, if the pulse width is long such that the thermal time constant of the die reaches a self-destruct temperature before the module case heat sink lowers the temperature of the power amplifier, the module might fail independently of the average temperature calculated by integrating the temperature profile.

Note that failure trigger points are in general statistical in nature. Reaching a failure trigger point usually means there is a 50% chance of failure. The probability of failure when the trigger point is exceeded is a function of the amount exceeded and the standard deviation of the failure mechanism.

19.5.3   Example: Maximum FET Channel Temperature as a Function of RF Pulse Width

This example shows the channel temperature rise as a function of RF pulse width (PW). The baseline conditions are ambient temperature of 50°C, peak pulse power of 150W, the thermal resistance of the die of 2°C/W, pulse width of ≤ 1 thermal time constant, and duty factor of 1%. The thermal time constant of the module case and metal chassis are much longer than the thermal time constant of the die.

Since the thermal time constants of the module case are much larger than the thermal time constant of the die and the pulse width is less than one thermal time constant, the module case has almost no effect on the thermal analysis of the peak power signal. The module case and the metal chassis absorb average heat from the die [i.e., heat generated by 1% of the peak power (150W) or 1.5W].

Figure 19.10 shows that the channel temperature rises to 240°C when the pulse width is equal to one thermal time constant. When the pulse width is equal to 0.3 thermal time constant, the channel temperature rises to 128°C as shown in Figure 19.11.

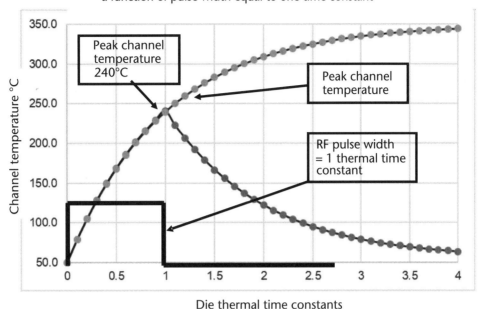

Figure 19.10   Channel temperature increases from a steady state 50°C during the RF pulse width (PW) equal to one thermal time constant and decreasing after the trailing edge of the pulse.

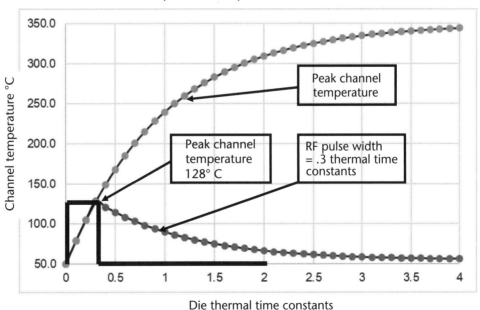

Figure 19.11   Channel temperature increase from a steady state 50°C during the RF pulse width (PW) equal to 0.3 thermal time constant and decreasing after the trailing edge of the pulse.

19.6 Conclusions

With all other factors the same, the average power and the average channel temperature rise under both pulse width conditions (Figures 19.8 and 19.9) are equal. Peak channel powers are significantly different. The exponential nature of the reliability equation, which relates to the channel temperature rise to the device MTTF, could be degraded under the longer pulse conditions. If the both pulse widths in this example were much less than the thermal time constant of the die, the peak temperature differential would be small. Note that the maximum width of a pulse-modulated RF signal should be much less than the thermal time constant of the power amplifier die to minimize the degradation of the reliability of the power amplifier module.

References

[1]   Bahl, S., "A Comprehensive Methodology to Qualify the Reliability of GaN Products," Texas Instruments, March 2015.

[2]   Vigrass, J., "Calculation of Semiconductor Failure Rates" Technical paper.

[3]   Bahl, A., D. Ruiz, and D. Lee, "Product-level Reliability of GaN Devices," *IEEE International Reliability Physics Symposium (IRPS)*, April 2016.

[4]   Ellerman, P., "Calculating Reliability Using FIT & MTTF: Arrhenius HTOL Model," Microsemi Corp., MicroNoteTM 1002, January 2012.

[5]   Qorvo Corp. "GaN on SiC 15 Years of Reliability & Producibility," 2015.

[6]   Çengal, Y. A., and A. J. Ghajar, *Heat and Mass Transfer Fundamentals & Applications Fifth Edition*, "Cooling of Electronic Equipment," Chapter 15, McGraw-Hill Education 2015.

[7]   Melito, M., A. Gaito, and G. Sorrentino, "Thermal Effects and Junction Temperature Evaluation of Power MOSFETs," STMicroelectronics, Application Note AN4783, November 2015.

[7]   Analog Devices, Inc., "Thermal Design Basics," MT-093 Tutorial, 2009.

[8]   Hienonen, R., M. Karjalainen, and R. Lankinen, "Verification of the Thermal Design of Electronic Equipment," VTT Automation, Technical Research Centre of Finland, Espoo, 1997.

[9]   Kumar, S., "Thermal Management of RF and Digital Electronic Assemblies using Optimized Materials and PCB Designs," American Standard Circuits, Inc.

[10]   Parker Hannifin Corporation, "Thermal Interface Materials for Electronics Cooling," Chomerics, Therm. Cat. 1002, November 2016.

[11]   Universal Science, "Thermal Conductivity of Materials."

[12]   Seshasayee, N., "Understanding Thermal Dissipation and Design of a Heatsink," Texas Instruments Application Report SLVA462, May 2011.

[13]   Wang, Y., "Electrical and Thermal Analysis of Gallium Nitride HEMTS," Thesis, Naval Postgraduate School, Monterey, California, June 2009.

[14]   Schwitter, B., et al., "Steady State and Transient Thermal Analyses of GaAs pHEMT Devices," AWR Corp.

[15]   Sattar, A., "Power MOSFET Basics," IXYS Corporation, Note IXAN0061.

CHAPTER 20

Electromagnetic Interference

20.1 Introduction

Electromagnetic interference (EMI) is a disturbance generated by an external source that affects the operational equipment or a disturbance generated by the operational equipment that effects external equipment. The transmission media can be conductive along wires or radiated emissions from or to the equipment of interest [1].

Radio frequency interference (RFI) and electromagnetic interference (EMI) are terms that are used interchangeably and are generally accepted to mean the same [2]. Conducted emissions are signals traveling through a physical conductor that may be along power cables or other interconnection wires or cabling. Radiated emissions are signals converted to electromagnetic energy propagating from or to a device.

EMI concerning a particular device is separated into four categories.

1. *Radiated emissions:* Signals radiated from a device that disrupts surrounding equipment.
2. *Conducted emissions:* Signals conducted from a device that disrupts surrounding connected equipment.
3. *Radiated susceptibility:* Signals radiated from surrounding devices that disrupts the equipment of concern.
4. *Conducted susceptibility:* Signals conducted from surrounding devices that disrupts the equipment of concern.

20.2 The Concept of Electromagnetic Compatibility

Electromagnetic comparability (EMC) is a concept that a device can function in a known externally sourced electromagnetic environment and simultaneously not introducing a disturbing number of electromagnetic signals so other devices operating in the same environment also operate satisfactorily [3].

There are two primary aspects to EMC:

1. The ability of electronic systems to operate without interfering with other systems.
2. The ability of such systems to operate as intended within a specified electromagnetic environment.

EMC or the elimination of EMI can be a major problem. Overcoming interference requires either the removal of the source or protect the destination circuit.

The ultimate goal, achieving EMC, is to have the effected equipment operating as intended [4]. EMC at the system level may have to satisfy certain emission standards, so that the equipment under concern does not adversely affect other equipment.

A simple EMI model consists of three elements: an EMI source, a coupling path, and a receptor. The source and the receptor are designed for specific functions and are therefore usually not amenable to major modifications to satisfy interference produce or interference received from another device.

For example, if an SSPA is designed to emit enough output to satisfy the signal-to-noise requirement of a communications link, it is not acceptable to reducing the output power to satisfy an EMC requirement. If that is done, the link may be nonoperational and therefore not an acceptable EMC solution. The primary focus solving the EMC issue is on the coupling path.

20.3 Mutual Coupling and Crosstalk on a Printed Circuit Board

There are primarily two types of radiated interference: capacitive coupling and inductive coupling. Both are considered near-field events [5] since the distance between the transmitting and receiving components are usually less than a wavelength.

20.3.1 Capacitive Coupling

There exists a mutual capacitance between two conductors separated by a dielectric. Figure 20.1 shows two parallel traces on a printed circuit board.

The actual amount of mutual capacitance of traces on a printed circuit board are a function of the geometry of the structure [6].

Note that on mutual capacitance the traces do not have to be perfectly parallel for a nonzero mutual capacitance. Mutual capacitance and the effective signal interference is a maximum when the parallel traces are close together (small values of the distance between the traces d) and the length of the parallel traces (L) increases.

To reduce mutual capacitance, decrease the length (L) of the parallel traces and increase the distance, d, between these parallel traces.

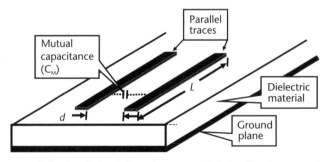

Figure 20.1 Two parallel printed circuit board traces and their effective mutual capacitance.

20.3.2 Inductive Coupling

Assume that a signal ($S1$) flows through an inductor L_1 (the transmitting inductor) producing a current that creates a creates a magnetic field. When a second inductor L_2 (the receiving inductor) is placed in the magnetic field, a current is induced in L_2 replicating the signal current in the transmitting inductor. This mechanism is called inductive coupling. The transmission path is modeled as a mutual inductance L_M between the transmitting loop ($S1$, L_1, and RL) and the receiving loop ($R1$, L_2, and $R2$) [7]; see Figure 20.2.

Inductive coupling is a relatively near-field event where the inductive transmitter (L_1) and the inductive receiver (L_2) are in proximity and mechanically aligned to receive the magnetic field. These issues are relatively easily avoided by doing the following.

To limit unintended mutual inductance:

1. Use a magnetic shield of high permeability material (e.g., mu-Metal) around the transmitting inductor (L_1) and/or the receiving inductor (L_2).
2. Orient the inductors orthogonally so the magnetic fields do not interact.
3. The magnetic field intensity is inversely proportional to the square of the distance from the source [8], so physically moving the transmitting and receiving inductors away from each other can have a great effect in reducing the mutual coupling.
4. Magnetic fields are minimized when the transmitting and receiving inductors are perpendicular to each other.

20.3.3 Crosstalk

The term crosstalk comes from early analog phones where one could hear voices due to electromagnetic coupling from nearby lines. In the context of printed circuit traces, crosstalk is when the energy from the electromagnetic fields of a signal on one trace couples to an adjacent trace (see Figure 20.3).

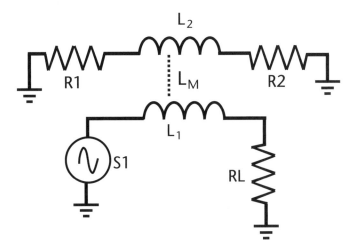

Figure 20.2 Inductively coupled coils with mutual inductance L_M.

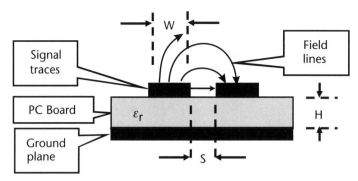

Figure 20.3   Crosstalk on a pair of microstrip traces.

Crosstalk increases as the physical dimension between the lines decrease [9]. To reduce crosstalk in microstrip circuits [4]:

1. Increase the spacing S as far as possible between signal traces.
2. Avoid parallel traces as much as possible. Orthogonal traces have the least amount of crosstalk.
3. Ground lines between parallel signal traces in some cases can increase crosstalk: ground traces have minimal advantage in microstrip circuits and sensitive parallel lines should be placed in a stripline configuration to minimize crosstalk.
4. Avoid bring traces closer than three times (S) the dielectric height (H).
5. The distance between the centers of two adjacent traces should be at least four times the trace width (W).
6. Design the transmission line so that the conductor is as close to the ground plane as possible (keep the dielectric height H as small as possible).

20.4 Far-Field Radiated Emissions and Radiated Susceptibility

Far-field interference is the reception of electromagnetic waves which is a combination of magnetic fields (H plane) at right angles to an electric field (E plane). The direction of travel is perpendicular to both the plane of the electric field and the magnetic field [10] (see Figure 20.4).

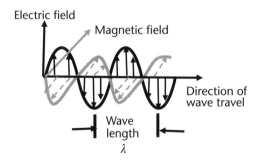

Figure 20.4   Propagation an electromagnetic wave.

20.4.1 Far-Field Approximation

The far-field begins far from the antenna such that the radiation pattern does not change shape with distance, that is, the wave front is approximately linear (i.e., the wave front from a point source is considered flat [11]). When the wave front is flat, the wave is considered planar (i.e., a plane wave) (see Figure 20.5).

The definition of where the far-field begins (i.e., distance R) is not well defined but is generally considered to be in the range of $2 \cdot \lambda$ to $10 \cdot \lambda$, where λ is the signal wavelength.

20.4.1.1   Example: Far-Field Approximation

If signal is radiated at a frequency of 10 GHz, the wavelength in air is approximately 3 cm. Assuming that the far-field approximation (R_{FAR}) begins at $5 \cdot \lambda$:

$$\left(R_{\mathrm{FAR}}\right) \approx 5 \cdot \lambda = 5 \cdot 3 \text{ cm} = 15 \text{ cm} (\approx 6 \text{ inches}) \tag{20.1}$$

The power density of the field decreases proportional to $1/R^2$ where R is the distance from the source.

20.4.2 Chassis Shielding

EMI mitigation focusing on the path of the electromagnetic wave relies primarily on containing the device in a metallically shielded enclosure (a metallic chassis).

SSPAs are usually enclosed in a metallic chassis that inherently provides some attenuation to radiation emitted from the unit and reciprocally attenuates externally sourced radiation from entering the unit. Since most SSPAs are relatively high gain

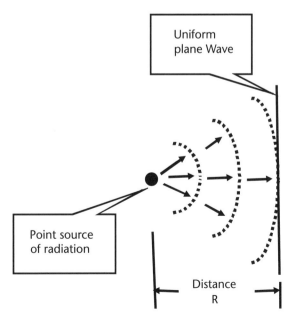

Figure 20.5   Far-field (plane wave) approximation.

units, they may be susceptible to internally generated signals feeding back from the output to the input. This internal radiation could provide sufficient feedback to make the unit unstable (oscillate).

Summarizing the main radiation issues, external radiation and susceptibility and internal radiation and susceptibility, this issue is typically mitigated by internally shielding the amplifier stages as well as providing a metal shield around the entire unit as shown in Figure 20.6. The SSPA has four amplification stages (Amp #1 through Amp #4). The output stage consists of two parallel amplifiers (Amp #4-1 and Amp #4-2). An internal septum separates the two input amplifiers (Amp #1 and Amp #3) from the two output amplifiers (Amp #1 and Amp #3) with a metallic wall. The opening through the wall is much less than a wavelength (λ) limiting the radiation between the two compartments.

20.4.3 Chassis and Wall Thickness

Most of time, the chassis thickness is determined by the structural integrity of the device and sufficiently thick to shield all of the radiated microwave signals. The actual minimum thickness is related to the frequency of the signals to be contain or shielded and the respective skin depth (δ_S) [3].

$$\delta_S = \sqrt{\frac{1}{\left(\pi \cdot F \cdot \mu_0 \cdot \mu_r \cdot \sigma\right)}} \qquad (20.2)$$

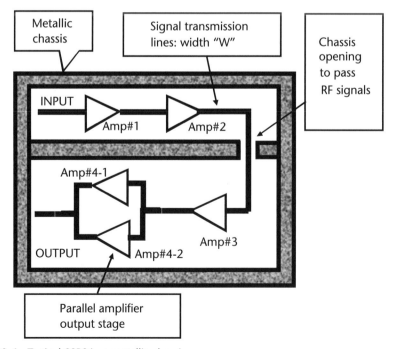

Figure 20.6 Typical SSPS in a metallic chassis.

where δ_S is the skin depth, F is the frequency of interest (Hz), μ_0 is the permeability of free space ($1.25664 \cdot 10^{-6}$ H/m), μ_r is the relative permeability of the material, and σ is the conductivity of the material in Siemens per meter (S/m).

20.4.3.1   Example of Isolation as a Function of Skin Depth

Assume an aluminum chassis shielding a signal at 1,500 MHz. F is 1,500 MHz, $\mu \approx 1$, and $\sigma \approx 3.65 \cdot 10^7$ S/m. The skin depth is $8.46826 \cdot 10^{-05}$ inches (0.002150939 mm).

Table 20.1 shows the signal penetration in the aluminum chassis as a function number (N) of skin depths ($N \cdot \delta_S$). The residual signal $10 \cdot \text{Log} [(1 - \text{Penetration})]$ is the radiation attenuation due to the chassis.

From Table 20.1, it is obvious that the thickness is not an issue for a chassis that requires a structural metal (11 skin depths providing -95.54 dB of attenuation is 0.000932 inch thick).

Note that on signal isolation for nonmetallic material using metallic plated surfaces some high-volume chassis are made from nonmetallic materials that are metal plated to shield the radiated signals. A skin depth calculation is necessary to ensure that the plating thickness is sufficient to provide the necessary shielding. As noted for aluminum, a plating of seven skin depths is needed for $\approx$ 60 dB of isolation and 11 skin depths for $\approx$ 95 dB of isolation. Plating thickness should be calculated at the lowest signal frequency requiring isolation.

20.4.4   Radiation Through Chassis Openings

Openings in a chassis are often required to accommodate adjustments and the passing signal between amplifier compartments (see Figure 20.6). These openings may compromise shielding effectiveness by providing paths for high-frequency interference.

The length of an opening in the metallic chassis or compartment in the chassis is used to evaluate the isolation of fields entering or leaving. The evaluation of the

Table 20.1   Isolation as a Function of the Number of Skin Depths for Aluminum at 1,500 MHz

Number of Skin Depths (δ_S)	Skin Depth Penetration	Penetration Thickness (Mils)	Attenuation (dB)
1	63.2%	0.085	−8.69
2	86.5%	0.169	−17.37
3	95.0%	0.254	−26.06
4	98.2%	0.339	−34.74
5	99.3%	0.423	−43.43
6	99.8%	0.508	−52.12
7	99.9%	0.593	−60.80
8	99.966%	0.677	−69.49
9	99.988%	0.762	−78.17
10	99.995%	0.847	−86.86
11	99.998%	0.932	−95.54

signal isolation is similar to the behavior of a signal through a slot antenna. The signal isolation or shielding effectiveness ($Slot_{\text{Isolation}}$) calculation as a function of the length of the opening in an enclosure [3] is:

$$Slot_{\text{Isolation}}(\text{dB}) = 20 \cdot \text{Log}\left(\frac{\lambda}{2 \cdot L}\right) \tag{20.3}$$

where λ is the wavelength of the interference and L is the maximum dimension of the opening.

Maximum radiation through an opening occurs when the longest dimension of the opening is equal to one half-wavelength (0-dB isolation).

Note that on isolation through a slot in the chassis the calculations in Table 20.2 calculate the signal loss through a slot in a chassis. Other considerations are

Table 20.2 Isolation as a Function of the Fraction of Wavelengths (1/N) of a Slot in the Chassis

Slot Length $L = \lambda/N$	Slot Length Wavelengths	Isolation dB
N	L	
2	0.5	0.00
5	0.2	7.96
10	0.1	13.98
20	0.05	20.00

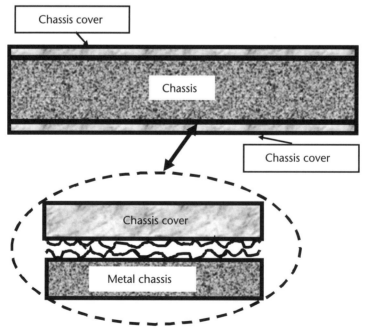

Figure 20.7 Expanded view of a chassis cover connection.

that the signal loss from the PC trace to the slot is not considered in this calculation, the signal loss from the slot to the receiver is not calculated, and the longest dimension of the slot is used for the isolation calculation.

20.4.5   Radiation Through Chassis Covers

Chassis covers are metallic-to-metallic connections. As shown in Figure 20.7, the expanded view of the cover to metal chassis contact shows that the interface is not perfectly smooth.

The actual signal isolation between the chassis and the cover is related to the spaces between the metal contacts similarly to that calculated for a slotted antenna [3]. Table 20.3 is an expansion of Table 20.2 to show the isolation to a slot length of $\lambda/5,000$, which calculates to 67.96 dB.

As shown in Table 20.3, the roughness between the metal-to-metal contact acting as a slotted antenna can significantly degrade the cover to chassis isolation [12]. An EMI gasket fills in the gaps between the metal-to-metal contact and significantly decreases the maximum slot length which in turn significantly increases the isolation provided by the gasketed chassis cover.

20.4.6   Limitations on Chassis Walls and Cover Separation from the Signal Path

From an EMI point of view, there is an advantage to separate each function into small self-contained cavities of the main chassis. Figure 20.6 shows the SSPA constructed in two cavities. This construction techniques limits external EMI and feedback from the output to the input that could cause amplifier instabilities. The issue with placing each module in a small compartment is the proximity of the walls and cover to the microstrip transmission line.

Table 20.3   Isolation as a Function of the Number Wavelengths for Very Small Slots That May Be Encountered Through Direct Metal-to-Metal Contacts

Slot Length $L = \lambda/N$ N	Slot Length Wavelengths L	Isolation dB
2	0.5	0.00
5	0.2	7.96
10	0.1	13.98
20	0.05	20.00
50	0.02	27.96
100	0.01	33.98
200	0.005	40.00
500	0.002	47.96
1,000	0.001	53.98
2,000	0.0005	60.00
5,000	0.0002	67.96

In microstrip circuits, the calculation of effective dielectric constant is based on waves emanating from the top of the transmission line being radiated into free space. When the microstrip circuit is placed in a chassis, some of the waves are terminated on the sides by metals walls and on the top by a metal cover.

20.4.6.1 Cover Effects on Microstrip Circuits

Figure 20.8 shows a microstrip transmission line with width (W) on a dielectric material (ε_r) of thickness (H). The cover is at a height Hc above the dielectric material.

A metallic cover in proximity to the transmission line raises the capacitance per length, and therefore lowers the impedance.

Equations analyzing the decrease impedance (ΔZ) of a nominal 50Ω microstrip transmission line radiating into free space as a function of cover height (Hc) [13] are

$$\Delta Z = \begin{cases} A & \text{for } \dfrac{W}{H} \leq 1 \\[2ex] A \cdot B & \text{for } \dfrac{W}{H} \geq 1 \end{cases} \tag{20.4}$$

$$A = 270\left[1 - \tanh\left(0.28 + 1.2\sqrt{\dfrac{Hc}{H}}\right)\right] \tag{20.5}$$

$$B = 1 - \tanh^{-1}\left(\dfrac{0.48\sqrt{\dfrac{W}{H} - 1}}{\left(1 + \dfrac{Hc}{H}\right)^2}\right) \tag{20.6}$$

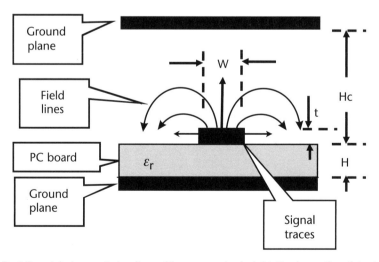

Figure 20.8 Microstrip transmission line with a cover at a height Hc above the dielectric material.

20.4.6.2   Example of the Cover Effect on the Impedance of a Microstrip Transmission Line

Assume that a 50Ω microstrip transmission line has no cover (i.e., that is the cover is an infinite height above the dielectric substrate). Microstrip has the following characteristics:

- Substrate height: $H = 20$ mils;
- Relative dielectric constant: $\varepsilon_r = 4.3$;
- Transmission-line width: $W = 38.16$ mils;
- Transmission-line thickness: $t = 0.685$ mil;
- Frequency (F): $F = 5{,}000$ MHz;
- Calculated impedance (Z_0): $Z_0 = 50\Omega$;
- Change in impedance (ΔZ) in ohms;
- Effective impedance (Z_{eff}) in ohms.

Table 20.4 shows the calculated results for cover height (Hc) three to five times the dielectric height (H).

20.4.6.3   Typically Accepted Practices for Cover and Wall Separation

It is generally accepted that cover heights and wall separations having a minimal effect on the performance of the microstrip transmission line are:

1. The distance from the cover to the dielectric material should be at least five times the height of the dielectric material

$$\frac{Hc}{H} \geq 5 \tag{20.7}$$

2. The distance from the edge of the conductor to the wall (D_W) should be at least four times the width of the conductor

$$D_W \geq 4W \tag{20.8}$$

Table 20.4   Change in Impedance of a 50Ω Microstrip Line as a Function of Cover Height Above the Dielectric Material

Hc/H	Hc mils	ΔZ Ohms	Z_{eff} Ohms
2	40	9.643	40.36
3	60	4.649	45.35
4	80	2.48	47.52
5	100	1.418	48.58

20.4.7   Notes on Radiated Emissions and Radiated Susceptibility

The mechanism for attenuating radiated emissions from an external source entering the unit of concern or attenuating internal radiation emitted externally interfering with other devices is passive and reciprocal. The resulting attenuation is the same for signals leaving the device or entering the device so the focus on the design to meet the requirements are the most difficult high-level signals.

20.4.7.1   Radiated Signal-Level Suppression

Attenuation of radiated signals is frequency-dependent. The radiated suppression must function over the frequency of operation. Many times, the EMI requirements cover the harmonics of the signals so the design may be required to function over a much higher range than the signal frequencies.

20.4.7.2   Digital Signal Frequencies

The highest frequency of interest with respect to interference of digital signals is related to the rise time of the signal not necessarily the pulse width. For a single pole filter, the rise time (RT) and effective frequency (Fr) are related by the following equation:

$$Fr = \frac{0.35}{RT} \tag{20.9}$$

High-frequency interference signals are present even if the system clock rates are slow. In the digital signal calculation example, a signal with a 3.5-ns rise time has a related frequency with respect to calculating interference of 100 MHz.

$$Fr = \frac{0.35}{RT} = \frac{0.35}{3.5}\,\text{nS} = 100 \text{ MHz} \tag{20.10}$$

20.5 Conducted Emissions and Conducted Susceptibility

Conducted emissions and conducted susceptibility are unintentional signals (spurious signals) propagated along power lines or signal conductors that interfere with other connected devices or signals that come from other devices that interfere with the operation of the SSPA, respectively.

In conducted emissions, the maximum conducted emitted levels are usually stated by the system designer and are set such that the emissions will not interfere with the operation of other devices in the system. In conducted susceptibility, conducted susceptibility states the limits of spurious, and noise on any wire DC power connection or cable conducted to a device. The device must meet all of its specification while being subjected to this interference signals.

20.5.1   Conducted Emission and Susceptibility on Signal Lines

Wires or cables subjected to conducted emissions and susceptibility carry RF, digital, or analog signals.

1. *RF signals:* The RF signal chain can be bandwidth-limited to the spectrum of interest may not be directly subjected to extraneous signals. Typically, some form of RF filtering between amplifier stages rejects spurious and baseband signals to below the level of concern.
2. *Digital signals:* Digital signal lines are designed to have a high immunity from interferences. Filtering of the digital DC power lines tends to prevent the digital signal from interacting and interfering with other connected functions in a device. Note that on digital logic speeds it is generally considered to be a good practice to avoid using digital logic faster than necessary. Logic speeds on the order of the RF signal spectrum can create spurious emissions in the RF amplifier chain.
3. *Analog signals:* Analog signals are generally band-limited and should be filtered to limit out-of-band signal emissions and susceptibility. Low-level analog signals should be avoided if possible because low signal (S) to noise (N) plus interference (I) [$S/(N + I)$] levels could mean a high rate of errors when the signal is processed.

20.5.2   Conducted Emission and Susceptibility on DC Power Lines

DC-to-DC converters (voltage regulators) are a common source of interference conducted emission and a thoroughfare for external signals to enter the RF signal path (i.e., conducted susceptibility).

20.5.2.1   Conducted Emissions on the DC Power Lines

DC-to-DC converters are an efficient means of converting an input DC voltage to a level compatible with the DC requirements of an RF amplifier. One means of accomplishing this this task is by chopping the input DC signal at a fixed frequency with a variable duty factor such that the average signal is at the new desired DC voltage.

20.5.2.2   DC-to-DC Converter Switching Speed

Between the minimum and maximum switching voltage, the DC-to-DC converters could be operating in the linear mode dissipating a considerable amount of power. If the switching speeds are very fast, the voltage regulator is only in the linear mode for short period and the efficiency of the regulator is improved. Note that, on the conflicting requirements relating to DC-to-DC converter rise and fall times, fast switch times (short rise and fall times) result in a more efficient DC-to-DC converter and short rise and fall times result in higher-frequency emissions, which tend to be more difficult to reject.

20.5.2.3   Voltage Ripple on the Output of the DC Converter

An inductor and a large capacitor set the switching frequency and the level of the output voltage ripple, respectively [14]. The frequency of the ripple could result in conducted emissions. The level of the output voltage ripple relates directly to the level of conducted emissions from the DC converter.

20.5.2.4   Controlling the Externally Conducted Emissions from the DC-to-DC Converter

The negative aspect of this type of regulator is the very fast switching of substantial currents, which can possibly generate high-level signals at frequencies much higher than the voltage regulator switching frequency. A portion of these signals can be found at the input of the voltage regulator and conducted through the input DC voltage input path. Voltage regulator emissions on the DC power lines can be controlled by filtering these signals with a passive lowpass filter.

20.5.2.5   Controlling the Internally Conducted Emissions from the DC-to-DC Converter

Internally, the voltage regulators are designed to reject low-frequency signals from DC (zero frequency) to typically a few hundred kilohertz [15] but provide little protection at higher frequencies. Higher-frequency emissions are attenuated with a lowpass filter similar to the lowpass filter on the DC-to-DC converter input.

20.5.2.6   Conducted Susceptibility (Interference) on the DC Power Lines

The conducted susceptibility is a requirement that the device of concern operate to within all its requirements when the DC input voltage lines are contaminated with a given level of spurious signals and noise. The most common solution to this problem is resolved by isolating and filtering the signal and power lines such that the spurious and noise are attenuated to a level where the impact will be negligible.

20.5.2.7   RF Signal Degradation Caused by DC Power Line Susceptibility (Interference)

Interference is typically from signals conducted or radiated onto system common DC power lines. Once in the system on the DC power lines, there are paths to the amplifier modules where the signals could modulate the RF signals or add to the RF signals as spurious components. Modulating the RF signal with noise or spurious signals could be the most detrimental because the interference may not be eliminated by filtering.

20.5.2.8   Isolating and Filtering Incoming Spurious and Noise

A common technique for cleaning up incoming signals is to isolate the signals in an enclosed section of the chassis. A passive filter in this enclosed section attenuates the noise and spurious components before they can contaminate the main functions of the assembly (see Figure 20.9).

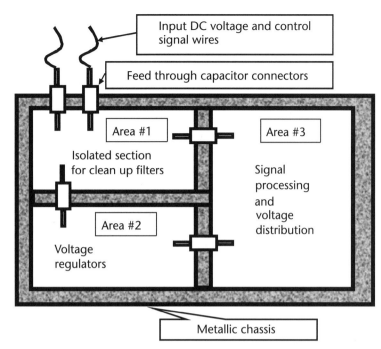

Figure 20.9   Example of a chassis layout used to clean up incoming noise and spurious signals.

The chassis shown has three compartments for isolating the DC and low-frequency control signals:

- *Area #1:* The Isolated section for clean-up filters section brings the outside signals possibly contaminated with noise and other unwanted components into the main chassis. The figure shows incoming wires connected to feed-through capacitors. The incoming signals are filtered in this area to attenuate the unwanted frequency components. Note that on input passive filters the filters are bidirectional and therefore attenuate unwanted signals from entering or leaving the assembly. These filters therefore also contribute to limiting the level of conducted emissions.

- *Area #2:* In voltage regulators, the cleaned-up input DC voltage is passed to the on-board voltage regulators where the necessary DC voltages for the SSPA modules are created and distributed. The voltage regulators creating the various voltages required for operation are typically efficient DC-to-DC converters using high-power chopped signals. The feed-through capacitors help to attenuate the high-frequency components of the spurious signals created by the voltage regulators. In addition to the feed-through capacitors, a parallel combination of high-value capacitors is usually necessary to bring the voltage ripple created in the regulators to an acceptable level.

- *Area #3:* In the signal processing and voltage distribution section, it is necessary that all signals entered this section are sufficiently devoid of noise and spurious signals such that the SSPA can operate within its required specification. This section may consist of control signals or amplifier modules.

20.5.2.9 Feed-Through Capacitors

Whenever possible, it is a good practice for non-RF signals to pass from compartment to compartment within the chassis through the chassis walls via a feed-through capacitor shown in Figure 20.10.

Feed-through capacitors are designed to pass DC and low-frequency signals while blocking RF signals by effectively shorting them to ground. The unit is through the chassis wall connector that contains a low-inductance disc capacitor (high resonant frequency) that is embedded in a metallic assembly (sometimes screwed into the chassis). This device is preventing the RF signals from entering or escaping the cavity.

20.6 DC Power-Line Filtering

Conducted emissions are at known frequencies and can be readily contained with proper filtering. The conducted susceptibility frequencies and levels are typically from an accumulation of many sources and devices in the system and therefore

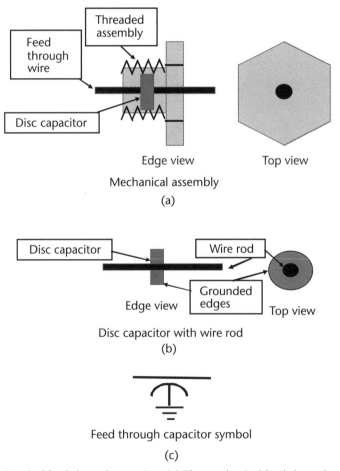

Figure 20.10 A typical feed-through capacitor. (a) The mechanical feed-through connection, (b) the internal disc capacitor with a hole in the center to pass the signal wire, and (c) the electrical symbol.

usually assumed to exist over a wide-frequency spectrum with a stated maximum spectral density.

The primary filter reducing the effects of DC power-line interference is located in Area #1 shown in Figure 20.9. This section of the chassis is the first interface between the unit of concern (the SSPA as an example) and the outside world (the remainder of the system excluding the SSPA).

20.6.1   Wideband Power Line Filtering (Area #1)

The filtering requirements for Area #1 is the difference between the specified maximum DC power line interference [I_{max} (dBV)] at the input to the SSPA and the acceptable level of interference at the input to the amplifier module (the output of the voltage regulators [I(dBV)] (i.e., the output of Area #2) (see Figure 20.11).

The voltage regulators (Area #2) typically attenuate input low-frequency signals from DC to usually around a few hundred kilohertz (Fv). Power-line signals in the RF and microwave range are filtered at the power and control inputs of the amplifier modules, but it is still a good idea to filter interference signals in Area #1. The input passive filters are typically designed to begin attenuating signals at frequency Fc (see Figure 20.12) even if it coincides with the local filtering at the amplifier level. RF-conducted interference can easily turn into radiated interference if they are not contained in Area #1.

The cutoff frequency of the low pass filter in Area #1 (Fc) is the frequency where the interference rejection in the voltage regulation reach the maximum allowable level (dBV) such that the combination of rejection from the power-line filters in Area #1 and the voltage regulator (Area #2) is such that the maximum allowable interference level [I(dBV)] is never exceeded.

20.6.2   Notes on Wideband Power-Line Filtering

On susceptibility notation, dBV is dB below 1-V RMS and susceptibility is typically specified in volts RMS. Note on the voltage regulator frequency response that the rejection frequency of a voltage regulator (Fv) varies with the regulator design and decoupling capacitors used. Check the specifications for the voltage regulators before designing the input passive filter.

The susceptibility specification is usually spread over a wide-frequency range (i.e., multiple decades). The filter design at the input must satisfy all the requirements

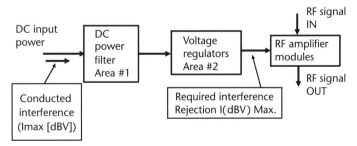

Figure 20.11   Conducted interference from the DC voltage input to the RF amplifier module DC input.

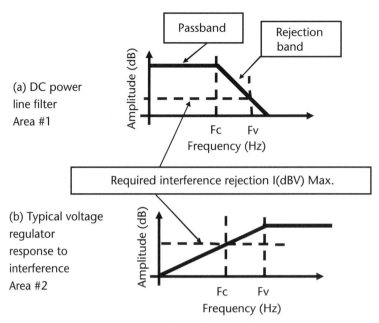

Figure 20.12   Interference filter rejection in input filter Area #1 and voltage regulator Area #2.

at all frequencies. Many times, this requires multiple lowpass filters used in tandem to cover the entire specified frequency range because the component (capacitor and inductor) self-resonance limits the rejection range of a single lowpass filter.

20.7   DC-DC Converter Efficiency and Effect on Spurious Emissions

DC-DC converters (Area #2) are used to provide the required regulated voltage to the amplifier modules from the raw DC power going into the SSPA module (Area #1).

20.7.1   Regulator Efficiency

The efficiency of DC-DC converter is a critical requirement and very much dependent on the output current, output waveform ripple, and switching time rise and fall time of the switching network.

Current losses are losses as a function of output current occur in the series FET switch (on resistance) and the inductor ESR. A ripple waveform dissipates power in a capacitor's ESR [14]. The switching frequency rise and fall time is the finite time that a DC-to-DC converter waveform is between states (rise time and fall time) and is a period that the device is operating in the linear mode dissipating power.

20.7.2   Conflicting Requirements Designing a DC-to-DC Converter

The conflicting requirements for DC-to-DC converter design are: the reservoir capacitor, the switching frequency [16], and high-value inductors.

20.7.2.1   Reservoir Capacitor

The reservoir capacitor reduces the voltage regulator output ripple; the higher the capacitance, the lower the ripple. Most of time, the limitation on this device is the size and the resonant frequency of the component. Usually, the resonant frequency issue can be resolved with multiple parallel capacitors of smaller capacitance and size (lower value and smaller size have higher resonant frequencies).

20.7.2.2   Switching Frequency

The design of the switching circuit in a DC-DC converter is a compromise between the frequency and the rise time of the switch. Faster switching frequencies require smaller coils, but during the transition time between states, the DC-DC converter is operating in the linear mode. During this time, power is being dissipated. The rise and fall times are related to the components, not the switching frequency period, so the higher the switching frequency, the more time per period the DC voltage converter is operating in the linear mode. Higher switching frequency can result in more power dissipated and lower efficiencies.

20.7.2.3   High-Value Inductors

Designing lower switching frequencies to increase DC-to-DC converter efficiencies result in high-value inductors. Higher-value inductors are larger and have a higher loss due to an increase in the ESR. The lower frequencies also require larger reservoir capacitors.

References

[1]   MIL-STD-451G, "Requirements for the Control of Electromagnetic Interference Characteristics of Subsystems and Equipment," Department of Defense Interface Standard, March 2015.

[2]   Automation Direct, "Applied EMI/RFI Techniques," August 2015.

[3]   Analog Devices, Inc., "EMI, RFI, and Shielding Concepts," MT-095 Tutorial, 2009.

[4]   Lun, T., "Designing for Board Level Electromagnetic Compatibility," Freescale Semiconductor Application Note AN2321, October 2005.

[5]   Schneider, M., "Design Considerations to Reduce Conducted and Radiated EMI," Purdue University, April 2010.

[6]   Baker, B., "PCB Signal Coupling Can Be a Problem," *EDN Network*, December 2013.

[7]   Getz, R., and B. Moeckel, "Understanding and Eliminating EMI in Microcontroller Applications," National Semiconductor, Application Note 1050, August 1996.

[8]   Duarte, R., and G. Felic, "Analysis of the Coupling Coefficient in Inductive Energy Transfer Systems," Hindawi Publishing Corporation, Active and Passive Electronic Components, 2014.

[9]   Intel, "Si Bootcamp Boot Camp: Crosstalk," DEG Training, 00030687, March 2007.

[10] Staff, "The Dark Force of Evil in Electronics: Electromagnetic Interference," *Electronic Design*, June 2009.

[11] Ulaby, F., and D. Long, *Microwave Radar and Radiometric Remote Sensing*, The University of Michigan Press, 2014.

[12] Clupper, H., and J. Wheeler, "The Effects of Gaps Introduced into a Continuous EMI Gasket," W. L. Gore and Associates, Inc.

[13] Gupta, K., *Microwave Integrated Circuits*, New York: Wiley, 1974.

[14] Texas Instruments Incorporated, "Switching Regulator Fundamentals," Application Report SNVA559A, September 2016.

[15] Maxim Integrated Products, "Improved Power-Supply Rejection for Linear Regulators," Application Note 883, October 2002.

[16] Martin, A., "AN-2162 Simple Success with Conducted EMI from DC to DC Converters," Texas Instruments, Application Report SNVA489C, April 2013.

Table of Constants

Physical Constants Useful in Microwave Engineering

Constant	Description	Value	Units
k	Boltzmann's constant	1.38065030E-23	joule/kelvin
c	Velocity of light in free space	2.99792458E+08	meters/second
E	Base number of natural logarithms	2.71828183E+00	none
H	Planck's constant	6.62606876E-34	joule-second
Z_0	Impedance of free space	3.76730313E+02	Ohms
ε_0	Permittivity of free space	8.85418782E-12	farad/meter
m_e	Electron mass	9.10938188E-31	kilogram
π	Pi	3.14159265E+00	none
q	Electron charge	1.60217646E-19	coulombs
μ_0	Permeability of free space (exact value is $4 \cdot \pi \times 10^{-7}$)	1.25663706E-06	henry/meter

Adapted from NIST website and the *CRC Handbook of Chemistry and Physics*.

Table of Dielectric Constants and Loss Tangents of Typical Microwave Materials

Substance	Dielectric Constant (Relative to the Dielectric Constant in a Vacuum)	Loss Tangent
Air	1.00054	0.0002 at 1 GHz
Alumina: 96%	10	0.0002 at 100 MHz
Alumina: 99.5%	9.6	0.0003 at 10 GHz
Beryllium oxide	6.7	0.006 at 10 GHz
Diamond	5.5 to 10	
Epoxy glass PCB	5.2	
FR-4 (G-10): low resin	4.9	0.008 at 100 MHz
Fused silica (glass)	3.8	
Gallium arsenide (GaAs)	13.1	0.0016 at 10 GHz
Germanium	16	
Mylar	3.2	
Nylon	3.2 to 5	
Polyamide	2.5 to 2.6	
Quartz (fused)	4.2	
Silicon	11.7 to 12.9	0.005 at 1 GHz, 0.015 at 10 GHz
Silicone rubber	3.6	
Teflon (polytetra fluoroethylene [PTFE])	2.0 to 2.1	0.00028 at 3 GHz
Vacuum (free space)	1	

Table of Printed Circuit Board Standard Copper Thickness

PCB Copper Thickness

1/2 oz.	0.689 mil
1 oz.	1.378 mils
2 oz.	2.756 mils

Common Frequency Bands

IEEE Standard Letter Designations for Radar Bands

Band Designation	Nominal Frequency Range
HF	3–30 MHz
VHF	30–300 MHz
UHF	300–3,000 MHz
SHF	3–30 GHz
EHF	30–300 GHz
L	1–2 GHz
S	2–4 GHz
C	4–8 GHz
X	8–2 GHz
Ku	12–18 GHz
K	18–27 GHz
Ka	27–40 GHz
Q	33–50 GHz (common usage)
V	40–75 GHz
W	75–110 GHz

List of Acronyms

AC	Alternating current
AWGN	Additive white Gaussian noise
BER	Bit error rate
CMOS	Complementary metal oxide semiconductor
CW	Continuous-wave RF signal (as opposed to a pulsed signal)
dB	Decibel
dBc	Decibel with respect to a carrier
dBm	Decibel with respect to 1 mW
dBW	Decibel with respect to 1W
DC	Direct current
DF	Duty factor
EM	Electromagnetic
EMC	Electromagnetic compatibility
EMI	Electromagnetic interference
EW	Electronic warfare
FET	Field-effect transistor
GaAs	Gallium arsenide
GaN	Gallium nitride
HEMT	High electron mobility transistor
HPA	High-power amplifier
IF	Intermediate frequency
IM	Intermodulation distortion
MMIC	Monolithic microwave integrated circuit
MTBF	Mean time between failures
MTTF	Mean time to failure
P1dB	Power level that the gain compresses by 1 dB
PA	Power amplifier
PAE	Power-added efficiency
PC	Printed circuit
PCB	Printed circuit board

PPM	Parts per million
PTFE	Polytetrafluoroethylene (Teflon)
RF	Radio frequency
RFI	Radio frequency interference
SMT	Surface mount technology
SSPA	Solid state power amplifier
TTL	Transistor-transistor logic
VSWR	Voltage standing-wave ratio

About the Author

 **Howard Hausman** received his BSEE and MSEE degrees from Polytechnic University/New York University and is the president and CEO of RF Microwave Consulting Service. Mr. Hausman is also an adjunct professor at Hofstra University, School of Engineering. Mr. Hausman was the president and CEO of MITEQ, Inc., an industry leader in microwave engineering, for 7 years until the company was sold to L3 Corp. Prior to being appointed the president and CEO of MITEQ and continuing after the L3 acquisition, Mr. Hausman was and is a microwave engineering consultant, chief scientist, senior principal engineer, chief technology officer, and vice president of engineering. He designed microwave low noise amplifiers, power amplifiers, passive devices, frequency converters, oscillators, synthesizers, receivers, and transmitters. The microwave devices that he designed are used in ground-based satellite communications systems, spaceborne microwave systems, missile guidance systems, radar systems, reconnaissance systems, commercial aircraft Wi-Fi systems, and defense electronics systems.

Mr. Hausman was also an adjunct professor at Polytechnic University, now NYU Tandon School of Engineering, where he taught graduate and undergraduate courses in electronic engineering. He is also a recipient of the New York University/Polytech Distinguished Alumni Award, the IEEE LI Section Alex Gruenwald Award "For outstanding contributions to enhance the knowledge of the IEEE LI Section members in Satellite Communications and Microwave Theory," and a 2014 Award from NASA for work on the Mars Landing Systems. He has lectured around the world and authored many papers relating to microwave systems, microwave power amplifiers, microwave components, satellite communications, radar, and reconnaissance systems. Mr. Hausman was awarded a patent titled "Measuring Satellite Linearity from Earth Using a Low Duty Cycle Pulsed Microwave Signal" (U.S. Patent 8,391,781B2).

Index

Artech House Microwave Library

Behavioral Modeling and Linearization of RF Power Amplifiers, John Wood

Chipless RFID Reader Architecture, Nemai Chandra Karmakar, Prasanna Kalansuriya, Randika Koswatta, and Rubayet E-Azim

Control Components Using Si, GaAs, and GaN Technologies, Inder J. Bahl

Design of Linear RF Outphasing Power Amplifiers, Xuejun Zhang, Lawrence E. Larson, and Peter M. Asbeck

Design Methodology for RF CMOS Phase Locked Loops, Carlos Quemada, Guillermo Bistué, and Iñigo Adin

Design of CMOS Operational Amplifiers, Rasoul Dehghani

Design of RF and Microwave Amplifiers and Oscillators, Second Edition, Pieter L. D. Abrie

Digital Filter Design Solutions, Jolyon M. De Freitas

Discrete Oscillator Design Linear, Nonlinear, Transient, and Noise Domains, Randall W. Rhea

Distortion in RF Power Amplifiers, Joel Vuolevi and Timo Rahkonen

Distributed Power Amplifiers for RF and Microwave Communications, Narendra Kumar and Andrei Grebennikov

Electric Circuits: A Primer, J. C. Olivier

Electronics for Microwave Backhaul, Vittorio Camarchia, Roberto Quaglia, and Marco Pirola, editors

EMPLAN: Electromagnetic Analysis of Printed Structures in Planarly Layered Media, Software and User's Manual, Noyan Kinayman and M. I. Aksun

An Engineer's Guide to Automated Testing of High-Speed Interfaces, Second Edition, José Moreira and Hubert Werkmann

Envelope Tracking Power Amplifiers for Wireless Communications, Zhancang Wang

Microwave Radio Transmission Design Guide, Second Edition, Trevor Manning

Microwave and RF Semiconductor Control Device Modeling, Robert H. Caverly

Microwave Transmission Line Circuits, William T. Joines, W. Devereux Palmer, and Jennifer T. Bernhard

Microwaves and Wireless Simplified, Third Edition, Thomas S. Laverghetta

Modern Microwave Circuits, Noyan Kinayman and M. I. Aksun

Modern Microwave Measurements and Techniques, Second Edition, Thomas S. Laverghetta

Modern RF and Microwave Filter Design, Protap Pramanick and Prakash Bhartia

Neural Networks for RF and Microwave Design, Q. J. Zhang and K. C. Gupta

Noise in Linear and Nonlinear Circuits, Stephen A. Maas

Nonlinear Microwave and RF Circuits, Second Edition, Stephen A. Maas

On-Wafer Microwave Measurements and De-Embedding, Errikos Lourandakis

Passive RF Component Technology: Materials, Techniques, and Applications, Guoan Wang and Bo Pan, editors

Practical Analog and Digital Filter Design, Les Thede

Practical Microstrip Design and Applications, Günter Kompa

Practical Microwave Circuits, Stephen Maas

Practical RF Circuit Design for Modern Wireless Systems, Volume I: Passive Circuits and Systems, Les Besser and Rowan Gilmore

Practical RF Circuit Design for Modern Wireless Systems, Volume II: Active Circuits and Systems, Rowan Gilmore and Les Besser

Production Testing of RF and System-on-a-Chip Devices for Wireless Communications, Keith B. Schaub and Joe Kelly

Q Factor Measurements Using MATLAB®, Darko Kajfez

Technologies for RF Systems, Terry Edwards

Terahertz Metrology, Mira Naftaly, editor

TRAVIS 2.0: Transmission Line Visualization Software and User's Guide, Version 2.0, Robert G. Kaires and Barton T. Hickman

Understanding Microwave Heating Cavities, Tse V. Chow Ting Chan and Howard C. Reader

Understanding Quartz Crystals and Oscillators, Ramón M. Cerda

Vertical GaN and SiC Power Devices, Kazuhiro Mochizuki

For further information on these and other Artech House titles, including previously considered out-of-print books now available through our In-Print-Forever® (IPF®) program, contact:

Artech House Publishers	Artech House Books
685 Canton Street	16 Sussex Street
Norwood, MA 02062	London SW1V 4RW UK
Phone: 781-769-9750	Phone: +44 (0)20 7596 8750
Fax: 781-769-6334	Fax: +44 (0)20 7630 0166
e-mail: artech@artechhouse.com	e-mail: artech-uk@artechhouse.com

Find us on the World Wide Web at: www.artechhouse.com